Keller · Wavelets in Geodesy and Geodynamics

Wolfgang Keller

Wavelets in Geodesy and Geodynamics

Walter de Gruyter · Berlin · New York

Author
Wolfgang Keller, Prof. Dr. Sc-techn.
Geodätisches Institut
Universität Stuttgart
Geschwister-Scholl-Str. 24/D
70174 Stuttgart, Germany
E-mail: Wolfgang.Keller@gis.uni-stuttgart.de

This book contains 130 figures.

♾ Printed on acid-free paper which falls within the guidelines of the ANSI to ensure permanence and durability.

Library of Congress Cataloging-in-Publication Data

Keller, Wolfgang.
Wavelets in geodesy and geodynamics / Wolfgang Keller.
p. cm.
Includes bibliographical references and index.
ISBN 3-11-017546-0 (alk. paper)
1. Wavelets (Mathematics) 2. Geology – Mathematics. I. Title.
QE33.2.W38K45 2004
2004045458

ISBN 3-11-017546-0

Bibliographic information published by Die Deutsche Bibliothek

Die Deutsche Bibliothek lists this publication in the Deutsche Nationalbibliografie; detailed bibliographic data is available in the Internet at <http://dnb.ddb.de>.

Printed in Germany
Cover design: Rudolf Hübler, Berlin
Typeset using the authors TEX files: I. Zimmermann, Freiburg
Printing and binding: Hubert & Co. GmbH & Co. Kg, Göttingen

Preface

For many years, digital signal processing has been completely governed by the theory of Fourier transformation and its numerical implementation – the Fast Fourier Transform (FFT).

The main disadvantage of the Fourier theory is the underlying assumption that the signals to be processed have time-wise or space-wise invariant statistical properties. In many applications the deviation from such a stationary behavior is exactly the information to be extracted from the signal. Such patterns occur for instance in seismic reflection signals. Hence one of the first contributions to a field, which is now called wavelet analysis, comes from the analysis of seismic signals. Without mentioning the name wavelet, the paper [43] marks somehow the beginning of the signal processing by wavelets. Even more, one of the most frequently used wavelet is named after Morlet – one of the authors of this paper.

The simplest thinkable wavelet is derived from the Haar base which has already been known since 1910. The disadvantage of this Haar wavelet is its low degree of smoothness. The French mathematician Y. Meyer [73] succeeded in the construction of a family of orthogonal smooth wavelets. With the concept of multi-resolution analysis (MRA), introduced by Mallat and Meyer, the gap between wavelet analysis and filtering was bridged. Unfortunately, the wavelets were either irregular and with a finite support – as the Haar wavelet – or smooth and with an infinite support. Due to the contribution of I. Daubechies [15] a method was found to construct wavelets which are both smooth and of finite support. An extensive historic overview of the development of wavelet theory can be found in [48].

Despite the fact that the first applications of wavelet transformation were in geo-sciences, it was not before the nineties that the geo-science community started to use wavelets. Meanwhile numerous applications of wavelet theory in geo-sciences can be found. In this book we present the mathematical foundations of wavelet theory as well as its applications in geo-sciences. The presentation of the material is somehow a compromise between a popular and a strictly mathematical presentation: The definitions and results will be given in a rigorous mathematical way. Proofs of the results will be included only if the respective proof is comparatively short and helps to understand the subject. In other cases the reader is referred to the literature and illustrating examples and remarks are given instead.

Since nowadays wavelet theory is a wide field, the selection of material which is included here is naturally subjective. It is governed by the personal experience and taste of the author.

The book has developed from a graduate course on wavelets held at the Department of Geomatics Engineering at The University of Calgary (Alberta), Canada. It is addressed to graduate students who are interested in digital signal processing. The reader is assumed to have a mathematical background at graduate level. Mathematical tools which are necessary for the understanding and which are beyond this level are included in the appendices.

The book consists of three main chapters. Since wavelet theory developed from Fourier theory on the one side and filter theory on the other side, both subjects are shortly sketched in the first chapter. The second chapter is devoted to the basics of wavelet theory. It includes the continuous as well as the discrete wavelet transform, both in the one- and in the two-dimensional case. A special emphasis is laid on orthogonal wavelets with finite support, as they play an important role in most applications. Other wavelets, such as spline wavelets or Meyer wavelets are excluded, which does not mean they are less important.

Since geo-sciences deal with data given on a sphere, spherical wavelets are extensively discussed. They come in two forms: harmonic wavelets, which have a special spherical harmonics expansion, and triangulation based wavelets, which are wavelets on a polyhedral approximation of the sphere.

In the last chapter some applications of wavelets in geo-sciences are reviewed. Also here the selection of the material presented is not claimed to be representative. A more important criterion for the inclusion of a subject is that its geophysical background can be explained with comparatively little effort. In some examples the original data material was not available. Therefore the wavelet analysis was carried out with synthetical data which reflect the main features of the missing original data.

I would like to thank the Walter de Gruyter Publishing Company for including this book into its Geodesy series, Dr. Manfred Karbe for his support, and Dr. Irene Zimmermann for her advice in the LaTeX typesetting of this textbook.

Stuttgart, January 2004 *Wolfgang Keller*

Contents

Notation

$\langle f, g\rangle$	Scalar product
$\|f\|$	Norm
$\boldsymbol{A}^\top$	Transposed matrix
$\boldsymbol{A}^{-\top}$	Inverse of the transposed matrix
$\boldsymbol{A}^{-1}$	Inverse matrix
A^*	Adjoint operator
A^{-1}	Inverse operator
$\bar{z}$	Complex conjugate of z
σ	Unit sphere

Sets

$\mathbb{N}$	Natural numbers
$\mathbb{Z}$	Integers
$\mathbb{R}$	Real numbers
$\mathbb{C}$	Complex numbers

Special functions

P_n	Legendre polynomial
$\bar{Y}_{nm}$	Fully normalized surface spherical harmonic
H_{nm}	Spherical harmonic
J_0	Bessel function of first kind and order 0

Signals

$f(t)$, $f(x)$	Continuous time or space dependent signal
$\{x_n\}$	Discrete signal sequence
$\delta(t)$	Delta distribution
δ_{ij}	Kronecker symbol
$\chi_{[a,b]}$	Characteristic function of the interval $[a, b]$
$\Pi = \chi_{[-1,1]}$	Characteristic function of the interval $[-1, 1]$

Spaces

$C_0^\infty(\mathbb{R})$	Space of all infinitely often differentiable function having a compact support
$L_2(\mathbb{R})$	Space of square integrable functions on $\mathbb{R}$
$L_2([a, b])$	Space of square integrable functions on $[a, b]$
$l_2(\mathbb{C})$	Space of square summable complex sequences

$\mathrm{Harm}_{p,\dots,q}$	Span of all spherical surface harmonics of degree p up to degree q
$\mathcal{H}(\{A_n\},\sigma)$	Sobolev space of all functions harmonic outside the unit sphere σ which have degree variances decaying faster than A_n^{-1}
$U \oplus V$	Direct sum of the spaces U and V

Operators

$\frac{\partial^{\|\alpha\|} f(x_1,\dots,x_n)}{\partial x_1^{\alpha_1} \cdot \partial x_n^{\alpha_n}}$	Partial derivative of f
∇f	Gradient of f
$f * g$	Convolution of f and g

Transforms

$\hat{f}$, $\mathcal{F}\{f\}$	Fourier transform of f
$\mathcal{F}^{-1}\{g\}$	Inverse Fourier transform of g
$H(z)$	z-transform of the sequence $\{h_n\}$
$\mathcal{G}\{f\}$	Garbor transform of f
$\mathcal{W}\{f\}$	Continuous wavelet transform
$\mathcal{W}^{eu}\{f\}$	Two-dimensional continuous wavelet transform

1 Fourier analysis and filtering

1.1 Fourier analysis

Let f be a periodic function with the period 2π. It is well known that f has the following series expansion:

$$f(x) = \sum_{n \in \mathbb{Z}} c_n e^{\imath nx}, \tag{1.1}$$

$$c_n = \frac{1}{2\pi} \int_{-\pi}^{\pi} f(x) e^{-\imath nx} \, dx. \tag{1.2}$$

This series representation is called Fourier series of f. It can be interpreted as a superposition of oscillations of different wavelengths $\lambda_n = 2\pi/n$, having the amplitudes c_n. If the period changes from 2π to $2T$, the Fourier series changes to

$$f(x) = \sum_{n \in \mathbb{Z}} c_n e^{\imath \frac{2\pi}{2T} nx}, \tag{1.3}$$

$$c_n = \frac{1}{2T} \int_{-T}^{T} f(x) e^{-\imath \frac{2\pi}{2T} nx} \, dx. \tag{1.4}$$

The spectral lines are now at the frequencies $\omega_n = \frac{2\pi}{2T} n$. This means the distance between the spectral lines gets smaller if the period $2T$ gets larger than 2π. If the assumption of periodicity is dropped by increasing T to infinity, the line spectrum changes to a continuous spectrum. For this limit process the following notations have to be introduced:

$$\Delta\omega = \frac{2\pi}{2T}, \quad \omega_n = n\Delta\omega, \quad c(\omega_n) = \frac{c_n}{\Delta\omega}.$$

With these notations the Fourier series can be written as

$$f(x) = \sum_{n \in \mathbb{Z}} c(\omega_n) e^{\imath \omega_n x} \Delta\omega, \tag{1.5}$$

$$c(\omega_n) = \frac{1}{2\pi} \int_{-T}^{T} f(x) e^{-\imath \omega_n x} \, dx. \tag{1.6}$$

The series (1.5) can be considered as Riemann's partial sum of the integral

$$\int_{-\infty}^{\infty} c(\omega) e^{\imath \omega x} \, d\omega.$$

Hence in the limit $T \to \infty$ the following representation of the non-periodic function f is obtained:

$$f(x) = \int_{-\infty}^{\infty} c(\omega) e^{\imath \omega x} \, d\omega, \tag{1.7}$$

$$c(\omega) = \frac{1}{2\pi} \int_{-\infty}^{\infty} f(x) e^{-\imath \omega x} \, dx. \tag{1.8}$$

The interpretation of these relations is that f can be represented as a superposition of oscillations having all frequencies between $-\infty$ and ∞. Of course, every frequency has a different intensity in this superposition. The function $c(\omega)$ indicates the energy in the infinitesimal frequency-band $(\omega, \omega + d\omega)$. The function $c(\omega)$ is usually called the amplitude spectrum of f and the transition from f to c is called Fourier transform.

For a detailed mathematical discussion the Fourier transform of functions, defined on $\mathbb{R}^n$, will be introduced.

Definition 1.1.1. Let f, g be functions which are defined on $\mathbb{R}^n$ and fulfill

$$\int_{\mathbb{R}^n} |f| \, d\boldsymbol{x} < \infty, \quad \int_{\mathbb{R}^n} |g| \, d\boldsymbol{x} < \infty. \tag{1.9}$$

The following functions are called *Fourier transform* of f and *inverse Fourier transform* of g, respectively:

$$^n\mathcal{F}\{f\}(\omega) = \hat{f}(\omega) = (2\pi)^{-n/2} \int_{\mathbb{R}^n} f(\boldsymbol{x}) e^{-\imath \boldsymbol{x}^\top \omega} \, d\boldsymbol{x}, \tag{1.10}$$

$$^n\mathcal{F}^{-1}\{g\}(\boldsymbol{x}) = (2\pi)^{-n/2} \int_{\mathbb{R}^n} g(\omega) e^{\imath \boldsymbol{x}^\top \omega} \, d\omega \tag{1.11}$$

with $\boldsymbol{x}, \omega \in \mathbb{R}^n$.

Obviously, the following relation holds:

$$^n\mathcal{F}^{-1}\{^n\mathcal{F}\{f\}\} = f.$$

Additionally, the Fourier transform is an isometric operation, because

$$\int_{\mathbb{R}^n} f(\boldsymbol{x}) \overline{g(\boldsymbol{x})} \, d\boldsymbol{x} = \int_{\mathbb{R}^n} \hat{f}(\omega) \overline{\hat{g}(\omega)} \, d\omega \tag{1.12}$$

is valid. The special case of $f = g$ is known as Parseval's identity:

Lemma 1.1.1 (Parseval's identity). *Let f fulfill* (1.9). *Then*

$$\int_{\mathbb{R}^n} f^2(\boldsymbol{x}) \, d\boldsymbol{x} = \int_{\mathbb{R}^n} \hat{f}^2(\omega) \, d\omega \tag{1.13}$$

is true.

The Fourier transform has some interesting properties:

Lemma 1.1.2 (Shift relation). *For a function f fulfilling* (1.9) *the following relation holds:*

$$ {}^n\mathcal{F}\{f(\bullet - \boldsymbol{b})\} = e^{-\imath \boldsymbol{b}^\top \omega} \cdot {}^n\mathcal{F}\{f\}(\omega). \tag{1.14} $$

This means, the shift of the signal by a vector $\boldsymbol{b}$ produces a modulation of its spectrum by $e^{-\imath \boldsymbol{b}^\top \omega}$.

Lemma 1.1.3 (Scaling relation). *For a function f fulfilling* (1.9) *the following relation holds:*

$$ {}^n\mathcal{F}\{f(a \cdot \bullet)\}(\omega) = \frac{1}{a}\, {}^n\mathcal{F}\{f\}\left(\frac{\omega}{a}\right). \tag{1.15} $$

In other words, compressing the signal f by the factor a results in a dilatation of the spectrum by the same factor.

The most important properties of the Fourier transform are the convolution theorem and the differentiation theorem. These two theorems reduce the infinitesimal operations convolution and differentiation to algebraic operations.

Definition 1.1.2. Let the functions f, g be sufficiently regular. The new function $(f * g)$, defined by

$$ (f * g)(\boldsymbol{x}) := \int_{\mathbb{R}^n} f(\boldsymbol{y}) g(\boldsymbol{x} - \boldsymbol{y})\, d\boldsymbol{y} = \int_{\mathbb{R}^n} f(\boldsymbol{x} - \boldsymbol{y}) g(\boldsymbol{y})\, d\boldsymbol{y} \tag{1.16} $$

is called the *convolution* of f and g.

By means of the Fourier transform the integral operation convolution is transformed into the algebraic operation multiplication of the corresponding spectra.

Theorem 1.1.1 (Convolution theorem). *Let the functions f, g be sufficiently regular. Then*

$$ {}^n\mathcal{F}\{f * g\} = (2\pi)^{n/2} \cdot {}^n\mathcal{F}\{f\} \cdot {}^n\mathcal{F}\{g\} \tag{1.17} $$

holds.

Proof.

$$ \begin{aligned} {}^n\mathcal{F}\{f * g\}(\omega) &= (2\pi)^{-n/2} \int_{\mathbb{R}^n} \int_{\mathbb{R}^n} f(\mathbf{y}) g(\mathbf{x} - \mathbf{y})\, d\mathbf{y} e^{-\imath \mathbf{x}^\top \omega}\, d\mathbf{x} \\ &= (2\pi)^{-n/2} \int_{\mathbb{R}^n} \int_{\mathbb{R}^n} g(\mathbf{x} - \mathbf{y}) e^{-\imath \mathbf{x}^\top \omega}\, d\mathbf{x}\, f(\mathbf{y})\, d\mathbf{y} \\ &= (2\pi)^{-n/2} \int_{\mathbb{R}^n} \int_{\mathbb{R}^n} g(\mathbf{z}) e^{-\imath (\mathbf{z}+\mathbf{y})^\top \omega}\, d\mathbf{z}\, f(\mathbf{y})\, d\mathbf{y} \\ &= (2\pi)^{-n/2} \int_{\mathbb{R}^n} g(\mathbf{z}) e^{-\imath \mathbf{z}^\top \omega}\, d\mathbf{z} \cdot \int_{\mathbb{R}^n} f(\mathbf{y}) e^{-\imath \mathbf{y}^\top \omega}\, d\mathbf{y} \\ &= (2\pi)^{n/2} \cdot {}^n\mathcal{F}\{f\} \cdot {}^n\mathcal{F}\{g\}. \end{aligned} $$

□

Many integral transformations of Physical Geodesy are convolutions. Perhaps the most important one is the Stokes formula in planar approximation:

$$N = \frac{1}{4\pi\gamma}\left(\frac{1}{\|x\|} * \Delta g\right). \tag{1.18}$$

It computes geoid undulations N from given gravity anomalies Δg. The numerical evaluation of the convolution integral is very time-consuming. Thanks to very efficient numerical algorithms for the Fourier transform and the inverse Fourier transform, this direct evaluation of the convolution integral can be bypassed:

$$N = (2\pi)^{-n/2} \cdot {}^n\mathcal{F}^{-1}\left\{{}^n\mathcal{F}\left\{\frac{1}{4\pi\gamma}\frac{1}{\|x\|}\right\} \cdot {}^n\mathcal{F}\{\Delta g\}\right\},$$

which is given as a schematic view in Figure 1.1.

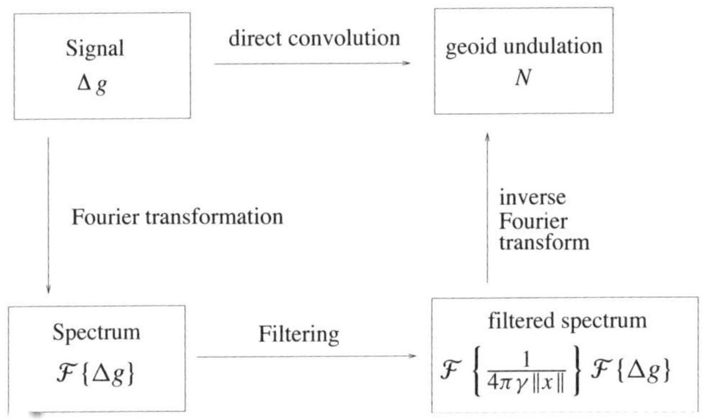

Figure 1.1. Schematic view of by-passed convolution

Let $\alpha \in (\mathbb{N} \cup \{0\})^n$ be a so-called multi-index. With $|\alpha| = \sum_{i=1}^n \alpha_i$ the order of the multi-index is denoted. With the help of this multi-index a compact notation for partial derivatives is introduced:

$$D^\alpha f := \frac{\partial^{|\alpha|} f}{\partial x_1^{\alpha_1} \partial x_2^{\alpha_2} \dots \partial x_n^{\alpha_n}}. \tag{1.19}$$

Theorem 1.1.2. *For sufficiently regular f the equality*

$$^n\mathcal{F}\{D^\alpha f\} = \imath^{|\alpha|}\omega^\alpha \cdot {}^n\mathcal{F}\{f\}(\omega) \tag{1.20}$$

holds with ω^α defined by

$$\omega^\alpha := \omega_1^{\alpha_1}\omega_2^{\alpha_2}\dots\omega_n^{\alpha_n}.$$

If the signal f has certain symmetry properties, these properties propagate to its Fourier transform. Let $f : \mathbb{R}^2 \to \mathbb{R}^1$ be sufficiently regular. Suppose f is of the following structure:

$$f(\boldsymbol{x}) = \bar{f}(r), \quad r := \|\boldsymbol{x}\|.$$

Then the following relations are valid:

$$\hat{f}(|\omega|) = {}^2\mathcal{F}\{f\}(\omega) = \int_0^\infty \bar{f}(r) J_0(r|\omega|) r\, dr, \tag{1.21}$$

$$\bar{f}(r) = {}^2\mathcal{F}^{-1}\{\hat{f}(|\omega|)\} = \int_0^\infty \hat{f}(|\omega|) J_0(r|\omega|)|\omega|\, d|\omega|. \tag{1.22}$$

In these formulae, J_0 is the Bessel function of first kind and order zero.

The two formulae above indicate that for a so-called *isotropic* function, i.e. a function which is invariant with respect to rotations, the two-dimensional Fourier transform can be replaced by a one-dimensional integral transformation. This integral transformation is called Hankel transform.

Recalling the interpretation of the Fourier transform $\hat{f}$ of a signal f as its spectrum, one has to realize that signals having spectra with a non-compact support are pure mathematical fiction. Real signals are always recorded by some measuring device. No measuring device is capable to follow arbitrarily short-wavelength oscillations. Therefore every recorded signal f is somehow band-limited:

$$\hat{f}(\omega) = 0, \quad |\omega| > \nu.$$

Band-limited signals have very interesting properties:

Theorem 1.1.3 (Sampling theorem). *Let $f : \mathbb{R} \to \mathbb{R}$ be a one-dimensional band-limited signal,*

$$\hat{f}(\omega) = 0, \quad |\omega| > \nu.$$

Then f can be exactly reconstructed from its sampling values:

$$f(x) = \sum_{n\in\mathbb{Z}} f(x_n) \frac{\sin(\nu(x - x_n))}{\nu(x - x_n)}, \quad x_n = \frac{n\pi}{\nu}. \tag{1.23}$$

Proof. Since $\hat{f}$ is band-limited, it can be represented by a Fourier series

$$\hat{f}(\omega) = \sum_{n\in\mathbb{Z}} f_n e^{\imath n \frac{2\pi}{2\nu}\omega},$$

$$f_n = \frac{1}{2\nu} \int_{-\nu}^{\nu} \hat{f}(\omega) e^{-\imath n \frac{2\pi}{2\nu}\omega}\, d\omega.$$

Because $\hat{f}(\omega) = 0$, $|\omega| > \nu$, the following relation holds:

$$\begin{aligned} f_n &= \frac{1}{2\nu}\int_{-\infty}^{\infty} \hat{f}(\omega) e^{-\imath n \frac{2\pi}{2\nu}\omega}\, d\omega \\ &= \frac{\sqrt{2\pi}}{2\nu} f(-n\frac{\pi}{\nu}) \\ &= \frac{\sqrt{2\pi}}{2\nu} f(-x_n). \end{aligned}$$

Consequently, the spectrum $\hat{f}$ can be determined completely by the sampled values of the signal

$$\hat{f}(\omega) = \frac{\sqrt{2\pi}}{2\nu}\sum_{n\in\mathbb{Z}} f(-x_n) e^{\imath x_n \omega}.$$

Applying the inverse Fourier transform one obtains

$$\begin{aligned} f(x) &= \frac{1}{\sqrt{2\pi}}\int_{\mathbb{R}} \hat{f}(\omega) e^{\imath x\omega}\, d\omega \\ &= \frac{1}{\sqrt{2\pi}}\int_{-\nu}^{\nu} \hat{f}(\omega) e^{\imath x\omega}\, d\omega \\ &= \frac{1}{\sqrt{2\pi}}\frac{\sqrt{2\pi}}{2\nu}\sum_{n\in\mathbb{Z}} f(-x_n)\int_{-\nu}^{\nu} e^{\imath\omega(x+x_n)}\, d\omega \\ &= \sum_{n\in\mathbb{Z}} f(x_n)\frac{1}{2\nu}\int_{-\nu}^{\nu} e^{\imath\omega(x-x_n)}\, d\omega \\ &= \sum_{n\in\mathbb{Z}} f(x_n)\frac{1}{2\nu\imath(x-x_n)}\left(e^{\imath\nu(x-x_n)} - e^{-\imath\nu(x-x_n)}\right) \\ &= \sum_{n\in\mathbb{Z}} f(x_n)\frac{\sin(\nu(x-x_n))}{\nu(x-x_n)}. \end{aligned}$$

□

Sampling step-size Δx and limiting frequency ν are in a close relationship

$$\nu = \frac{\pi}{\Delta x}. \tag{1.24}$$

The limiting frequency ν is called *Nyquist frequency*. The wavelength λ belonging to the Nyquist frequency is

$$\lambda = \frac{2\pi}{\nu}. \tag{1.25}$$

Ergo the sampling step-size Δx has to be chosen accordingly

$$\Delta x = \frac{\pi}{\nu} = \frac{\lambda}{2}, \tag{1.26}$$

to be able to resolve the shortest wavelength occurring in the signal. In other words: It is impossible to resolve wavelengths shorter than $\lambda = 2\Delta x$ from data sampled at the sampling rate Δx.

In many cases the sampling rate is given by the measuring device. On the other hand, the signal f to be measured could contain wavelengths smaller than twice the sampling rate. If in this case the spectrum $\hat{f}$ is determined from the sampled values $f(x_n)$, the frequencies above the Nyquist frequency ν corrupt the frequencies below ν.

Theorem 1.1.4 (Aliasing). *Let f be band-limited $\hat{f}(\omega) = 0$, $|\omega| > \nu$. If f is undersampled by the sampling rate*

$$\Delta x = \frac{\pi}{\nu(1-\alpha)}, \quad 0 < \alpha < \frac{2}{3}, \tag{1.27}$$

the function

$$\bar{f}(x) := \sum_{n\in\mathbb{Z}} f(x_n)\frac{\sin(\nu(1-\alpha)(x-x_n))}{\nu(1-\alpha)(x-x_n)}, \quad x_n = \frac{n\pi}{\nu(1-\alpha)}, \tag{1.28}$$

reconstructed from the sampling values, has the spectrum

$$\hat{\bar{f}}(\omega) = \hat{f}(\omega) + \hat{f}(\omega + 2\nu(1-\alpha)) + \hat{f}(\omega - 2\nu(1-\alpha)). \tag{1.29}$$

Proof.

$$\begin{aligned}
&\bar{f}(x_n)\\
&= f(x_n) = \frac{1}{2\pi}\int_{-\nu}^{\nu} \hat{f}(\omega)e^{\imath\omega\frac{n\pi}{\nu(1-\alpha)}}\,d\omega\\
&= \frac{1}{2\pi}\int_{-\nu}^{-\nu(1-\alpha)} \hat{f}(\omega)e^{\imath\omega\frac{n\pi}{\nu(1-\alpha)}}\,d\omega + \frac{1}{2\pi}\int_{-\nu(1-\alpha)}^{\nu(1-\alpha)} \hat{f}(\omega)e^{\imath\omega\frac{n\pi}{\nu(1-\alpha)}}\,d\omega\\
&\quad + \frac{1}{2\pi}\int_{\nu(1-\alpha)}^{\nu} \hat{f}(\omega)e^{\imath\omega\frac{n\pi}{\nu(1-\alpha)}}\,d\omega\\
&= \frac{1}{2\pi}\int_{-3\nu(1-\alpha)}^{-\nu(1-\alpha)} \hat{f}(\omega)e^{\imath\omega\frac{n\pi}{\nu(1-\alpha)}}\,d\omega + \frac{1}{2\pi}\int_{-\nu(1-\alpha)}^{\nu(1-\alpha)} \hat{f}(\omega)e^{\imath\omega\frac{n\pi}{\nu(1-\alpha)}}\,d\omega\\
&\quad + \frac{1}{2\pi}\int_{\nu(1-\alpha)}^{3\nu(1-\alpha)} \hat{f}(\omega)e^{\imath\omega\frac{n\pi}{\nu(1-\alpha)}}\,d\omega\\
&= \frac{1}{2\pi}\int_{-\nu(1-\alpha)}^{\nu(1-\alpha)} \hat{f}(\omega-2\nu(1-\alpha))e^{\imath\omega\frac{n\pi}{\nu(1-\alpha)}}\,d\omega + \frac{1}{2\pi}\int_{-\nu(1-\alpha)}^{\nu(1-\alpha)} \hat{f}(\omega)e^{\imath\omega\frac{n\pi}{\nu(1-\alpha)}}\,d\omega\\
&\quad + \frac{1}{2\pi}\int_{-\nu(1-\alpha)}^{\nu(1-\alpha)} \hat{f}(\omega+2\nu(1-\alpha))e^{\imath\omega\frac{n\pi}{\nu(1-\alpha)}}\,d\omega\\
&= \frac{1}{2\pi}\int_{-\nu(1-\alpha)}^{\nu(1-\alpha)} [\hat{f}(\omega)+\hat{f}(\omega-2\nu(1-\alpha))+\hat{f}(\omega+2\nu(1-\alpha))]e^{\imath\omega\frac{n\pi}{\nu(1-\alpha)}}\,d\omega.
\end{aligned}$$

Since $\hat{\bar{f}}$ is band-limited it has the Fourier expansion

$$\hat{\bar{f}}(\omega) = \sum_{n\in\mathbb{Z}} \bar{f}_n e^{\imath n \frac{2\pi}{2\nu(1-\alpha)}\omega}, \quad \bar{f}_n = \frac{1}{2\nu(1-\alpha)} \int_{-\nu(1-\alpha)}^{\nu(1-\alpha)} \hat{\bar{f}}(\omega) e^{\imath\omega\frac{n\pi}{\nu(1-\alpha)}}\, d\omega.$$

As in the previous lemma it can be shown that

$$\begin{aligned}\bar{f}_n &= \frac{\sqrt{2\pi}}{2\nu(1-\alpha)} \bar{f}(x_n) \\ &= \int_{-\nu(1-\alpha)}^{\nu(1-\alpha)} \left[\hat{f}(\omega) + \hat{f}(\omega - 2\nu(1-\alpha)) + \hat{f}(\omega + 2\nu(1-\alpha))\right] e^{\imath\omega\frac{n\pi}{\nu(1-\alpha)}}\, d\omega\end{aligned}$$

holds. This means the coefficients $\bar{f}_n$ are simultaneously the *Fourier* coefficients of $\hat{\bar{f}}$ and of $\left[\hat{f}(\omega) + \hat{f}(\omega - 2\nu(1-\alpha)) + \hat{f}(\omega + 2\nu(1-\alpha))\right]$. Since there is a one-to-one relationship between a function and its Fourier coefficients, the relation

$$\hat{\bar{f}} = \left[\hat{f}(\omega) + \hat{f}(\omega - 2\nu(1-\alpha)) + \hat{f}(\omega + 2\nu(1-\alpha))\right]$$

follows. □

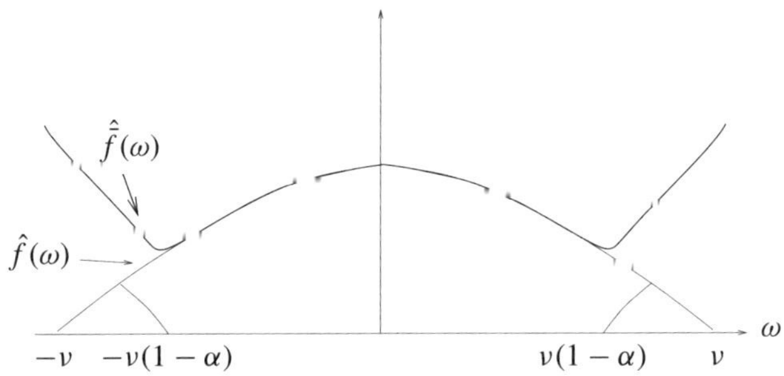

Figure 1.2. Aliasing

Remark. The theorem has several consequences.

First of all it is clear that due to the under-sampling the spectrum $\hat{f}(\omega)$ cannot be resolved up to the Nyquist frequency ν. As a consequence of the sampling theorem only a resolution up to $\bar{\nu} = \nu(1-\alpha)$ can be expected.

Additionally, one has to keep in mind that the sampling theorem can only be applied if the signal f is band-limited to the frequency-band $[-\nu(1-\alpha), \nu(1-\alpha)]$. But this is not the case. The signal is band-limited to the larger frequency band $[-\nu, \nu]$. Hence the resolved spectrum does not only contain the influence of $\hat{f}$ between $-\nu(1-\alpha)$ and $\nu(1-\alpha)$, but also the influence of $\hat{f}$ in the frequency bands $[-\nu, -\nu(1-\alpha))$

and $(\nu(1-\alpha), \nu]$. The theorem describes how these frequency-bands interact with the resolved spectrum: The frequencies $|\omega| > \nu(1-\alpha)$ outside the resolution limit are aliased by the frequencies $\nu(1-2\alpha) \leq |\omega| \leq \nu(1-\alpha)$. Only the small frequency band $|\omega| \leq \nu(1-2\alpha)$ can be resolved exactly. For this reason the theorem is frequently called *aliasing theorem.*

The discrete sampling is not the only difference between the theoretical *Fourier* spectrum and the spectrum computed from the data. The second difference is due to the finite sampling time $t \in [-T, T]$. Considering the finite sampling length, the question has to be discussed how the spectrum, computed from a finite sampling length, differs from the theoretical Fourier spectrum. Basically, there are two possibilities to continue a signal sampled in $[-T, T]$ to the whole $\mathbb{R}^1$:

1. continuation by zero values and
2. periodic continuation.

Lemma 1.1.4 (Zero padding).

$$\mathcal{F}\left\{f \cdot \Pi\left(\frac{\bullet}{T}\right)\right\}(\omega) = \frac{1}{\sqrt{2\pi}}\mathcal{F}\{f\} * (T\,\mathrm{sinc}(T\bullet))(\omega). \tag{1.30}$$

Proof.

$$\begin{aligned}
\mathcal{F}\left\{f \cdot \Pi\left(\frac{\bullet}{T}\right)\right\}(\omega) &= \mathcal{F}^{-1}\left\{f \cdot \Pi\left(\frac{\bullet}{T}\right)\right\}(-\omega) \\
&= \mathcal{F}^{-1}\left\{\mathcal{F}\{\mathcal{F}^{-1}\{f\}\} \cdot \mathcal{F}\left\{\mathcal{F}^{-1}\left\{\Pi\left(\frac{\bullet}{T}\right)\right\}\right\}\right\}(-\omega) \\
&= \frac{1}{\sqrt{2\pi}}\left(\mathcal{F}^{-1}\{f\} * \mathcal{F}^{-1}\left\{\Pi\left(\frac{\bullet}{T}\right)\right\}\right)(-\omega) \\
&= \frac{1}{\sqrt{2\pi}}\int_{\mathbb{R}} \mathcal{F}^{-1}\{f\}(-\omega - y)\mathcal{F}^{-1}\left\{\Pi\left(\frac{\bullet}{T}\right)\right\}(y)\,dy \\
&= \frac{1}{\sqrt{2\pi}}\int_{\mathbb{R}} \mathcal{F}\{f\}(\omega + y)\mathcal{F}\left\{\Pi\left(\frac{\bullet}{T}\right)\right\}(-y)\,dy \\
&= \frac{1}{\sqrt{2\pi}}\int_{\mathbb{R}} \mathcal{F}\{f\}(\omega - y)\mathcal{F}\left\{\Pi\left(\frac{\bullet}{T}\right)\right\}(y)\,dy \\
&= \frac{1}{\sqrt{2\pi}}\mathcal{F}\{f\} * (T\,\mathrm{sinc}(T\bullet))(\omega).
\end{aligned}$$

□

The lemma states that a continuation by zero values maintains the continuous nature of the spectrum but causes a blurring by the convolution with the sinc function (sine cardinal or sampling function). The bigger T, the smaller is this effect, which is frequently called *spectral leakage*.

The other possibility is the periodic continuation

$$\tilde{f}(x) = f(x - 2kT), \quad t \in [(2k-1)T, (2k+1)T]. \tag{1.31}$$

In this case the periodic continued function has a discrete line-spectrum. The spectral lines can be computed as Fourier coefficients

$$\tilde{f}(x) = \sum_{n\in\mathbb{Z}} c_n e^{\imath n \frac{\pi}{T} x}, \quad c_n = \frac{1}{2T}\int_{-T}^{T} \tilde{f}(x) e^{-\imath n \frac{\pi}{T} x}\, dx. \tag{1.32}$$

The question arises, how the Fourier coefficients c_n can be related to the Fourier spectrum $\hat{f}(\omega)$.

Lemma 1.1.5 (Periodic continuation). *Let $\tilde{f}$ be the periodic continuation of f corresponding to* (1.31). *Then*

$$c_n = \frac{1}{2}\left(\hat{f} * \operatorname{sinc}(T\bullet)\right)\left(n\frac{\pi}{T}\right) \tag{1.33}$$

holds.

Proof.

$$\begin{aligned} c_n &= \frac{1}{2T}\int_{-T}^{T} \tilde{f}(x) e^{-\imath n \frac{\pi}{T} x}\, dx = \frac{\sqrt{2\pi}}{2T}\frac{1}{\sqrt{2\pi}}\int_{-\infty}^{\infty} \tilde{f}(x) \Pi\left(\frac{x}{T}\right) e^{-\imath n \frac{\pi}{T} x}\, dx \\ &= \frac{\sqrt{2\pi}}{2T}\mathcal{F}\left\{f\cdot\Pi\left(\frac{\bullet}{T}\right)\right\}\left(n\frac{\pi}{T}\right) = \frac{1}{2}\left(\hat{f} * \operatorname{sinc}(T\bullet)\right)\left(n\frac{\pi}{T}\right). \end{aligned}$$

□

The Fourier coefficients of the periodically continued function $\tilde{f}$ are the values of the spectrum $\hat{f}$ of f, sampled at the discrete places $n\frac{\pi}{T}$ and blurred by the finite sampling length.

The next question which has to be dealt with is the numerical computation of the Fourier spectrum of a sampled signal. Since there are only a finite number of sampled values, only a finite number of spectral values can be computed. Hence the numerically computed spectrum is implicitly the spectrum of the periodic continuation. Consequently, the integrals

$$c_n = \frac{1}{2T}\int_{-T}^{T} \tilde{f}(x) e^{-\imath n \frac{\pi}{T} x}\, dx \tag{1.34}$$

have to be approximated by some numerical quadrature formula. The simplest quadrature formula one could think of is the *trapezian rule*:

$$\begin{aligned} c_n &= \frac{1}{2T}\int_{-T}^{T} \tilde{f}(x) e^{-\imath n \frac{\pi}{T} x}\, dx \\ &\approx \frac{1}{2T}\frac{\Delta x}{2}\left(f(-T)e^{\imath n\pi} + 2\sum_{i=-N+1}^{N-1} f(i\Delta x) e^{-\imath n \frac{\pi}{T} i \Delta x} + f(T) e^{-\imath n \pi}\right), \\ \Delta x &= \frac{T}{N}, \quad n = -N, \ldots, N-1, \end{aligned}$$

which approximates the integral by the area of a polygon. Due to the periodic continuation $f(T) = f(-T)$ holds. This leads to the following simplification

$$c_n \approx \frac{1}{2N}\frac{1}{2}\Big(f(-T)e^{-\iota n\pi} + 2\sum_{i=-N+1}^{N-1} f(i\Delta x)e^{-\iota n\frac{\pi}{T}i\Delta x} + f(-T)e^{\iota n\pi}\Big)$$
$$= \frac{1}{2N}\sum_{i=-N}^{N-1} f(i\Delta x)e^{-\iota ni\frac{\pi}{N}}.$$

Definition 1.1.3. The coefficients

$$\bar{c}_n := \frac{1}{2N}\sum_{j=-N}^{N-1} f(j\Delta x)e^{-\iota nj\frac{\pi}{N}} \tag{1.35}$$

are called the *discrete* Fourier transform (DFT) of the function f sampled at the places $j\Delta x,\ j = -N, \dots, N-1$.

In the general case the DFT coefficients $\bar{c}_n$ are more or less accurate approximations of the Fourier coefficients c_n of the periodic continuation of f. Nevertheless, if f is band-limited the DFT even gives the exact Fourier-coefficients.

Lemma 1.1.6. *Let f be periodic on $[-T, T]$. Furthermore, let f be band-limited, i.e., let $c_n = 0$, $|n| > N$ hold for the Fourier coefficients*

$$c_n := \frac{1}{2T}\int_{-T}^{T} f(t)e^{-\iota\frac{n\pi}{T}t}\,dt. \tag{1.36}$$

Finally, let $\{\bar{c}_n\}$ be the DFT of f. Then

$$c_n = \bar{c}_n \tag{1.37}$$

holds.

Proof.

$$\bar{c}_n = \frac{1}{2N}\sum_{j=-N}^{N-1} f(j\Delta x)e^{-\iota j\frac{n\pi}{N}}$$
$$= \frac{1}{2N}\sum_{j=-N}^{N-1}\Big(\sum_{k=-N}^{N-1} c_k e^{\iota k\frac{j\pi}{T}\Delta x}\Big)e^{-\iota j\frac{n\pi}{N}}$$
$$= \frac{1}{2N}\sum_{j=-N}^{N-1}\Big(\sum_{k=-N}^{N-1} c_k e^{\iota kj\frac{\pi}{N}}\Big)e^{-\iota jn\frac{\pi}{N}} =$$
$$= \frac{1}{2N}\sum_{k=-N}^{N-1} c_k \sum_{j=-N}^{N-1} e^{\iota j\frac{\pi}{N}(k-n)} = \sum_{k=-N}^{N-1} c_k\delta_{kn} = c_n.$$

□

As a preliminary summary of the results obtained so far, it can be said that

- the DFT gives the exact Fourier coefficients c_n, if the signal f is periodic and band-limited,
- these Fourier coefficients are the sampling values of the *Fourier* spectrum $c_n = \hat{f}(n\frac{\pi}{T})$ of the periodic continuation of a band-limited non-periodic signal and that
- these coefficients are the sampling values of the aliased spectrum of an arbitrary but sufficiently regular signal.

This means that the result of the DFT is a reasonable approximation of the spectrum if both the sampling length T and the sampling step-size Δx are properly chosen.

The most important disadvantage of the DFT is its computational inefficiency. Suppose that there are $2N$ sampling values. Then it needs $O(4N^2)$ operations to compute the $2N$ spectral values. If the periodicity of the factors $e^{-\imath jn\frac{\pi}{N}}$ is exploited, the computation sequence of the DFT can be rearranged in such a way that only $O(4N(\mathrm{ld}(N)+1))$ operations are necessary.

This rearrangement of the computation sequence is called *Fast Fourier Transform* (FFT) and it is only due to the dramatic reduction of the computational load by the FFT that spectral methods found their way into many applications.

The basic idea which leads to the FFT is to compute the DFT separately for the even and for the odd sampling points and to combine the results: In this way the DFT can be rewritten as

$$\begin{aligned} c_n &= \frac{1}{2N}\sum_{j=0}^{2N-1} f(j\Delta x)e^{-\imath jn\frac{\pi}{N}} \\ &= \frac{1}{2N}\sum_{j=0}^{N-1} f(2j\Delta x)e^{-\imath 2jn\frac{\pi}{N}} + \frac{1}{2N}\sum_{j=0}^{N-1} f((2j+1)\Delta x)e^{-\imath(2j+1)n\frac{\pi}{N}} \\ &= \frac{1}{2N}\sum_{j=0}^{N-1} f(2j\Delta x)\big(e^{-\imath jn\frac{\pi}{N}}\big)^2 + e^{-\imath n\frac{\pi}{N}}\frac{1}{2N}\sum_{j=0}^{N-1} f((2j+1)\Delta x)\big(e^{-\imath jn\frac{\pi}{N}}\big)^2. \end{aligned} \tag{1.38}$$

Now two cases have to be distinguished.

a) $n < N$:

$$\begin{aligned} c_n = \frac{1}{2N}\sum_{j=0}^{N-1} f(2j\Delta x)\big(e^{-\imath jn\frac{\pi}{N}}\big)^2 \\ + e^{-\imath n\frac{\pi}{N}}\frac{1}{2N}\sum_{j=0}^{N-1} f((2j+1)\Delta x)\big(e^{-\imath jn\frac{\pi}{N}}\big)^2. \end{aligned} \tag{1.39}$$

b) $n = N + k \geq N$:

$$\begin{aligned} c_{N+k} &= \frac{1}{2N} \sum_{j=0}^{N-1} f(2j\Delta x)\big(e^{-\imath jk\frac{\pi}{N}}\big)^2\big(e^{-\imath j\pi}\big)^2 \\ &\quad - e^{-\imath k\frac{\pi}{N}} \frac{1}{2N} \sum_{j=0}^{N-1} f((2j+1)\Delta x)\big(e^{-\imath jk\frac{\pi}{N}}\big)^2\big(e^{-\imath j\pi}\big)^2 \\ &= \frac{1}{2N} \sum_{j=0}^{N-1} f(2j\Delta x)\big(e^{-\imath jk\frac{\pi}{N}}\big)^2 - e^{-\imath k\frac{\pi}{N}} \frac{1}{2N} \sum_{j=0}^{N-1} f((2j+1)\Delta x)\big(e^{-\imath jk\frac{\pi}{N}}\big)^2. \end{aligned} \tag{1.40}$$

If now the coefficients c_n^0, c_n^1 are defined by

$$c_n^0 = \frac{1}{2N} \sum_{j=0}^{N-1} f(2j\Delta x)\big(e^{-\imath jk\frac{\pi}{N}}\big)^2 \tag{1.41}$$

$$c_n^1 = \frac{1}{2N} \sum_{j=0}^{N-1} f((2j+1)\Delta x)\big(e^{-\imath jk\frac{\pi}{N}}\big)^2, \tag{1.42}$$

the DFT reads

$$c_n = c_n^0 + \big(e^{-\imath n\frac{\pi}{N}}\big)c_n^1, \quad n = 0, \dots, N-1 \tag{1.43}$$

$$c_{N+k} = c_k^0 - \big(e^{-\imath k\frac{\pi}{N}}\big)c_k^1, \quad k = 0, \dots, N-1. \tag{1.44}$$

The computation of the c_n^0 and the c_n^1 coefficients is basically a DFT for the even and for the odd sampling points. Therefore, for this rearrangement of the computation sequence, $2O(N^2)$ instead of $4O(N^2)$ operations are necessary. This reduces the computational load by 50 %.

This idea can be repeated by the rearrangement of the computation of the c_n^0 and c_n^1 and this iterated rearrangement leads to the FFT with $O(4N\,\mathrm{ld}(N))$ necessary operations.

Finally it should be mentioned once again that the results of the FFT completely coincide with the results of the DFT. But the computational efficiency of the FFT is much higher.

One of the classical textbooks about Fourier transform and its applications is [9]. Since 1984 FFT methods have been applied to geoid computation and terrain reduction. Representative for the many publications about this item we only mention the papers [94], [96], [98] and [99].

1.2 Linear filters

A function $f \in L_2(\mathbb{R})$ is uniquely determined by its Fourier spectrum. Frequently it is useful to split the signal f into several frequency channels:

$$\hat{f}(\omega) = \sum_{j=1}^{n} \hat{f}_j(\omega), \tag{1.45}$$

$$f(x) = \sum_{j=0}^{n} f_j(x). \tag{1.46}$$

If these channels are orthogonal, i.e. if

$$\int_{\mathbb{R}} f_i(x)\overline{f_j}\,dx = \int_{\mathbb{R}} \hat{f}_i(\omega)\overline{\hat{f}_j}(\omega)\,d\omega = \delta_{ij} \tag{1.47}$$

holds, the frequency decomposition is called *orthogonal.*

The simplest way to construct an orthogonal frequency decomposition would be the splitting of the frequency domain into n non-overlapping frequency bins, and $\hat{f}_j$ could be defined as the restriction of $\hat{f}$ to the j-th bin. Since every channel becomes an ideal bandpass such a frequency decomposition is practically not realizable. Therefore the question arises how to construct realizable orthogonal frequency decompositions. This question leads to the theory of filter banks.

In filter theory, instead of the signal $f \in L_2(\mathbb{R})$ itself the series of its values $\{f(n)\} \in l_2(\mathbb{Z})$, sampled at the integers is considered. Please keep in mind that for a band limited signal the sequence of its sampling values is equivalent to the continuous signal (sampling theorem).

Definition 1.2.1. A *digital convolution filter* H is a mapping

$$\begin{aligned} H\colon l_2(\mathbb{Z}) &\to l_2(\mathbb{Z}) \\ x &\mapsto \Big\{(Hx)_n := \sum_{k\in\mathbb{Z}} h_k x_{n-k}\Big\}. \end{aligned} \tag{1.48}$$

Many properties of a digital convolution filter can be described by the z-transform of the sequence $\{h_k\}$ of its filter coefficients.

Definition 1.2.2. The *z-transform* of the sequence of the filter coefficients of a convolution filter H

$$H(z) := \sum_{k\in\mathbb{Z}} h_k z^{-k} \tag{1.49}$$

is called its *transfer function.*

Remark. The transfer function H describes how the corresponding filter amplifies or attenuates a certain frequency $e^{\imath 2\pi\omega x}$: Let f be the signal $e^{\imath 2\pi\omega x}$ sampled at the integers, i.e.

$$f(n) = e^{\imath 2\pi\omega n}. \tag{1.50}$$

The filtered signal is the sequence

$$g(n) = \sum_{k\in\mathbb{Z}} h_k e^{\imath(n-k)2\pi\omega} = e^{\imath n 2\pi\omega} \sum_{k\in\mathbb{Z}} h_k e^{-\imath k 2\pi\omega} = H(e^{\imath 2\pi\omega}) f(n). \tag{1.51}$$

Therefore the result of the filtering of a periodic signal with the frequency ω is again a periodic signal of the same frequency ω, but with the amplitude multiplied by the factor $H(e^{\imath 2\pi\omega})$. Since every signal can be represented by a superposition of periodic signals, the transfer function describes the attenuation or amplification of each frequency, contained in the signal.

The most important property of the z-transform is that it maps the convolution of two signals into the product of their z-transforms:

Theorem 1.2.1 (Convolution theorem). *Let $u, v \in l_2(\mathbb{Z})$ be two signals with the z-transforms*

$$U(z) = \sum_{k\in\mathbb{Z}} u_k z^{-k}, \quad V(z) = \sum_{k\in\mathbb{Z}} v_k z^{-k}. \tag{1.52}$$

The convolution $w = u * v$ defined by

$$w_n := \sum_{k\in\mathbb{Z}} u_k v_{n-k} \tag{1.53}$$

has the z-transform

$$W(z) = U(z) \cdot V(z). \tag{1.54}$$

Proof.

$$\begin{aligned}
W(z) &= \sum_{n\in\mathbb{Z}} w_n z^{-n} \\
&= \sum_{n\in\mathbb{Z}} z^{-n} \sum_{k\in\mathbb{Z}} u_k v_{n-k} \\
&= \sum_{k\in\mathbb{Z}} u_k \sum_{n\in\mathbb{Z}} v_{n-k} z^{-n} \\
&= \sum_{k\in\mathbb{Z}} u_k \sum_{l\in\mathbb{Z}} v_l z^{-(k+l)} \\
&= \sum_{k\in\mathbb{Z}} u_k z^{-k} \sum_{l\in\mathbb{Z}} v_l z^{-l} \\
&= U(z) \cdot V(z).
\end{aligned}$$

□

A filter bank is an array of filters, which are used in order to decompose a given signal $\{x(n)\}$ into several frequency channels and to reconstruct the signal from the individual channel information. In its simplest version it is a two-channel filter bank:

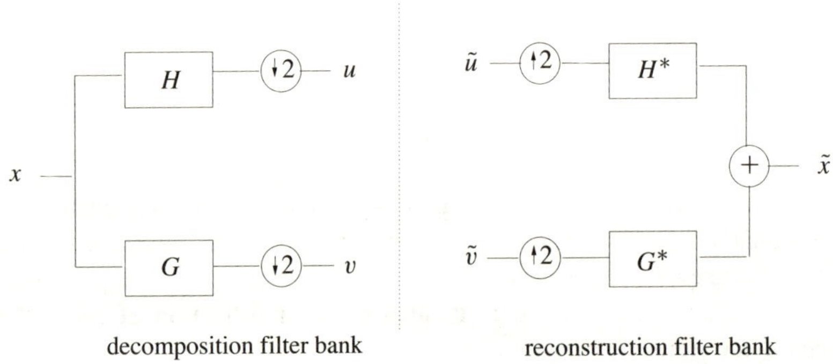

Figure 1.3. Two channel filter bank

In the analysis filter bank the signal passes a high-pass filter G and a low-pass filter H. After passing these filters it will be down-sampled by the factor 2.

$$\ldots, x(-2), x(-1), x(0), x(1), x(2), \cdots \xrightarrow{\downarrow 2} \ldots, x(-2), x(0), x(2), \ldots \tag{1.55}$$

In order to guarantee that the reconstructed signal has the same number of samples as the original signal, the signals in the individual channels have to be up-sampled by the factor 2 before passing through the reconstruction filters H^*, G^*.

$$\ldots, x(-2), x(0), x(2), \cdots \xrightarrow{\uparrow 2} \ldots, x(-2), 0, x(0), 0, x(2), \ldots \tag{1.56}$$

Now the question arises under which conditions the reconstructed signal $\tilde{x}$ coincides with the original signal.

Lemma 1.2.1. *Let H, G, H^*, G^* be the z-transforms of the decomposition and reconstruction filters H, G, H^*, G^*. If the reconstructed signal coincides with the original signal*

$$x = \tilde{x},$$

then the following conditions are fulfilled:

$$H(-z)H^*(z) + G(-z)G^*(z) = 0, \tag{1.57}$$

$$H(z)H^*(z) + G(z)G^*(z) = 2. \tag{1.58}$$

Proof. Consider a signal y, which is obtained from a signal x, first by down-sampling and then by up-sampling it:

$$\ldots, y(-2), y(-1), y(0), y(1), y(2), \cdots = \cdots, x(-2), 0, x(0), 0, x(2), \ldots$$

Hence for the z-transform of the signal y we have

$$\begin{aligned}Y(z) &= \sum_{k\in\mathbb{Z}} y_k z^{-k} = \sum_{k\in\mathbb{Z}} x_{2k} z^{-2k}\\ &= \sum_{k\in\mathbb{Z}} \frac{1}{2}(x(k)z^{-k} + x(k)(-z)^{-k})\\ &= \frac{1}{2}(X(z) + X(-z)).\end{aligned}$$

Consequently, the z-transform of the reconstructed signal $\tilde{x}$ is

$$\begin{aligned}\tilde{X}(z) &= H^*(z)\cdot\frac{1}{2}(H(z)X(z) + H(-z)X(-z))\\ &\quad + G^*(z)\cdot\frac{1}{2}(G(z)X(z) + G(-z)X(-z))\\ &= \frac{1}{2}((H(z)H^*(z) + G(z)G^*(z))X(z)\\ &\quad + (H(-z)H^*(z) + G(-z)G^*(z))X(-z)).\end{aligned}$$

From the condition $\tilde{X}(z) = X(z)$ follows

$$\begin{aligned}H(-z)H^*(z) + G(-z)G^*(z) &= 0,\\ H(z)H^*(z) + G(z)G^*(z) &= 2.\end{aligned}$$

□

The conditions (1.57) and (1.58) are only necessary conditions. Condition (1.57) is already fulfilled for the special choice

$$G(-z) := z^l H^*(z), \quad G^*(z) := -z^{-l} H(-z), \quad l \in \mathbb{Z}. \tag{1.59}$$

The remaining condition (1.58) leads to

$$H(z)H^*(z) - (-1)^l H(-z)H^*(-z) = 2. \tag{1.60}$$

Consequently, the problem to construct an exact reconstructing filter bank is reduced to the problem to determine a *Laurent* polynomial, which satisfies (1.60). For the product M of the two filters H and H^*

$$M(z) := H(z)H^*(z)$$

the condition (1.58) leads to

$$2 = M(z) - (-1)^l M(-z) = \sum_{k\in\mathbb{Z}} m_k(z^{-k} - (-1)^l(-z)^{-k}). \tag{1.61}$$

Obviously, this condition can only be fulfilled for an odd l. Therefore one has to find a filter M with the property

$$2 = M(z) + M(-z) \tag{1.62}$$

and to decompose this filter afterwards according to (1.2). From the resulting low-pass filter H the corresponding high-pass filter can be derived using (1.59).

Example 1.2.1. In order to illustrate the construction ideas for a filter bank, here a small example is given.

Starting from condition (1.62) for the coefficients of the filter M follows:

$$2 = \sum_{k\in\mathbb{Z}} m_k(z^{-k} + (-z)^{-k}) = \sum_{k\in\mathbb{Z}} m_k(1 + (-1)^k)z^{-k}.$$

Consequently, $m_0 = 1$ and $m_{2k} = 0$, $k \neq 0$ holds. The shortest possible filter which fulfills (1.62) has the filter coefficients $m_0 = 1$ and $m_{-1}, m_1 \neq 0$. Hence

$$M(z) = (m_{-1}z + 1 + m_1 z^{-1})$$

must hold. According to (1.2) the relation

$$(m_{-1}z + 1 + m_1 z^{-1}) = M(z) = H(z)H^*(z)$$

has to be valid. Obviously

$$\begin{aligned} M(z) &= m_{-1}z\left(\frac{m_1}{m_{-1}}z^{-2} + \frac{1}{m_{-1}}z^{-1} + 1\right) \\ &= m_{-1}z\left[\left(\sqrt{\frac{m_1}{m_{-1}}}z^{-1} + \frac{1}{2\sqrt{m_1 m_{-1}}}\right)^2 + \left(1 - \frac{1}{4m_1 m_{-1}}\right)\right] \end{aligned}$$

is valid. Hence the filter M is of the product form (1.2), if $m_1 m_{-1} = \frac{1}{4}$ holds. The coefficient m_{-1} can be chosen arbitrarily. For the sake of simplicity $m_{-1} = m_1 = 1/2$ is set. This yields

$$M(z) = \frac{1}{2}(z + 2 + z^{-1}) = \frac{1}{2}(1 + z^{-1})^2 \cdot z.$$

Immediately one can conclude

$$H^*(z) = \frac{1}{\sqrt{2}}(1 + z^{-1}), \quad H(z) = \frac{1}{\sqrt{2}}(1 + z).$$

Using (1.59) this leads to

$$G(z) = \frac{1}{\sqrt{2}}(1 - z), \quad G^*(z) = \frac{1}{\sqrt{2}}(1 - z^{-1}).$$

Finally, the analysis filters

$$(Hx)_n := \frac{1}{\sqrt{2}}x_n + \frac{1}{\sqrt{2}}x_{n-1}, \quad (Gx)_n := \frac{1}{\sqrt{2}}x_n - \frac{1}{\sqrt{2}}x_{n-1}$$

and the reconstruction filters

$$(H^*x)_n := \frac{1}{\sqrt{2}}x_n + \frac{1}{\sqrt{2}}x_{n+1}, \quad (G^*x)_n := \frac{1}{\sqrt{2}}x_n - \frac{1}{\sqrt{2}}x_{n+1}.$$

are obtained.

The transfer functions of the low-pass filter H and of the high-pass filter G are given by

$$\begin{aligned}|H(e^{\imath 2\pi\omega})| &= \frac{1}{\sqrt{2}}|1+e^{\imath 2\pi\omega}| = \frac{1}{\sqrt{2}}(1+e^{\imath 2\pi\omega}+e^{-\imath 2\pi\omega})^{1/2} \\ &= \frac{1}{\sqrt{2}}(2+2\cos(2\pi\omega))^{1/2}\end{aligned}$$

and

$$\begin{aligned}|G(e^{\imath 2\pi\omega})| &= \frac{1}{\sqrt{2}}|1-e^{\imath 2\pi\omega}| = \frac{1}{\sqrt{2}}(2-e^{\imath 2\pi\omega}-e^{-\imath 2\pi\omega})^{1/2} \\ &= \frac{1}{\sqrt{2}}(2-2\cos(2\pi\omega))^{1/2}.\end{aligned}$$

For ω between minus and plus the *Nyquist frequency* $\nu = 1/2$, the transfer functions of both the high- and the low-pass filter, are displayed in Figure 1.4.

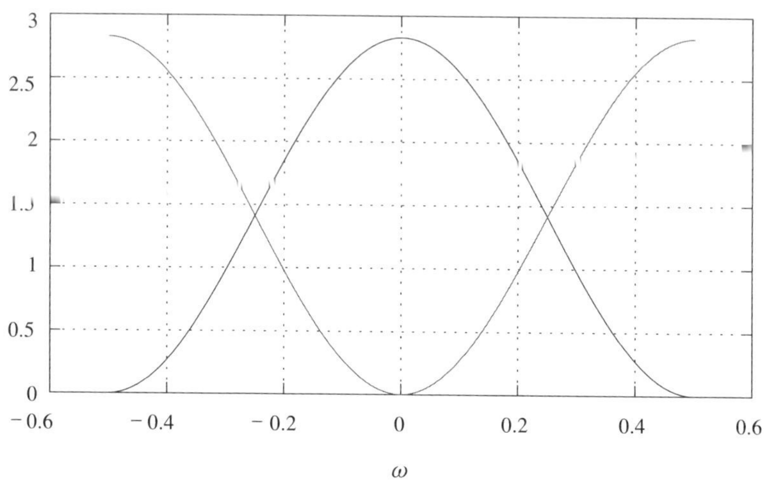

Figure 1.4. Transfer functions of high- and low-pass filter

It is clearly visible that these two simple filters do not separate the high- and the low-frequency channel very well.

Now we demonstrate for a simple example the effect of the decomposing and of the reconstructing filter bank. The following test signal will be used:

$$x(n) = e^{\imath \frac{n\pi}{2}}. \tag{1.63}$$

The results of the high- and low-pass filtering of the signal followed by a down-sampling are the signals u and v:

$$u_n = (Hx)_{2n} = \frac{1}{\sqrt{2}}\left(e^{\imath \frac{2n\pi}{2}} + e^{\imath \frac{(2n-1)\pi}{2}}\right) = \frac{1}{\sqrt{2}} e^{\imath n\pi}(1-\imath) \tag{1.64}$$

$$v_n = (Gx)_{2n} = \frac{1}{\sqrt{2}}\left(e^{\imath \frac{2n\pi}{2}} - e^{\imath \frac{(2n-1)\pi}{2}}\right) = \frac{1}{\sqrt{2}} e^{\imath n\pi}(1+\imath). \tag{1.65}$$

Let $\tilde{u}$ and $\tilde{v}$ denote the up-sampled versions of the signals u, v. Then

$$\tilde{u}_n = \begin{cases} u_{\frac{n}{2}}, & n \text{ even} \\ 0, & n \text{ odd}, \end{cases} \qquad \tilde{v}_n = \begin{cases} v_{\frac{n}{2}}, & n \text{ even} \\ 0, & n \text{ odd} \end{cases} \tag{1.66}$$

holds. Consequently

$$\begin{aligned}(H^*\tilde{u})_n + (G^*\tilde{v})_n &= \begin{cases} \frac{1}{\sqrt{2}}(\tilde{u}_{\frac{n}{2}} + \tilde{v}_{\frac{n}{2}}) = \frac{1}{2} e^{\imath \frac{n\pi}{2}} \cdot 2, & n \text{ even} \\ \frac{1}{\sqrt{2}}(\tilde{u}_{\frac{n+1}{2}} - \tilde{v}_{\frac{n+1}{2}}) = \frac{1}{2} e^{\imath \frac{n\pi}{2}} \cdot 2, & n \text{ odd} \end{cases} \\ &= x(n)\end{aligned} \tag{1.67}$$

is valid. Indeed, the reconstruction filters recover the previously decomposed signal exactly.

At the beginning of this section it has been already mentioned that the orthogonality of different frequency channels

$$\int_{\mathbb{R}} f_i(x) \bar{f}_j \, dx = \int_{\mathbb{R}} \hat{f}_i(\omega) \bar{\hat{f}}_j(\omega) \, d\omega = \delta_{ij} \tag{1.68}$$

is a desirable property of a filter bank. Now the question will be discussed, how the orthogonality of the filtered signal components reflects itself in the transfer functions of the filters.

A particular signal is the so-called unit pulse

$$\delta_n := \begin{cases} 1, & n = 0 \\ 0, & n \neq 0. \end{cases} \tag{1.69}$$

If this unit pulse is filtered by a filter H, the filter output are exactly the coefficients of the filter H:

$$(H\delta)_n = \sum_{k\in\mathbb{Z}} h_k \delta_{n-k} = h_k. \tag{1.70}$$

Therefore the sequence of filter coefficients is the so-called *impulse response* of the filter.

Definition 1.2.3. If the impulse response of a filter H contains only a finite number of non-vanishing filter coefficients h_n, the filter is called *finite impulse response filter* (FIR).

For an FIR filter the condition of perfect reconstruction of a filter bank can be simplified.

Theorem 1.2.2. *A perfect reconstruction filter* (*PR*) *satisfies*

$$H(z)H^*(z) + H(-z)H^*(-z) = 2. \tag{1.71}$$

For FIR filters, there exist numbers $a \in \mathbb{R}$ and $l \in \mathbb{Z}$, such that

$$G(z) = az^{-(2l+1)}H^*(-z), \quad G^*(z) = a^{-1}z^{-(2l+1)}H(-z) \tag{1.72}$$

Proof. The conditions (1.57) and (1.58) are a system of linear equations for the determination of H^*, G^* from H, G. Its solution is

$$\begin{bmatrix} H^*(z) \\ G^*(z) \end{bmatrix} = \frac{2}{D(z)} \begin{bmatrix} -G(-z) \\ H(-z) \end{bmatrix}$$

with the determinant

$$D(z) = H(-z)G(z) - H(z)G(-z).$$

Hence

$$G(z)G^*(z) = -\frac{D(-z)}{2}H^*(-z) \cdot \frac{2}{D(z)}H(-z) = -\frac{D(-z)}{D(z)}H^*(-z)H(-z)$$

is valid. Since

$$D(-z) = H(z)G(-z) - H(-z)G(z) = -D(z)$$

holds, by inserting this in equation (1.58) the condition

$$H(z)H^*(z) + H(-z)H^*(-z) = 2$$

follows.

If H is an FIR filter both $D(z)$ and $D^{-1}(z)$ must be a finite sum. This is only possible if both sums contain only one element $-2az^k$ and $-\frac{1}{2}a^{-1}z^{-k}$ respectively. Since $D(-z) = -D(z)$, $k = 2l + 1$ must hold. This proves

$$G(z) = az^{-(2l+1)}H^*(-z), \quad G^*(z) = a^{-1}z^{-(2l+1)}H(-z). \qquad \square$$

A filter bank with $H(z) = H^*(z)$ and $G(z) = G^*(z)$ is called a *conjugate mirror filter*. For a conjugate mirror filter condition (1.71) simplifies to

$$|H(z)|^2 + |H(-z)|^2 = 2. \tag{1.73}$$

Definition 1.2.4. A filter bank H, G is called *quadrature mirror filter* (QMF) if for each signal $x \in l_2(\mathbb{Z})$ the relation

$$\|Hx\|^2 + \|Gx\|^2 = \|x\|^2 \tag{1.74}$$

holds.

Interpreting condition (1.74), the filter bank splits the signal into two orthogonal components. There is no signal component which occurs both in the high-frequency *and* in the low-frequency channel. Indeed, in the case of orthogonal channels the following equation is true:

$$\begin{aligned} \|x\|^2 &= \langle Hx + Gx, Hx + Gx\rangle \\ &= \langle Hx, Hx\rangle + 2\langle Hx, Gx\rangle + \langle Gx, Gx\rangle \\ &= \|Hx\|^2 + \|Gx\|^2. \end{aligned} \tag{1.75}$$

We now discuss the question under which circumstances an FIR filter becomes a QMF filter.

Theorem 1.2.3. *A perfect reconstruction conjugate mirror filter is a QMF filter.*

Proof. For a PR filter the decomposition and the reconstruction filters are in the following relationship:

$$\begin{bmatrix} H^*(z) \\ G^*(z) \end{bmatrix} = \frac{2}{D(z)} \begin{bmatrix} -G(-z) \\ H(-z) \end{bmatrix}$$

with the determinant

$$D(z) = H(-z)G(z) - H(z)G(-z).$$

Inserting this into equation (1.58) the following equation is obtained:

$$0 = \frac{2}{D(z)}[-H(-z)G(-z) - H^*(z)G^*(z)].$$

For a conjugate mirror filter this relation simplifies to

$$H(-z)G(-z) + H(z)G(z) = 0.$$

Now let $u := Hx$, $v := Gx$ be the high- and the low-pass filtered components of a signal $x \in l_2(\mathbb{Z})$ and let U, V be their z-transforms. Using the convolution theorem for the z-transform, the scalar product between u and v can be expressed as follows:

$$\begin{aligned}\langle u, v\rangle &= (u * v)_0 = U(0) \cdot V(0)\\ &= H(0) \cdot G(0) \cdot (X(0))^2\\ &= \frac{1}{2}(H(-0)G(-0) + H(0)G(0)) \cdot (X(0))^2\\ &= 0\end{aligned}$$

Hence the filter bank is a QMF filter. □

2 Wavelets

2.1 Motivation

The theory and application of wavelets is a comparatively young branch in digital signal processing. It was developed in order to overcome some deficiencies of classical Fourier analysis. One of the disadvantages of Fourier analysis is the lack of time-localization. What has to be understood by *time-localization* will be shown by the following example.

Consider the following signal s:

$$s(t) = \begin{cases} \sin(440\pi t), & 0 \le t < 0.5 \\ 0, & 0.5 \le t < 1.0 \\ \sin(440\pi t), & 1.0 \le t < 1.5 \\ 0, & 1.5 \le t < 2.0 \\ \sin(440\pi t), & 2.0 \le t < 2.5 \\ 0, & 2.5 \le t < 3.0 \\ \sin(220\pi t), & 3.0 \le t < 4.0. \end{cases} \tag{2.1}$$

The graph of this signal is shown in Figure 2.1.

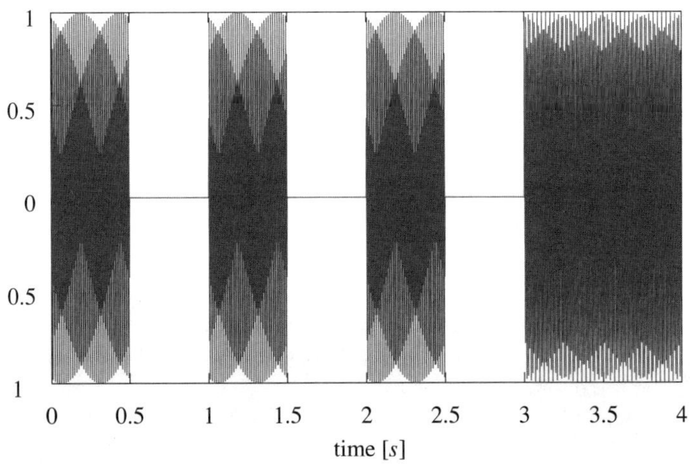

Figure 2.1. Four isolated notes

Obviously, the signal consists of four parts: Three tones of the frequency 220 Hz with a short duration and one tone of frequency 110 Hz with a long duration. They are separated from each other by a short break. The classical Fourier spectrum of this signal is displayed in the Figure 2.2.

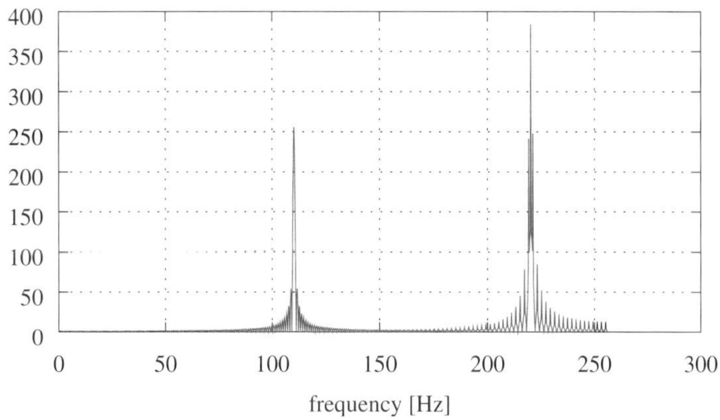

Figure 2.2. Fourier spectrum of the signal

The spectrum clearly shows the two frequencies occurring in the signal. But it does not contain any information on the time of occurrence of the individual frequencies. This is because the Fourier analysis always includes the whole time span of the signal. It would be much better to have a time-*and*-frequency analysis which resembles the notation in music: Not only the frequency of a tone but also the time span of its occurrence must be visible. Figure 2.3 shows the music notation of the signal under consideration.

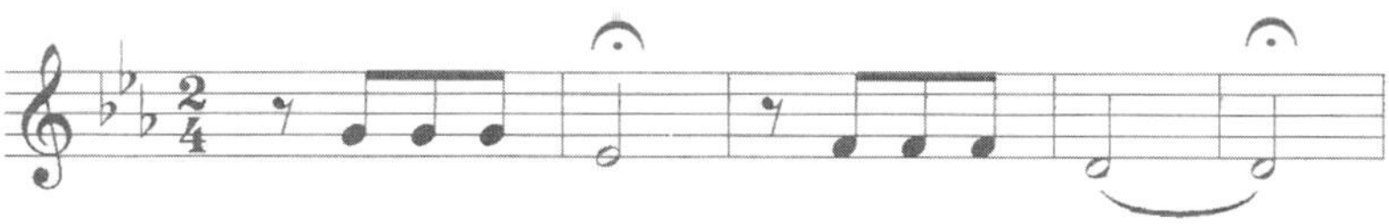

Figure 2.3. Music notation of the signal

In order to recover the time information, the windowed or short-time Fourier transform is used. The essence of the windowed Fourier transform is the masking of the signal by a moving window $g(t)$:

$$\mathcal{G}\{f\}(\omega, t) := \frac{1}{\sqrt{2\pi}} \int_{\infty}^{\infty} f(u)g(u-t)e^{-\imath u\omega}\, du \tag{2.2}$$

or in a different notation

$$\mathcal{G}\{f\}(\omega, t) = \frac{1}{\sqrt{2\pi}} \int_{-\infty}^{\infty} f(u) g_{\omega,t}(u)\, du, \quad g_{\omega,t}(u) := g(u-t)e^{-\imath\omega u}. \tag{2.3}$$

The window function g is usually a real, even function with the maximum concentration of energy in the low frequency band. A frequently used window function is

$$g(t) := \pi^{-1/4} e^{-\frac{t^2}{2}}. \tag{2.4}$$

For this choice of the window function the windowed Fourier transform is also called Gabor transform.

The short-time Fourier transform has some interesting properties. Similar to the usual Fourier transform the total energy of the signal equals the total energy of its short-time spectrum (Parseval's identity):

$$\int_{-\infty}^{\infty} |f(t)|^2\, dt = \frac{1}{2\pi} \int_{-\infty}^{\infty} \int_{-\infty}^{\infty} |\mathcal{G}\{f\}(\omega, t)|^2\, d\omega\, dt, \tag{2.5}$$

provided $\int_{-\infty}^{\infty} g^2(t)\, dt = 1$. The short-time Fourier transform has also an inverse transformation:

$$f(t) = \frac{1}{\sqrt{2\pi}} \int_{-\infty}^{\infty} \int_{-\infty}^{\infty} \mathcal{G}\{f\}(t, \omega) g(u-t) e^{\imath\omega t}\, d\omega\, dt. \tag{2.6}$$

Figure 2.4 shows the result of the Garbor transform of the signal (2.1).

It is clearly visible that in contrast to the classical Fourier transform the short-time transform has as well a time as a frequency resolution. The short-time spectrum resembles very much the music notation of the signal. Obviously, the time resolution of the Gabor transform is very good. The individual tones are clearly separated, but the frequency resolution is rather bad. Instead of a single frequency every tone is represented by a whole frequency band. The balance between time and frequency resolution is controlled by the length of the window g. The longer the window the better the frequency – but the worse the time resolution. This can be demonstrated by analyzing the same signal with an wider window. The result of the Garbor transform using a wider window is displayed in Figure 2.5. A wider window yields a better frequency but a worse time resolution.

The balance between time and frequency resolution can be quantified by the *uncertainty principle*: Since the window function is supposed to fulfill

$$\int_{-\infty}^{\infty} g^2(t)\, dt = 1,$$

its square can be considered as a probability density. Consequently, mean value and variance of this probability density can be computed:

$$\mu_g := \int_{-\infty}^{\infty} t g^2(t)\, dt = \int_{-\infty}^{\infty} u g_{\omega,t}^2(u) du - t, \tag{2.7}$$

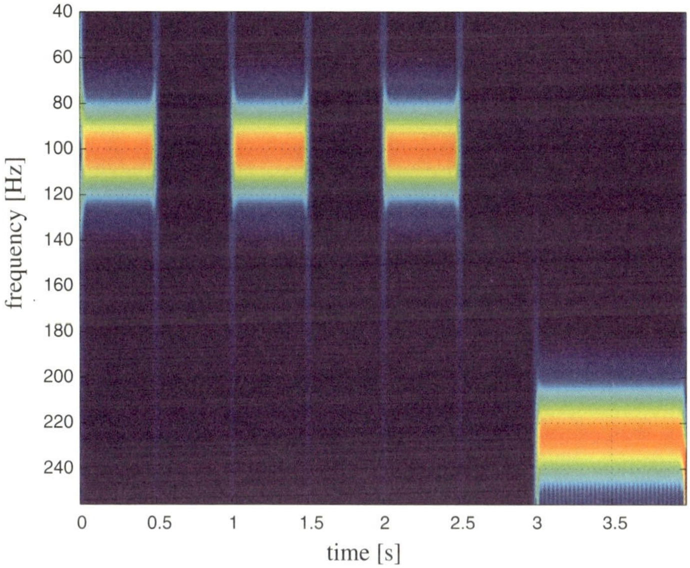

Figure 2.4. Short-time Fourier spectrum of the signal

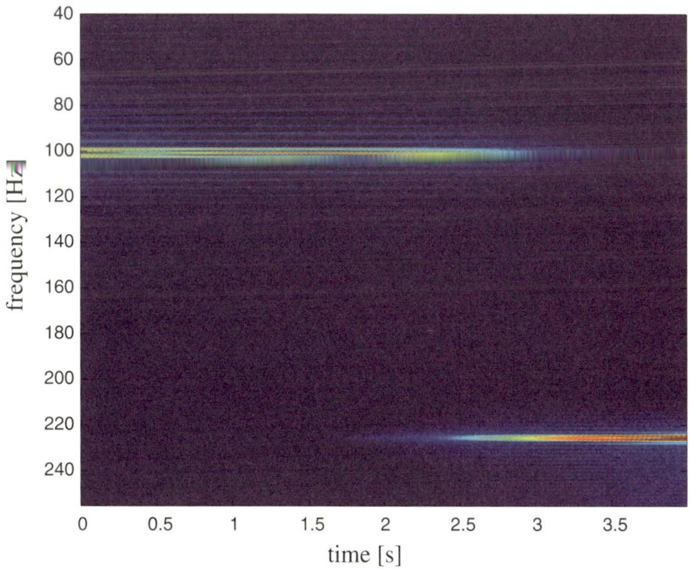

Figure 2.5. Short-time Fourier spectrum for a wider window

$$\sigma_g := \int_{-\infty}^{\infty} (t - \mu_g)^2 g^2(t)\, dt = \int_{-\infty}^{\infty} (u - (t + \mu_g))^2 |g_{\omega,t}|^2(u)\, du. \tag{2.8}$$

Following Parseval's identity, the square of the Fourier transform $\hat{g} := \mathcal{F}\{g\}$ of the window function is a probability density function too. Thus mean value and variance can be computed for $\hat{g}$ as well:

$$\mu_{\hat{g}} := \int_{-\infty}^{\infty} \omega \hat{g}^2(\omega)\, d\omega, \tag{2.9}$$

$$\sigma_{\hat{g}} := \int_{-\infty}^{\infty} (\omega - \mu_{\hat{g}})^2 \hat{g}^2(\omega)\, d\omega = \int_{-\infty}^{\infty} (\omega' - (\omega + \mu_{\hat{g}})^2 |\hat{g}_{\omega,t}|^2(\omega')\, d\omega'. \tag{2.10}$$

These parameters have an intuitively clear interpretation: The square of the window function $|g_{\omega,t}|^2$ is centered at $t + \mu_g$ and has a width of σ_g. The same is true for the spectrum: $|\hat{g}_{\omega,t}|^2$ is centered at $\omega + \mu_{\hat{g}}$. Figure 2.6 is to illustrate this situation.

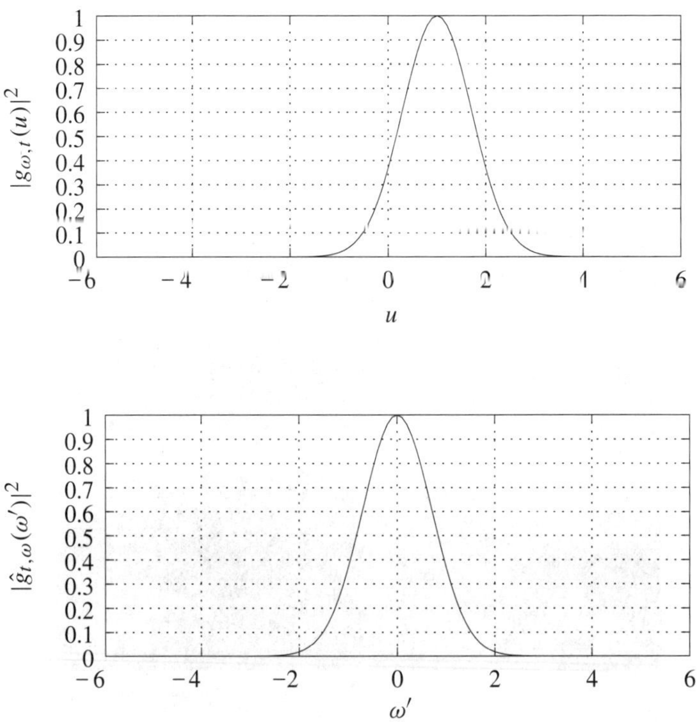

Figure 2.6. Window function and its spectrum

Using Parseval's identity the short-time Fourier transform can be expressed in two different ways:

$$\mathcal{G}\{f\}(\omega, t) = \frac{1}{\sqrt{2\pi}} \int_{-\infty}^{\infty} f(u) g_{\omega,t}(u)\, du \tag{2.11}$$

$$= \frac{1}{\sqrt{2\pi}} \int_{-\infty}^{\infty} \hat{f}(\omega') \overline{\hat{g}_{\omega,t}(\omega')}\, d\omega'. \tag{2.12}$$

The first equation states that $\mathcal{G}\{f\}$ is the signal looked through a window of width σ_g centered at t, and the second equation explains that the same quantity is also the spectrum of the signal looked through a window of width $\sigma_{\hat{g}}$. So, for every t the quantity $\mathcal{G}\{f\}(\omega, t)$ can be considered as the intensity of the signal f around the time t and at frequencies close to ω. The vague expressions *around* and *close to* can be quantified by the time resolution σ_g and the frequency resolution $\sigma_{\hat{g}}$.

Unfortunately, time and frequency resolution cannot be made arbitrarily small simultaneously. There is something like the *uncertainty principle* in quantum physics.

Theorem 2.1.1 (Uncertainty principle). *Let $g \in L_2(\mathbb{R})$, $\|g\| = 1$. Then*

$$\int_{-\infty}^{\infty} (t - t_0)^2 |g(t)|^2\, dt \cdot \int_{-\infty}^{\infty} (\omega - \omega_0)^2 |\hat{g}(\omega)|^2\, d\omega = \sigma_g^2 \sigma_{\hat{g}}^2 \geq \frac{1}{4} \tag{2.13}$$

holds for all $t_0, \omega_0 \in \mathbb{R}$.

Proof. The proof can be found in [75]. □

Numerical implementation of short-time Fourier transform. Let s_i, $i = 0, \dots, 2^N - 1$ be the sampled values of the signal to be analyzed. In order to reduce edge-effects an extended signal $\bar{s}_i$, $i = 0, \dots, 2^{N+1} - 1$ of the length 2^{N+1} is generated by the reflection

$$\bar{s}_i = \begin{cases} s_{2^{N-1}-i}, & i = 0, \dots, 2^{N-1} - 1 \\ s_{i-2^{N-1}}, & i = 2^{N-1}, \dots, 3 \cdot 2^{N-1} - 1 \\ s_{5 \cdot 2^{N-1}-i-1}, & i = 3 \cdot 2^{N-1}, \dots 2^{N+1} - 1. \end{cases} \tag{2.14}$$

In the next step a sequence of shifting parameters t_k, $k = 0, \dots, M$ has to be chosen. How this choice has to be made depends on the goal the short-time Fourier analysis is aiming at.

For each shifting parameter t_k the following operations have to be carried out:

- The window function $g_{t_k}(\bullet) := g(\bullet - t_k)$ has to be sampled with the same sampling width as the signal s yielding the sampled values $g_{k,i}$, $i = 0, \dots, 2^{N+1} - 1$.

- The signal and the shifted window have to be multiplied

$$h_{k,j} := \bar{s}_j \cdot g_{k,j}, \quad j = 0, \dots, 2^{N+1} - 1.$$

- $Gs_{j,k} := \mathcal{G}\{\bar{s}\}(\omega_j, t_k)$ can be computed as the FFT of $h_{k,j}$.

- The edges of the spectrum have to be stripped off

$$\mathcal{G}\{s\}(\omega_j, t_k) := Gs_{k,j+2^{N-1}}, \quad j = 0, \ldots, 2^N - 1.$$

The great disadvantage of the short-time Fourier transform is that the time resolution is the same for all frequencies. For high frequencies already a short time span is sufficient to resolve them; for lower frequencies a longer time span is necessary. This means the time and the frequency resolution should be coupled: for higher frequencies a better time resolution than for lower frequencies. This can be achieved by varying the window-width according to the frequency. This idea led to the invention of wavelets.

2.2 Continuous wavelet transformation

2.2.1 Concept

In the short-time Fourier transform the analyzing function $g_{\omega,t}$ was a hat-shaped contour filled with a sinusoidal as it is indicated in Figure 2.7.

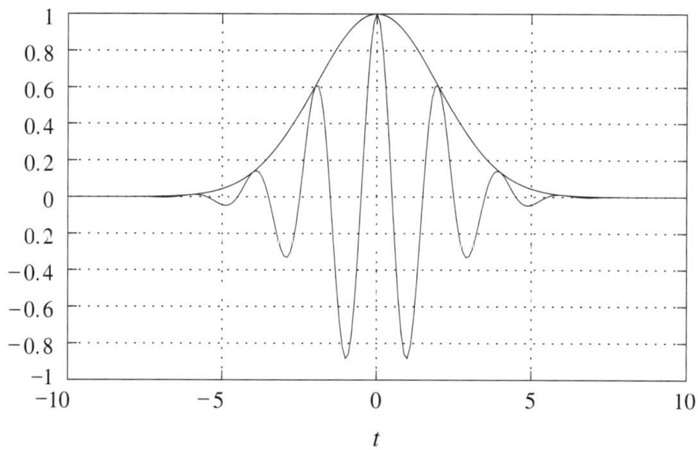

Figure 2.7. Analyzing function of short-time Fourier transform

Due to the fixed envelope g the time and frequency resolution is the same for every value of t, ω. It would be better to have an analyzing function with a high time resolution for high frequencies and a low time resolution for low frequencies. Such a function is a wavelet.

Definition 2.2.1. A function $\psi \in L_2(\mathbb{R})$ is called a *wavelet* if it fulfills the admissibility condition

$$0 < c_\psi := 2\pi \int_{-\infty}^{\infty} \frac{|\hat{\psi}(\omega)|^2}{|\omega|} d\omega < \infty, \tag{2.15}$$

and the continuous wavelet transform of a function $f \in L_2(\mathbb{R})$ with respect to this wavelet is given by

$$\mathcal{W}\{f\}(\lambda, t) := \frac{1}{\sqrt{c_\psi}} |\lambda|^{-1/2} \int_{-\infty}^{\infty} f(u)\psi\left(\frac{u-t}{\lambda}\right) du, \quad \lambda \in \mathbb{R}\setminus\{0\},\ t \in \mathbb{R}. \tag{2.16}$$

From the definition of a wavelet some important conclusions can be drawn. First, the admissibility condition implicates

$$0 = \hat{\psi}(0) = (2\pi)^{-1/2} \int_{-\infty}^{\infty} \psi(t)\, dt. \tag{2.17}$$

Therefore the wavelet has to have a vanishing mean value. It has to oscillate in some way. On the other hand there is the condition $\psi \in L_2(\mathbb{R})$. For this reason the function ψ has to decay in some way for $t \to \infty$. Hence the oscillation cannot last forever, it has to stop after a certain time span. The function ψ does not represent a wave but a small wave – a *wavelet*.

Figure 2.8 shows two examples of wavelets together with their spectra. On the left side the so-called *Mexican hat* wavelet and on the right side the Daubechies wavelet of order 4.

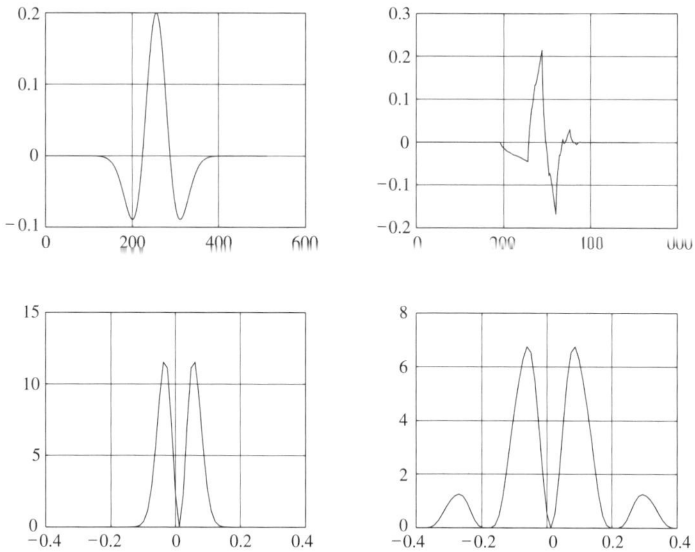

Figure 2.8. Mexican hat wavelet (top left) with its spectrum (bottom left) and Daubechies wavelet (top right) with its spectrum (bottom right)

In the continuous wavelet transformation (2.16) the analyzing function is $\psi(\frac{u-t}{\lambda})$. Similar to the short-time Fourier transform case, here the analyzing function is also a certain shape filled with oscillations. The big difference to the short-time Fourier transform is the change of the envelopes shape with the change of the scale λ.

- The shorter the scale λ is the more compressed is the envelope and the higher is the frequency of the oscillation.

Consequently, the time resolution increases with decreasing scale size and the frequency resolution decreases with decreasing scale size. In order to demonstrate the difference between continuous wavelet transformation and short-time Fourier transformation, an example similar to that given by I. Daubechies will be discussed. The following signal will be considered:

$$f(t) = \sin(1000\pi t) + \sin(2000\pi t) + [\delta(t - t_1) + \delta(t - t_2)], \tag{2.18}$$

$$t_1 = 0.075, \quad t_2 = 0.079.$$

The signal is displayed in Figure 2.9.

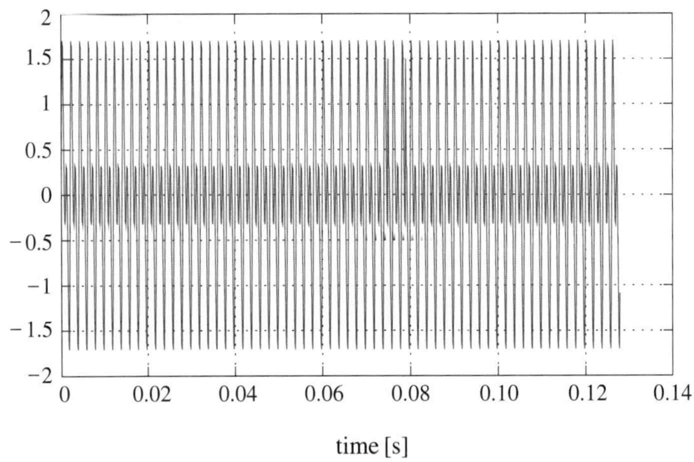

Figure 2.9. Signal with two frequencies and two impulses

If this signal is analyzed using the short-time Fourier transform then, depending on the window length, either the two frequencies can be resolved and the impulses get lost, or the impulses can be detected but the frequencies cannot be separated. This is visible in the two Figures 2.10 and 2.11.

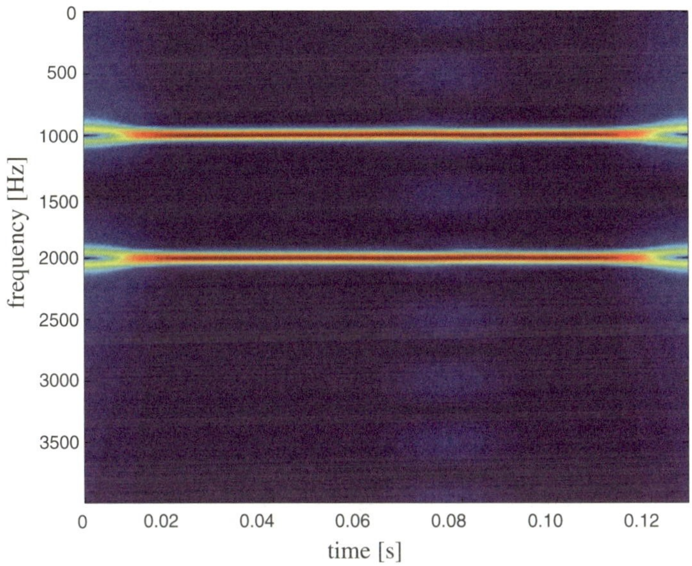

Figure 2.10. Short-time Fourier transform with wide window

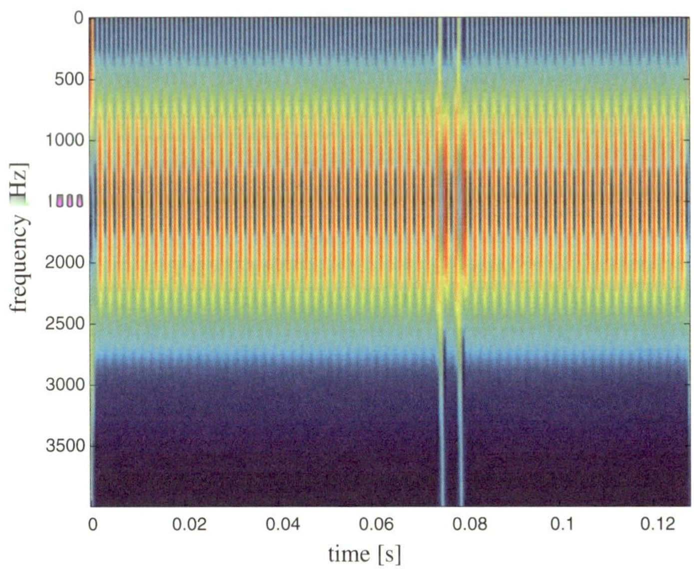

Figure 2.11. Short-time Fourier transform with small window

The great advantage of the continuous wavelet transformation is the ability to resolve both the two long-waved frequencies and the two extremely short impulses. This can be seen in Figure 2.12. The signal was analyzed by the so-called Morlet wavelet and this analysis shows both the two frequencies and the two impulses.

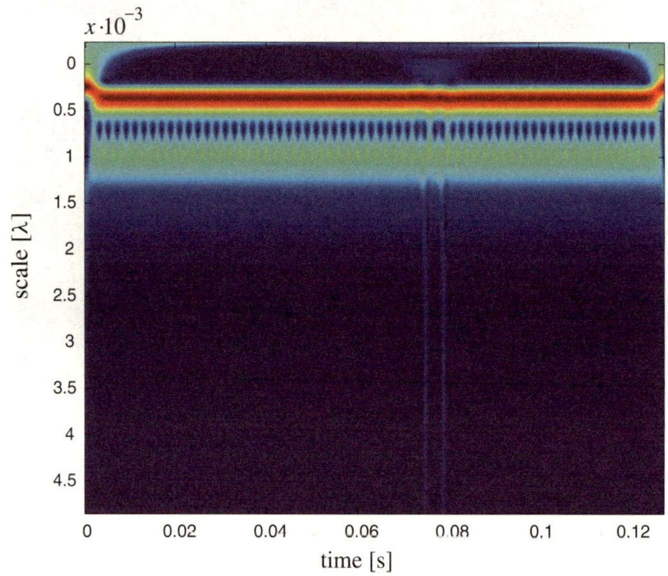

Figure 2.12. Continuous wavelet transformation

As in the case of the Fourier and the short-time Fourier transform also the continuous wavelet transformation has the isometry property and an uniquely determined inverse transformation.

Theorem 2.2.1. *The following equation is valid:*

$$\int_{-\infty}^{\infty} f^2(t)\,dt = \int_{-\infty}^{\infty}\int_{-\infty}^{\infty} |\mathcal{W}\{f\}(\lambda,t)|^2 \frac{d\lambda dt}{\lambda^2}. \tag{2.19}$$

Theorem 2.2.2. *The following equation is valid:*

$$f(u) = c_\psi^{-1/2} \int_{-\infty}^{\infty}\int_{-\infty}^{\infty} |\lambda|^{-1/2}\psi\left(\frac{u-t}{\lambda}\right) \mathcal{W}\{f\}(t,\lambda)\frac{d\lambda dt}{\lambda^2}. \tag{2.20}$$

The proofs of both theorems are given in [68].

Usually, Fourier transform is discussed in the context of filtering. In order to get a bit more familiar with wavelet transform it will be discussed now, how wavelet transform and filtering are related to each other. With the function

$$\psi_\lambda(t) := |\lambda|^{-1/2}\psi\left(-\frac{t}{\lambda}\right) \tag{2.21}$$

the continuous wavelet transform formally can be written as a convolution

$$\mathcal{W}\{f\}(t,\lambda) = c_\psi^{-1/2}\psi_\lambda * f. \tag{2.22}$$

Since $\hat{\psi}_\lambda(0) = 0$ and $\lim_{\omega\to\infty}\hat{\psi}_\lambda(\omega) = 0$ the wavelet transform is a band-pass filter. The spectrum of a wavelet ψ is concentrated around a certain frequency ω_0. Therefore $\hat{\psi}_\lambda$ is concentrated around ω_0/λ. Consequently, the wavelet transformation $\mathcal{W}\{f\}(\lambda,\bullet)$ contains only those Information about f, which are related to frequencies around ω_0/λ. The wavelet transformation is similar to a slicing of the signal f into separated frequency bands. In digital signal processing this decomposition of the signal is referred to as sub-band coding.

Numerical implementation of continuous wavelet transformation

- Let s_i, $i = 0,\dots,2^N - 1$ be the sampled values of the signal to be analyzed. In order to reduce edge-effects an extended signal $\bar{s}_i$, $i = 0,\dots,2^{N+1} - 1$ of the length 2^{N+1} is generated by reflection

$$\bar{s}_i = \begin{cases} s_{2^{N-1}-i}, & i = 0,\dots,2^{N-1}-1 \\ s_{i-2^{N-1}}, & i = 2^{N-1},\dots,3\cdot 2^{N-1}-1 \\ s_{5\cdot 2^{N-1}-i-1}, & i = 3\cdot 2^{N-1},\dots 2^{N+1}-1. \end{cases} \tag{2.23}$$

- The spectrum $\hat{s}_j$, $j = 0,\dots,2^{N+1}-1$ of the extended signal has to by computed by FFT.

- In the next step a sequence of scale parameters λ_k, $k = 0,\dots,M$ has to be chosen. This choice depends on the goal the continuous wavelet analysis is aiming at. For each scale parameter λ_k the following operations have to be carried out:

  - The wavelet $c_\psi^{-1/2}\psi_{\lambda_k}$ has to be sampled with the same sampling width as the signal s yielding the sampled values $\psi_{k,i}$.
  - The spectrum of the dilated wavelet $c_\psi^{-1/2}\psi_{\lambda_k}$ has to be computed from the samples $\psi_{k,i}$ using FFT, yielding the spectral values $\hat{\psi}_{k,j}$, $j = 0,\dots,2^{N+1} - 1$.
  - The spectra of the signal and of the dilated wavelet have to be multiplied

$$h_{k,j} := \hat{s}_j \cdot \hat{\psi}_{k,j}, \quad j = 0,\dots,2^{N+1} - 1.$$

  - The inverse FFT has to be applied to the sequence $h_{k,j}$ giving the values $Ws_{i,k}$
  - Stripping off the edges

$$\mathcal{W}\{s\}(\lambda_k, t_i) := Ws_{i-2^{N-1},k}, \quad i = 0,\dots,2^N - 1.$$

2.2.2 Time-frequency resolution

Let us introduce the function

$$\psi_{\lambda,t}(u) := |\lambda|^{-1/2}\psi\left(\frac{u-t}{\lambda}\right). \tag{2.24}$$

As in the case of the short-time Fourier transform, the continuous wavelet transform can also be written in two equivalent ways:

$$\mathcal{W}\{f\}(\lambda,t) = c_\psi^{-1/2}\int_{-\infty}^{\infty} f(u)\psi_{\lambda,t}(u)\,du = c_\psi^{-1/2}\int_{-\infty}^{\infty} \hat{f}(\omega)\overline{\hat{\psi}_{\lambda,t}(\omega)}\,d\omega. \tag{2.25}$$

If the wavelet ψ fulfills $\|\psi\| = 1$, both $|\psi_{\lambda,t}|^2$ and $|\hat{\psi}_{\lambda,t}|^2$ have the properties of probability densities. Consequently, expectations and variances can be computed:

$$t^{\lambda,t} := \int_{-\infty}^{\infty} u|\psi_{\lambda,t}(u)|^2\,du, \tag{2.26}$$

$$\sigma^2_{\psi_{\lambda,t}} = \int_{-\infty}^{\infty} (u - t^{\lambda,t})^2|\psi_{\lambda,t}(u)|^2\,du, \tag{2.27}$$

$$\omega^{\lambda,t} := \int_0^{\infty} \omega|\hat{\psi}_{\lambda,t}(\omega)|^2\,d\omega, \tag{2.28}$$

$$\sigma^2_{\hat{\psi}_{\lambda,t}} := \int_0^{\infty} (\omega - \omega^{\lambda,t})^2|\hat{\psi}_{\lambda,t}(\omega)|^2\,d\omega. \tag{2.29}$$

It is easy to verify that the following relations hold:

$$\sigma_{\psi_{\lambda,t}} = \lambda\sigma_{\psi_{1,0}}, \quad \sigma_{\hat{\psi}_{\lambda,t}} = \frac{\sigma_{\hat{\psi}_{1,0}}}{\lambda}. \tag{2.30}$$

Finally, also

$$\omega^{\lambda,t} = \frac{\omega^{1,0}}{\lambda} \tag{2.31}$$

is true.

If one recalls the wavelet transform as a band-pass filter, the center $\omega_0^{\lambda,t}$ of the filter characteristic changes with changing scale length λ. For increasing scale size λ the center of the passing band moves to the low frequencies, the window length $\sigma_{\hat{\psi}_{\lambda,t}}$ in the spectral domain decreases and the window length $\sigma_{\psi_{\lambda,t}}$ in the time domain increases. Therefore, the larger the scale size the better the frequency – and the worse is the time resolution.

In the (t,λ) phase-space the wavelet $\psi_{\lambda,b}$ cuts out a window which is centered at $(t^{\lambda,b}, \omega^{\lambda,b})$ and has a center-dependent size. The situation is displayed in Figure 2.13.

Quite similar to the short-time Fourier transform, also for the continuous wavelet transform time and frequency resolution cannot be improved simultaneously. Also here the uncertainty inequality

$$\sigma_{\psi_{\lambda,t}} \cdot \sigma_{\hat{\psi}_{\lambda,t}} \geq \frac{1}{4} \tag{2.32}$$

is valid.

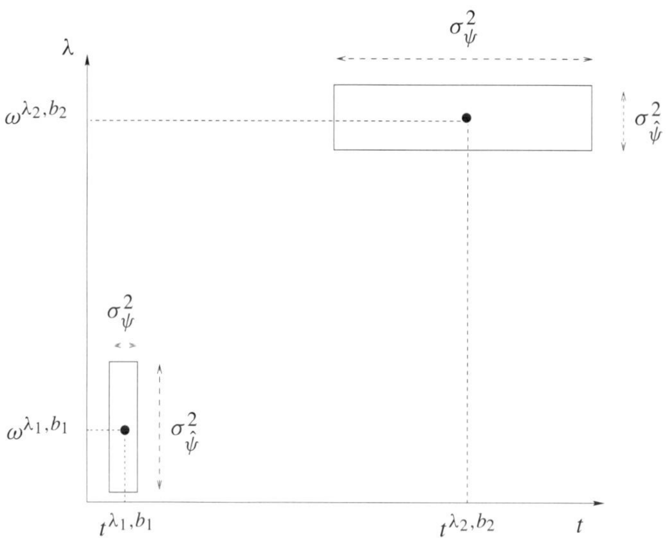

Figure 2.13. Phase-space window of $\psi_{\lambda,b}$ for different values of λ and b

2.2.3 Approximation properties

There is an infinite variety of wavelets. As a criterion to classify different types of wavelets and to give different wavelets certain meanings in terms of digital signal processing, the wavelets will be classified according to their behavior for decreasing scale sizes.

The basic idea is that the wavelet transform of a signal f approximates the derivative of f of a certain order. The wavelets can be classified according to the order of the derivative they approximate.

Definition 2.2.2. A wavelet ψ is called *of order* $N \in \mathbb{N}$, if the following relations hold:

1.
$$\int_{-\infty}^{\infty} t^k \psi(t)\, dt = 0, \quad 0 \le k < N, \tag{2.33}$$

2.
$$\int_{-\infty}^{\infty} t^N \psi(t)\, dt \neq 0. \tag{2.34}$$

As already mentioned, the order of a wavelet is in close relationship with the order of the derivative the wavelet transform approximates.

Theorem 2.2.3. *Let $\psi \in L_2(\mathbb{R})$ be a wavelet of order N and let f be sufficiently smooth. Define μ by*

$$\mu := (-1)^N \left(\int_{-\infty}^{\infty} t^N \psi(t)\, dt \right) / N!.$$

Then

$$\lim_{\lambda \to 0} \left\| \frac{\operatorname{sgn}^N(-\lambda)}{|\lambda|^{N+\frac{1}{2}}} \sqrt{c_\psi}\, \mathcal{W}\{f\}(\lambda, \bullet) - \mu f^{(N)}(\bullet) \right\| = 0. \tag{2.35}$$

Proof. [68]. □

In order to illustrate the content of this theorem, following trigonometric function will be used as the signal f:

$$f(u) := \sin(u). \tag{2.36}$$

As analyzing wavelet, the most simple wavelet – the Haar wavelet – will be applied:

$$\psi(u) := \begin{cases} 1, & 0 \le u < 0.5 \\ -1, & 0.5 \le u < 1 \\ 0, & \text{else.} \end{cases} \tag{2.37}$$

For this wavelet the quantity μ is

$$\mu = -\int_{-\infty}^{\infty} u \psi(u)\, du = \frac{1}{4}. \tag{2.38}$$

For c_ψ the following value is obtained:

$$c_\psi = 2\pi \int_{-\infty}^{\infty} \frac{|\hat{\psi}(\omega)|^2}{|\omega|}\, d\omega = \int_{-\infty}^{\infty} \frac{\sin^2(\omega/4)\operatorname{sinc}^2(\omega/4)}{|\omega|}\, d\omega = 2 \ln 2. \tag{2.39}$$

Then the continuous wavelet transformation of f with respect to the Haar wavelet is

$$\begin{aligned} \mathcal{W}\{f\}(\lambda, t) &= \frac{1}{\sqrt{\lambda c_\psi}} \int_{-\infty}^{\infty} f(u) \psi\left(\frac{u-t}{\lambda}\right) du \\ &= \frac{1}{\sqrt{\lambda c_\psi}} \left\{ \int_t^{t+\lambda/2} f(u)\, du - \int_{t+\lambda/2}^{t+\lambda} f(u)\, du \right\} \\ &= \frac{1}{\sqrt{\lambda c_\psi}} \left\{ -\cos(u)\Big|_t^{t+\lambda/2} + \cos(u)\Big|_{t+\lambda/2}^{t+\lambda} \right\} \\ &= \frac{1}{\sqrt{\lambda c_\psi}} \left\{ \cos(t) - 2\cos(t+\lambda/2) + \cos(t+\lambda) \right\}. \end{aligned} \tag{2.40}$$

Therefore one obtains

$$
\begin{aligned}
\frac{\operatorname{sgn}(-\lambda)\sqrt{c_\psi}}{|\lambda|^{3/2}} \mathcal{W}\{f\}(\lambda, t) &= -\frac{1}{\lambda^2}\{\cos(t) - 2\cos(t+\lambda/2) + \cos(t+\lambda)\} \\
&= -\frac{1}{\lambda^2}\{\cos t \\
&\quad - 2\cos t + 2\sin t\frac{\lambda}{2} + \cos t\frac{\lambda^2}{4} + O(\lambda^3) \\
&\quad + \cos t - \sin t\lambda - \frac{1}{2}\cos t\lambda^2 + O(\lambda^3)\} \\
&= -\left\{-\frac{1}{4}\cos t + O(\lambda)\right\} \\
&= \mu\left\{(\sin t)' + O(\lambda)\right\},
\end{aligned}
\tag{2.41}
$$

and hence

$$
\lim_{\lambda\to 0} \frac{\operatorname{sgn}(-\lambda)\sqrt{c_\psi}}{|\lambda|^{3/2}} \mathcal{W}\{f\}(\lambda, t) = \mu(\sin t)'. \tag{2.42}
$$

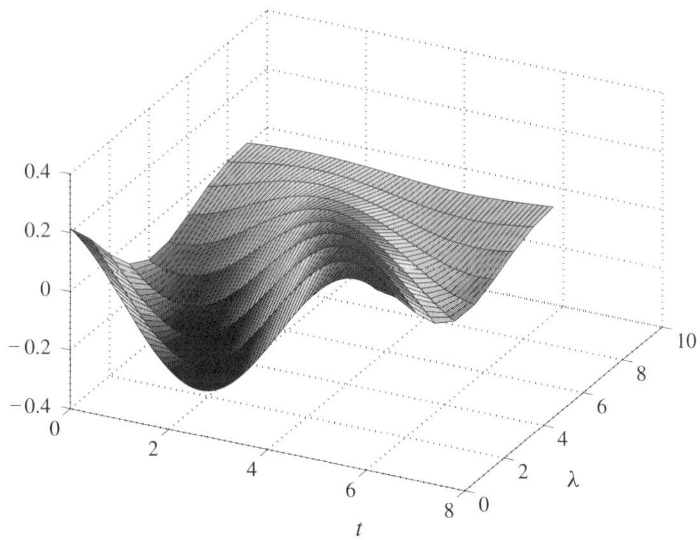

Figure 2.14. Wavelet transformation of sine function

This expression is displayed in Figure 2.14. A comparison with $F'(u) = \cos(u)$ as it is displayed in Figure 2.15 shows that for decreasing scales $\lambda \to 0$ the wavelet transform tends to the first derivative.

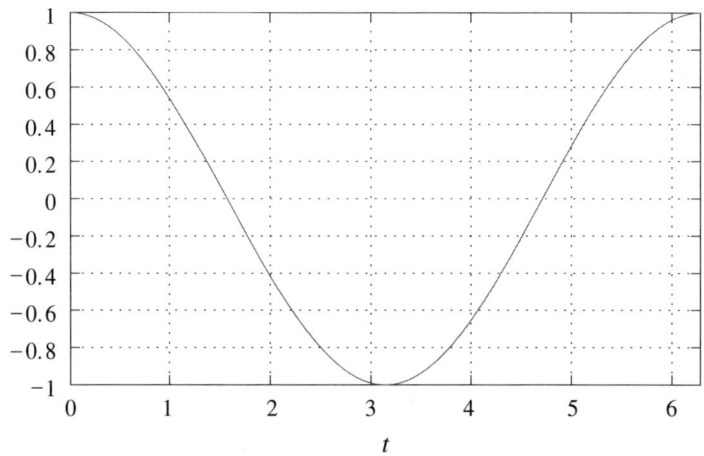

Figure 2.15. First derivative of the sine function

So far the order of a wavelet has been identified as one criterion for the classification of different types of wavelets. Now it will be shown that, depending of the order of the wavelet, the wavelet transformation of a function f can be used to characterize its smoothness. The situation is comparable to the Fourier spectrum. The decay of the Fourier spectrum is closely related to the regularity of the function. Something quite similar also holds for wavelets.

Theorem 2.2.4. *Let ψ be a wavelet of order N. For some $k \in \{1, \dots, N\}$ let hold $|f^{(k)}| < \infty$ then*

$$|\mathcal{W}\{f\}(\lambda, t)| = O(|\lambda|^{k+1/2}), \quad \lambda \to 0. \tag{2.43}$$

Proof. [68]. □

Quite similar to the case of Fourier transform the decay of the wavelet transform is determined by the order of differentiability. The difference is that the decay of the wavelet transform is determined not only by the order of differentiability but also by the order of the wavelet.

2.3 Discrete wavelet transformation

2.3.1 Frames

The computation of the wavelet transform of a signal and the reconstruction of a signal from its wavelet spectrums requires the evaluation of a single or a double integral respectively. For an efficient numerical implementation these integrals have to be discretized. The question arises, whether or not it is really necessary to know $\mathcal{W}\{f\}(\lambda, t)$ for all values of λ and t in order to reconstruct the signal.

Fortunately, it will turn out that the representation of a signal by its wavelet-spectrum is highly redundant. Therefore it will be possible to replace the integrals by sums. To be more precise, we consider the grid

$$\{(\lambda_0^m, nt_0) \mid m, n \in \mathbb{Z}\} \tag{2.44}$$

and ask ourselves, if it will be possible to reconstruct the signal f completely from the restriction of the wavelet spectrum to this grid. On the grid the following set of functions is considered:

$$\{\psi_{m,n}^{(\lambda_0,t_0)}(\bullet) := \lambda_0^{-m/2}\psi(\lambda_0^{-m}\bullet - nt_0) \mid m, n \in \mathbb{Z}\}. \tag{2.45}$$

The question is: Under which circumstances can a signal f be completely represented using this discrete set of functions only? The answer is: A complete representation is possible, if these functions form a so-called frame.

Definition 2.3.1. Let $\lambda_0 > 1$, $t_0 > 0$ and let $\psi \in L_2(\mathbb{R})$ be a wavelet. The set of functions $\{\psi_{m,n}^{(\lambda_0,t_0)} \mid m, n \in \mathbb{Z}\}$ is called a *wavelet frame* for $L_2(\mathbb{R})$ if there are positive constants A, B with

$$A\|f\|^2 \leq \sum_{m\in\mathbb{Z}}\sum_{n\in\mathbb{Z}} |\langle\psi_{m,n}^{(\lambda_0,t_0)}, f\rangle|^2 \leq B\|f\|^2. \tag{2.46}$$

The constants A, B are called the bounds of the frame. If $A = B$ holds, the frame is called *tight frame*.

The coefficients $\langle\psi_{m,n}^{(\lambda_0,t_0)}, f\rangle$ have a very instructive interpretation. Since

$$\langle\psi_{m,n}^{(\lambda_0,t_0)}, f\rangle = \mathcal{W}\{f\}(\lambda_0^m, nt_0) \tag{2.47}$$

holds, the values of the frame coefficients indicate the intensity of the signal at scales close to λ_0^m in the vicinity of the time nt_0.

Frames are important, since for them a unique relationship between the signal f and its coefficients $\langle\psi_{m,n}^{(\lambda_0,t_0)}, f\rangle$ can be established. To every frame $\{\psi_{m,n}^{(\lambda_0,t_0)} \mid m, n \in \mathbb{Z}\}$ a uniquely determined operator $T : L_2(\mathbb{R}) \to l_2(\mathbb{Z}^2)$ can be assigned in the following way:

$$(Tf)_{m,n} := \langle\psi_{m,n}^{(\lambda_0,t_0)}, f\rangle.$$

From equation (2.46) the following relation can be deduced

$$\sqrt{A}\|f\| \leq \|Tf\| \leq \sqrt{B}\|f\|.$$

Consequently, the operator T is bounded and has a bounded inverse operator. The existence of a bounded inverse is very important since it guarantees that the signal can be uniquely and stably reconstructed from its coefficients.

Hence, if the wavelet ψ together with the scale spacing λ_0 and the time spacing t_0 forms a frame, the signal is completely characterized by its wavelet spectrum on

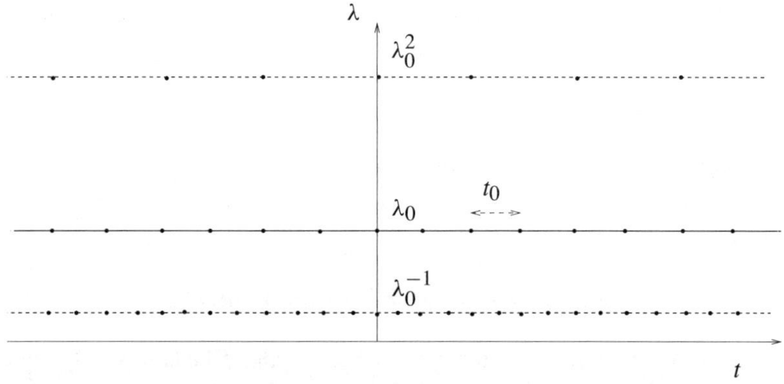

Figure 2.16. Grid for a wavelet frame

a discrete grid. This grid is not uniformly shaped but denser for smaller scales and wider for larger scales. The situation is displayed in Figure 2.16

The frame property is very important for the establishment of a unique relationship between a signal and its coefficients, but it does not answer the question of *how* the reconstruction has to be done. A simple answer is only possible for tight frames.

Theorem 2.3.1. *Let ψ be normalized, i.e. $\|\psi\| = 1$, and let $\{\psi_{m,n}^{(\lambda_0,t_0)} \mid m, n \in \mathbb{Z}\}$ be a tight frame with the bounds $A = B = 1$. Then the functions $\psi_{m,n}^{(\lambda_0,t_0)}$ form a complete orthonormal system in $L_2(\mathbb{R})$.*

Proof. In the first step it will be shown that for the system $\{\psi_{m,n}^{(\lambda_0,t_0)}\}$ Parseval's identity is valid: The relation

$$\|f\|^2 = \sum_{m\in\mathbb{Z}} \sum_{n\in\mathbb{Z}} |\langle \psi_{m,n}^{(\lambda_0,t_0)}, f\rangle|^2 \tag{2.48}$$

immediately follows from (2.46).

As a next step it will be shown that with ψ also the $\psi_{m,n}^{(\lambda_0,t_0)}$ are normalized:

$$\begin{aligned}
\|\psi_{m,n}^{(\lambda_0,t_0)}\|^2 &= \int_{\mathbb{R}} (\psi_{m,n}^{(\lambda_0,t_0)}(t))^2 \, dt \\
&= \int_{\mathbb{R}} (\lambda_0^{-m/2} \psi(\lambda_0^{-m} t - n t_0))^2 \, dt \\
&= \int_{\mathbb{R}} (\psi(x))^2 \, dx \\
&= 1.
\end{aligned}$$

In the last step the orthogonality will be proved: From (2.48) follows

$$\begin{aligned}1 &= \|\psi_{m,n}^{(\lambda_0,t_0)}\|^2 \\ &= \sum_{\mu,\nu\in\mathbb{Z}} |\langle\psi_{m,n}^{(\lambda_0,t_0)}, \psi_{\mu,\nu}^{(\lambda_0,t_0)}\rangle|^2 \\ &= \|\psi\|^2 + \sum_{(\mu,\nu)\neq(m,n)} |\langle\psi_{m,n}^{(\lambda_0,t_0)}, \psi_{\mu,\nu}^{(\lambda_0,t_0)}\rangle|^2.\end{aligned}$$

Hence

$$\sum_{(\mu,\nu)\neq(m,n)} |\langle\psi_{m,n}^{(\lambda_0,t_0)}, \psi_{\mu,\nu}^{(\lambda_0,t_0)}\rangle|^2 = 0$$

must hold. This is only possible for

$$\langle\psi_{m,n}^{(\lambda_0,t_0)}, \psi_{\mu,\nu}^{(\lambda_0,t_0)}\rangle = \delta_{n,\nu}\delta_{m,\mu}.$$

Orthonormality together with the validity of Parseval's identity guarantees completeness. □

For a complete orthonormal system the reconstruction of a signal from its coefficients is fairly easy, it is just the generalized Fourier series

$$f(t) = \sum_{m\in\mathbb{Z}}\sum_{n\in\mathbb{Z}} \langle\psi_{m,n}^{(\lambda_0,t_0)}, f\rangle \cdot \psi_{m,n}^{(\lambda_0,t_0)}(t) \tag{2.49}$$

For practical reasons dyadic frames, i.e. frames with $\lambda_0 = 2$, $t_0 = 1$ are preferred.

The question remains how a wavelets can be constructed in such a way that it forms a tight frame. This question will be discussed in the chapter on multi-resolution analysis.

2.4 Multi-resolution analysis

The concept of multi-resolution analysis is the key to the construction of orthogonal wavelet bases and for the fast decomposition of a signal into disjunct frequency bands. Before starting with a mathematical definition of the multi-resolution analysis a motivating example will be given. Let f be a signal from a certain subspace $V_{-1} \subset L_2(\mathbb{R})$. This signal is to be decomposed into a high- and a low-frequency part. The low-frequency part is obtained by an orthogonal projection $P_0 f$ into a smaller subspace $V_0 \subset V_{-1}$, which contains only the smooth functions of V_{-1}. The orthogonal complement of V_0 in V_{-1} will be denoted by W_0. The projection of the signal f into W_0 will be denoted by $Q_0 f$. In this way we have

$$f = P_0 f + Q_0 f, \tag{2.50}$$

$$V_{-1} = V_0 \oplus W_0. \tag{2.51}$$

The procedure can be repeated for $P_0 f$ by representing V_0 as the orthogonal sum of V_1, containing the smoother part, and of W_1, containing the rougher part of V_0. The corresponding projectors are called P_1 and Q_1.

$$P_0 f = P_1 P_0 f + Q_1 P_0 f \tag{2.52}$$

and therefore

$$\begin{aligned} f &= P_0 f + Q_0 f \\ &= P_1 P_0 f + Q_1 P_0 f + Q_0 f. \end{aligned} \tag{2.53}$$

The two-stage decomposition is illustrated in Figure 2.17. Iterating this technique,

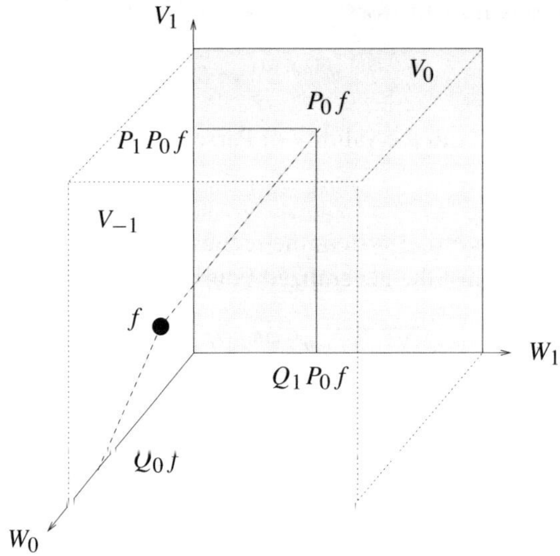

Figure 2.17. Two-stage decomposition

one obtains the relation

$$f = P_n P_{n-1} \dots P_0 f + Q_n P_{n-1} \dots P_0 f + Q_{n-1} P_{n-2} \dots P_0 f + \cdots + Q_0 f, \tag{2.54}$$

which is schematically displayed in Figure 2.18. This scheme corresponds to a decomposition of the signal f into different frequency-bands, as it is shown in Figure 2.19.

The decomposition of a signal into different frequency bands is the core of the multi-resolution analysis.

Definition 2.4.1. A nested sequence

$$\{0\} \subset \cdots \subset V_2 \subset V_1 \subset V_0 \subset V_{-1} \subset \cdots \subset L_2(\mathbb{R}) \tag{2.55}$$

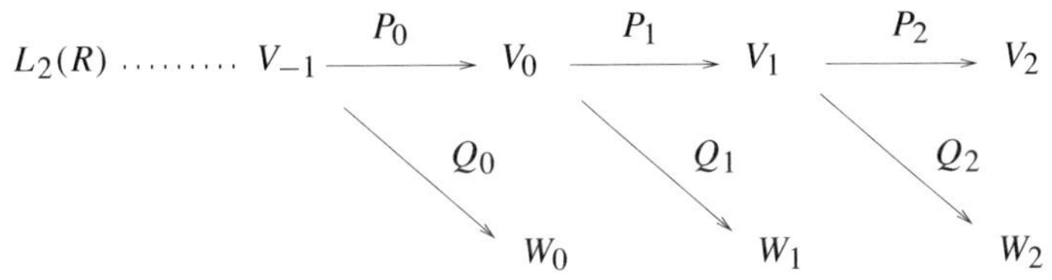

Figure 2.18. Multi-resolution analysis

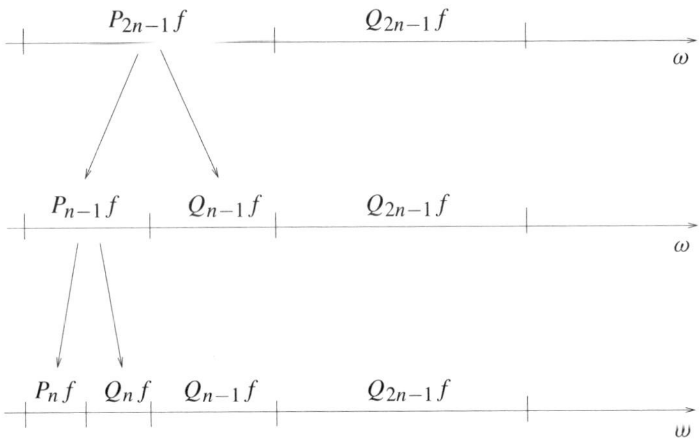

Figure 2.19. Decomposition into separate frequency bands

of closed subsets $V_m \subset L_2(\mathbb{R})$ is called a *multi-resolution analysis* (MRA) of $L_2(\mathbb{R})$, if the following four statements hold:

(i)

$$\overline{\bigcup_{m\in\mathbb{Z}} V_m} = L_2(\mathbb{R}). \tag{2.56}$$

(ii)

$$\bigcap_{m\in\mathbb{Z}} V_m = \{0\}. \tag{2.57}$$

(iii)

$$f(\bullet) \in V_m \Leftrightarrow f(2^m \bullet) \in V_0. \tag{2.58}$$

(iv) There is a function $\varphi \in L_2(\mathbb{R})$ with

$$V_0 = \overline{\operatorname{span}\{\varphi(\bullet - k) \mid k \in \mathbb{Z}\}} \tag{2.59}$$

and

$$A \sum_{k\in\mathbb{Z}} c_k^2 \le \|\sum_{k\in\mathbb{Z}} c_k \varphi(\bullet - k)\|_{L_2}^2 \le B \sum_{k\in\mathbb{Z}} c_k^2, \quad 0 < A, B \tag{2.60}$$

for all $\{c_k\}_{k\in\mathbb{Z}} \in l_2(\mathbb{Z})$.

Some of these properties are more of a technical nature. The essential property is (2.58). This equation expresses the fact that all spaces of an MRA are scaled versions of the base space V_0. The space V_0 itself is spanned by shifted versions of the so-called *scaling function* φ. For $m \to \infty$ the elements of V_m are getting more extended and dilated. For $m \to -\infty$ the spaces V_m contain increasingly finer structures.

The space V_0 is spanned by the shifted versions of the scaling function φ. Due to equation (2.58) the spaces V_m are spanned by the shifted versions of $\varphi_{m,k}(x) := 2^{-m/2}\varphi(2^{-m}x - k)$:

$$V_m = \overline{\operatorname{span}\{\varphi_{m,k} \mid k \in \mathbb{Z}\}}. \tag{2.61}$$

The following simple example is to illustrate the concept of MRA.

Example 2.4.1. Define the sequence of subspaces $\{V_m\}_{m\in\mathbb{Z}}$ as follows:

$$V_m := \{f \in L_2(\mathbb{R}) \mid \operatorname{supp} \mathcal{F}\{f\} \subset [-\pi 2^{-m}, \pi 2^{-m}]\}. \tag{2.62}$$

This means that all the subspaces consist of band-limited signals. Obviously,

$$\begin{aligned} f \in V_m &\Rightarrow \operatorname{supp} \mathcal{F}\{f\} \subset [-\pi 2^{-m}, \pi 2^{-m}] \subset [-\pi 2^{-(m-1)}, \pi 2^{-(m-1)}] \\ &\Rightarrow f \in V_{m-1} \end{aligned}$$

holds, which induces $V_m \subset V_{m-1}$. Since for $f \in V_m$ the relation

$$\begin{aligned} \operatorname{supp} \mathcal{F}\{f(2^m \bullet)\} &= \operatorname{supp} 2^{-m} \mathcal{F}\{f\}(2^{-m}\omega) = \{\omega \mid |2^{-m}\omega| \le \pi 2^{-m}\} \\ &= \{\omega \mid |\omega| \le \pi\} \end{aligned}$$

is valid,

$$f \in V_m \Rightarrow f(2^m \bullet) \in V_0 \tag{2.63}$$

can be concluded.

From equation (1.23) in the sampling theorem (p. 5) it is easy to see that

$$\varphi(x) = \frac{\sin(\pi x)}{\pi x} \tag{2.64}$$

is the scaling function of V_0.

Let $g \in \bigcap_{m\in\mathbb{Z}} V_m$. Hence $\operatorname{supp} \mathcal{F}\{g\} = \{0\}$ must hold. Consequently,

$$\mathcal{F}\{g\} = \begin{cases} a, & \omega = 0 \\ 0, & \text{else}, \end{cases} \tag{2.65}$$

i.e., the Fourier transform of g differs from zero in one single point only. Since the Lebesgue integral does not change if the integrant is changed on a set of zero measure one can conclude

$$g(x) = \frac{1}{\sqrt{2\pi}} \int_{\mathbb{R}} \mathcal{F}\{g\}(\omega) e^{\imath\omega x} \, d\omega = \frac{1}{\sqrt{2\pi}} \int_{\mathbb{R}} 0 e^{\imath\omega x} \, d\omega = 0. \tag{2.66}$$

Summarizing, one can say that the conditions (ii), (iii), (iv) and (v) in the definition of an MRA are fulfilled. The condition (i), which is of a bit more technical is not verified. We will suppose for the moment that this condition is fulfilled as well. Then the spaces (2.62) form an MRA of $L_2(\mathbb{R})$.

Now the question arises how the MRA is connected to the original goal, the construction of a tight wavelet frame. The answer is that the bases of the difference spaces W_m of an MRA will form a tight wavelet frame, provided the scaling function φ of the MRA has some nice properties. For this reason the properties of φ will be studied first.

Lemma 2.4.1. *Let φ be the scaling function of an MRA. There exist a uniquely determined sequence of real numbers $\{h_k\}_{k\in\mathbb{Z}}$ with*

$$\varphi(x) = \sqrt{2} \sum_{k\in\mathbb{Z}} h_k \varphi(2x - k). \tag{2.67}$$

Proof. The scaling function φ is an element of V_0. Since $V_0 \subset V_{-1}$ holds, $\varphi \in V_{-1}$ can be concluded. The functions $\{\sqrt{2}\varphi(2\bullet - k)\}_{k\in\mathbb{Z}}\}$ form a Riesz base of V_{-1}. Hence there are uniquely determined real numbers h_k with

$$\varphi(x) = \sqrt{2} \sum_{k\in\mathbb{Z}} h_k \varphi(2x - k).$$

□

Definition 2.4.2. Equation (2.67) is called *scaling equation* of the MRA and the coefficients h_k are called *scaling coefficients*.

As we remember, the shifted versions of the scale function φ form a Riesz base of V_0. Such a Riesz base is particularly nice if it is orthogonal:

$$\langle \varphi(\bullet - k), \varphi(\bullet - l) \rangle = \delta_{kl}. \tag{2.68}$$

Whether or not a scaling function forms an orthogonal Riesz base can be read from its scaling coefficients.

Lemma 2.4.2. *Let φ be the scaling function of an MRA. Let the shifted versions of φ form an orthogonal base of V_0. Then*

$$\sum_{k\in\mathbb{Z}} h_k h_{k+2m} = \delta_{0m} \tag{2.69}$$

holds.

Proof. A simple calculation yields

$$\begin{aligned}\delta_{0m} &= \langle\varphi(\bullet), \varphi(\bullet+m)\rangle \\ &= \Big\langle\sqrt{2}\sum_{k\in\mathbb{Z}} h_k\varphi(2x-k), \sqrt{2}\sum_{l\in\mathbb{Z}} h_l\varphi(2(x+m)-l)\Big\rangle \\ &= 2\sum_{k\in\mathbb{Z}}\sum_{l\in\mathbb{Z}} h_k h_l \langle\varphi(2x-k), \varphi(2x-l+2m)\rangle \\ &= 2\sum_{k\in\mathbb{Z}}\sum_{l\in\mathbb{Z}} h_k h_{l+2m} \langle\varphi(2x-k), \varphi(2x-l)\rangle \\ &= \sum_{k\in\mathbb{Z}}\sum_{l\in\mathbb{Z}} h_k h_{l+2m}\delta_{kl} \\ &= \sum_{k\in\mathbb{Z}} h_k h_{k+2m}.\end{aligned}$$

□

Lemma 2.4.3. *Let φ be the scaling function of an MRA. Then $\{\varphi(\bullet-k) \mid k\in\mathbb{Z}\}$ forms an orthogonal system if and only if*

$$\sum_{n\in\mathbb{Z}} |\hat{\varphi}(\omega+2\pi n)|^2 = \frac{1}{2\pi} \tag{2.70}$$

holds.

Proof. Suppose that $\{\varphi(\bullet-k) \mid k\in\mathbb{Z}\}$ is an orthogonal system. Then

$$\langle\varphi(\bullet-k), \varphi(\bullet-n)\rangle = \delta_{k,n}$$

holds. As a consequence one obtains

$$\begin{aligned}\delta_{0k} &= \langle\hat{\varphi(\bullet)}, \hat{\varphi}(\bullet)e^{-\imath k\bullet}\rangle \\ &= \int_{\mathbb{R}} |\hat{\varphi}(\omega)|^2 e^{\imath k\omega}\, d\omega \\ &= \int_0^{2\pi} \sum_{n\in\mathbb{Z}} |\hat{\varphi}(\omega+2\pi n)|^2 e^{\imath k\omega}\, d\omega.\end{aligned}$$

The equation can be interpreted as the computation of the k-th Fourier coefficient of $\sum_{n\in\mathbb{Z}} |\hat{\varphi}(\omega + 2\pi n)|^2$, which is shown to be $2\pi \cdot \delta_{0k}$. For $k = 0$ the result

$$\sum_{n\in\mathbb{Z}} |\hat{\varphi}(\omega + 2\pi n)|^2 = \frac{1}{2\pi}$$

follows immediately. □

For many purposes it will be convenient, to express the scaling equations in the Fourier domain. On this behalf a transfer function of the scaling coefficients is defined by

$$H(\omega) := \frac{\sqrt{2}}{2} \sum_{k\in\mathbb{Z}} h_k e^{-\imath k\omega}. \tag{2.71}$$

Consequently, the Fourier transformation of the scaling equation (2.67) results in the following simple product:

$$\hat{\varphi}(\omega) = H\left(\frac{\omega}{2}\right) \hat{\varphi}\left(\frac{\omega}{2}\right). \tag{2.72}$$

The orthogonality of the scaling function is characterized by the following theorem.

Theorem 2.4.1. *Let $\{V_m\}$ be an MRA of $L_2(\mathbb{R})$, which is generated by the orthogonal scaling function φ. Then the following relations are true:*

$$|H(\omega)|^2 + |H(\omega + \pi)|^2 = 1, \tag{2.73}$$

$$H(0) = 1, \quad H(\pi) = 0, \tag{2.74}$$

and

$$\sum_{k\in\mathbb{Z}} h_k = \sqrt{2}, \quad \sum_{k\in\mathbb{Z}} (-1)^k h_k = 0. \tag{2.75}$$

Proof. Obviously $\hat{\varphi}(0) \neq 0$ holds. With the help of (2.72) one obtains

$$H(0) = \frac{\hat{\varphi}(0)}{\hat{\varphi}(0)} = 1.$$

Let us now assume that the orthogonality condition (2.70) is fulfilled. Together with $H(0) = 1$ this leads to $H(\pi) = 0$. Since the scaling function φ is supposed to be orthogonal, equation (2.73) holds. If the relation (2.72) is inserted, the following result is obtained:

$$
\begin{aligned}
\frac{1}{2\pi} &= \sum_{n\in\mathbb{Z}} |\hat{\varphi}(\omega + 2\pi n)|^2 \\
&= \sum_{n\in\mathbb{Z}} \left| H\left(\frac{\omega}{2} + n\pi\right) \hat{\varphi}\left(\frac{\omega}{2} + n\pi\right)\right|^2 \\
&= \sum_{m\in\mathbb{Z}} \left| H\left(\frac{\omega}{2} + 2\pi m\right) \hat{\varphi}\left(\frac{\omega}{2} + 2\pi m\right)\right|^2 \\
&\quad + \sum_{m\in\mathbb{Z}} \left| H\left(\left(\frac{\omega}{2} + \pi\right) + 2\pi m\right) \hat{\varphi}\left(\left(\frac{\omega}{2} + \pi\right) + 2\pi m\right)\right|^2 \\
&= \left| H\left(\frac{\omega}{2}\right) \right|^2 \sum_{m\in\mathbb{Z}} \left|\hat{\varphi}\left(\frac{\omega}{2} + 2\pi m\right)\right|^2 + \left| H\left(\frac{\omega}{2} + \pi\right)\right|^2 \sum_{m\in\mathbb{Z}} \left|\hat{\varphi}\left(\left(\frac{\omega}{2} + \pi\right) + 2\pi m\right)\right|^2 \\
&= \left| H\left(\frac{\omega}{2}\right)\right|^2 \frac{1}{2\pi} + \left| H\left(\frac{\omega}{2} + \pi\right)\right|^2 \frac{1}{2\pi}.
\end{aligned}
$$

Hence

$$1 = |H(\omega)|^2 + |H(\omega + \pi)|^2$$

holds. □

At the moment we have the decomposition $V_{-1} = V_0 \oplus W_0$ and we have a base of V_0. What is still needed is a base of W_0. Since W_0 is the orthogonal complement of V_0 in V_{-1}, any base function ψ of W_0 has to be orthogonal to every element of V_0. This is certainly true, if ψ is orthogonal to every base function $\varphi(\bullet - k)$ of V_0. Due to (2.67) the base function has the representation

$$\varphi(x - k) = \sqrt{2} \sum_{l\in\mathbb{Z}} h_l \varphi(2x - 2k - l).$$

Hence the orthogonality condition reads

$$0 = \left\langle \psi, \sqrt{2} \sum_{l\in\mathbb{Z}} h_l \varphi(2x - 2k - l) \right\rangle. \tag{2.76}$$

Since $\psi \in W_0 \subset V_{-1}$, there are coefficients g_m with

$$\psi(x) = \sqrt{2} \sum_{m\in\mathbb{Z}} g_m \varphi(2x - m). \tag{2.77}$$

Inserting this in (2.76) one obtains

$$
\begin{aligned}
0 &= \left\langle \sqrt{2} \sum_{m\in\mathbb{Z}} g_m \varphi(2x - m), \sqrt{2} \sum_{l\in\mathbb{Z}} h_l \varphi(2x - 2k - l) \right\rangle \\
&= 2 \sum_{m\in\mathbb{Z}} \sum_{l\in\mathbb{Z}} g_m h_l \langle \varphi(2x - m), \varphi(2x - 2k - l)\rangle =
\end{aligned}
$$

$$= 2\sum_{m\in\mathbb{Z}}\sum_{l\in\mathbb{Z}} g_m h_{l-2k}\langle\varphi(2x-m), \varphi(2x-l)\rangle$$
$$= \sum_{m\in\mathbb{Z}} h_{m-2k} g_m.$$

This relation is an infinite-dimensional homogeneous linear system of equations for the coefficients g_m of the base function $\psi \in W_0$. Of course the system has several solutions. One solution is

$$g_m = (-1)^m h_{1-m}, \tag{2.78}$$

which can easily be verified:

$$\begin{aligned}
\sum_{m\in\mathbb{Z}} h_{m-2k} g_m &= \sum_{m\in\mathbb{Z}} h_{m-2k}(-1)^m h_{1-m} \\
&= \sum_{m\in\mathbb{Z}} h_m(-1)^m h_{1-m-2k} \\
&= \sum_{m\in\mathbb{Z}} h_{2m} h_{1-2(m+k)} - \sum_{m\in\mathbb{Z}} h_{2m+1} h_{-2(m+k)} \\
&= \sum_{\mu\in\mathbb{Z}} h_{-2(\mu+k)} h_{2\mu+1} - \sum_{m\in\mathbb{Z}} h_{2m+1} h_{-2(m+k)} \quad (\mu = -(m+k)) \\
&= 0.
\end{aligned}$$

In this way we have found a special element $\psi \in W_0$. But this particular element has additional properties:

- ψ is a wavelet,
- the shifted versions of ψ form a Riesz base of W_0.

Theorem 2.4.2. *Let $\{V_m\}_{\in\mathbb{Z}}$ be an MRA of $L_2(\mathbb{R})$, generated by the orthogonal scaling function φ. Let $\{h_k\}_{k\in\mathbb{Z}}$ be the scaling coefficients of φ. The function $\psi \in V_{-1}$ defined by*

$$\psi(x) := \sqrt{2}\sum_{k\in\mathbb{Z}} g_k\varphi(2x-k), \quad g_k := (-1)^k h_{1-k} \tag{2.79}$$

has the following properties:

(i) *ψ is a wavelet with $c_\psi = 2\ln 2$.*

(ii) *$\{\psi_{m,k}(x) := 2^{-m/2}\psi(2^{-m}x-k) \mid k \in \mathbb{Z}\}$ is a CONS[1] of W_m.*

(iii) *$\{\psi_{m,k} \mid m, k \in \mathbb{Z}\}$ is a tight wavelet frame in $L_2(\mathbb{R})$.*

[1] Complete orthonormal system, see Definition A.2.5 on p. 224

Proof. The proof can be found in [68]. □

The theorem shows the way for the systematic construction of tight wavelet frames

1. Construction of an MRA with an orthogonal scaling function φ.
2. Determination of the scaling coefficients $\{h_k\}_{k\in\mathbb{Z}}$.
3. Construction of the mother-wavelet ψ by (2.79).

Example 2.4.2. The earlier discussed example of an MRA spanned by the sinc function is not directly suited for the construction of a tight wavelet frame. Since

$$\begin{aligned}\left\langle \frac{\sin(\pi(x-k))}{\pi(x-k)}, \frac{\sin(\pi(x-l))}{\pi(x-l)} \right\rangle &= \left\langle \mathcal{F}\left\{\frac{\sin(\pi(x-k))}{\pi(x-k)}\right\}, \overline{\mathcal{F}\left\{\frac{\sin(\pi(x-l))}{\pi(x-l)}\right\}} \right\rangle \\ &= \left\langle \Pi\left(\frac{\omega}{\pi}\right) e^{-\imath\pi k\omega}, \Pi\left(\frac{\omega}{\pi}\right) e^{\imath\pi l\omega} \right\rangle \\ &= \int_{-\pi}^{\pi} e^{-\imath(l-k)\omega}\, d\omega \\ &= 2\pi\,\delta_{lk}\end{aligned}$$

holds, the scaling function φ is orthogonal. From (1.23) in the sampling theorem one can conclude

$$\begin{aligned}\varphi(x) &= \frac{\sin \pi x}{\pi x} \\ &= \sum_{n\in\mathbb{Z}} \frac{\sin(\pi n/2)}{\pi n/2} \frac{\sin(2\pi(x-n/2))}{2\pi(x-n/2)} \\ &= \sum_{n\in\mathbb{Z}} \frac{\sin(\pi n/2)}{\pi n/2} \frac{\sin(\pi(2x-n))}{\pi(2x-n)} \\ &= \sum_{n\in\mathbb{Z}} \frac{\sin(\pi n/2)}{\pi n/2} \varphi(2x-n).\end{aligned}$$

This leads to the scaling coefficients

$$h_n = \frac{\sin(\pi n/2)}{\pi n/2} \tag{2.80}$$

and to the mother wavelet

$$\psi(x) = \sum_{n\in\mathbb{Z}} (-1)^n \frac{\sin(\pi(1-n)/2)}{\pi(1-n)/2} \frac{\sin(\pi(2x-n))}{\pi(2x-n)}. \tag{2.81}$$

The resulting Shannon wavelet is displayed in Figure 2.20.

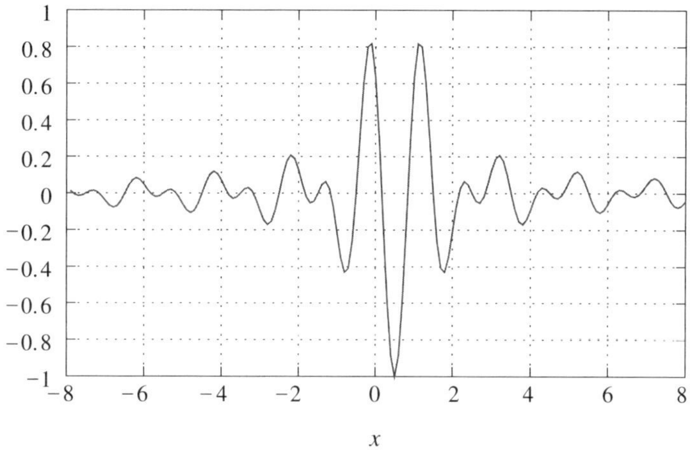

Figure 2.20. Shannon wavelet

Nevertheless, for practical applications the Shannon wavelet is not particularly suited, since it has an infinite number of non-vanishing scaling coefficients. Wavelets with a small number of non-vanishing scaling coefficients are much better to handle. For this reason in the next example a wavelet with the smallest possible number of non-vanishing scaling coefficients will be discussed.

Example 2.4.3 (Haar wavelet). Following our previously mentioned algorithm, let us start with the construction of an MRA of $L_2(\mathbb{R})$. The spaces V_m are defined by

$$V_m := \{f \in L_2(\mathbb{R}) \mid f(x) = \text{const for } x \in [2^m k, 2^m(k+1)[,\ k \in \mathbb{Z}\}. \tag{2.82}$$

It is easy to show that

$$\begin{aligned} f \in V_m &\Rightarrow f(x) = \text{const for } x \in [2^m k, 2^m(k+1)[\\ &\Rightarrow f(x) = \text{const for } x \in [2^{(m-1)}k, 2^{(m-1)}(k+1)[\\ &\Rightarrow f \in V_{m-1} \end{aligned} \tag{2.83}$$

holds. Hence

$$\{0\} \subset \cdots \subset V_1 \subset V_0 \subset V_{-1} \subset \cdots \subset L_2(\mathbb{R}) \tag{2.84}$$

is valid. Furthermore,

$$\begin{aligned} f \in V_m &\Rightarrow f(x) = \text{const for } x \in [2^m k, 2^m(k+1)[\\ &\Rightarrow f(2^m x) = \text{const for } x \in [k, k+1[\\ &\Rightarrow f(2^m \bullet) \in V_0 \end{aligned} \tag{2.85}$$

is valid. Let

$$\varphi(x) := \begin{cases} 1, & 0 \le x < 1 \\ 0, & \text{else}, \end{cases} \tag{2.86}$$

then obviously

$$V_0 = \overline{\text{span}\{\varphi(\bullet - k) \mid k \in \mathbb{Z}\}} \tag{2.87}$$

holds. Consequently, the spaces V_m form an MRA of $L_2(\mathbb{R})$ with the scaling function φ, the so-called Haar scaling function. The scaling function is displayed in Figure 2.21.

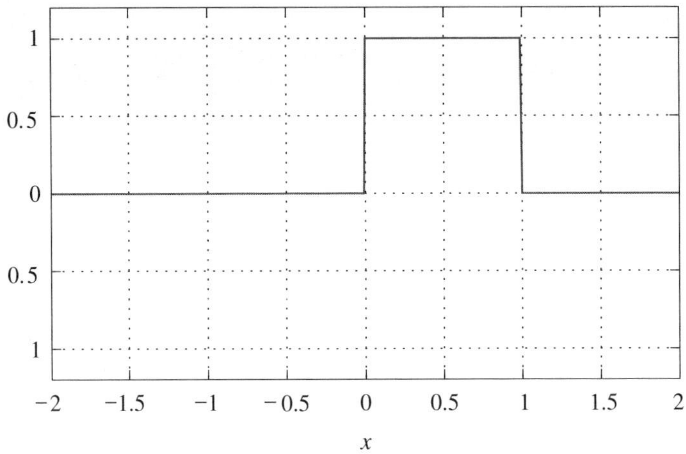

Figure 2.21. Haar scaling function

The next step is the determination of the scaling coefficients. Obviously,

$$\varphi_{1,k}(x) = \sqrt{2}\varphi(2x - k) = \sqrt{2} \begin{cases} 1, & \frac{k}{2} \le x < \frac{k+1}{2} \\ 0, & \text{else} \end{cases} \tag{2.88}$$

holds. Hence

$$\varphi(x) = \left(\frac{1}{\sqrt{2}} \varphi_{1,0}(x) + \frac{1}{\sqrt{2}} \varphi_{1,1}(x) \right) \tag{2.89}$$

is valid. Consequently, the scaling coefficients of the Haar scaling function are

$$h_0 = h_1 = \frac{1}{\sqrt{2}}, \qquad h_k = 0 \text{ else.} \tag{2.90}$$

The last step is the determination of the Haar wavelet

$$\psi(x) = (h_1 \varphi_{1,0}(x) - h_0 \varphi_{1,1}(x)) = \begin{cases} 1, & 0 \le x < 0.5 \\ -1, & 0.5 \le x < 1 \\ 0, & \text{else.} \end{cases} \tag{2.91}$$

The graph of the Haar wavelet ψ is plotted in Figure 2.22.

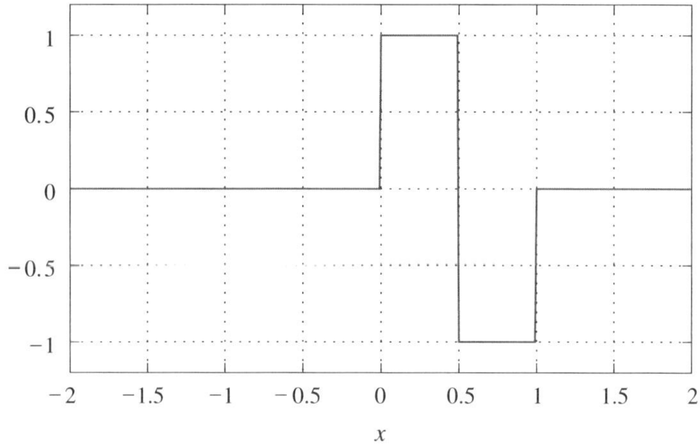

Figure 2.22. Haar wavelet

So far a systematic tool for the construction of a tight wavelet frame was developed. Now the question has to be addressed, how the Fourier coefficients of a given signal f with respect to a wavelet frame can be computed. This will lead to the famous Mallat algorithm.

2.5 Mallat algorithm

Using the technique of MRA, it was possible to construct a tight wavelet frame $\{\psi_{m,n}\}_{m,n\in\mathbb{Z}}$ in $L_2(\mathbb{R})$. Therefore every signal $f \in L_2(\mathbb{R})$ can be represented by

$$f(x) = \sum_{m\in\mathbb{Z}} \sum_{n\in\mathbb{Z}} d_n^{(m)} \psi_{m,n}(x), \quad d_n^{(m)} = \langle f, \psi_{m,n} \rangle. \tag{2.92}$$

Of course it does not make much sense to compute the coefficients $d_n^{(m)}$ by the numerical evaluation of the scalar products. Especially for large values of m the support of $\psi_{m,n}$ will grow very large and the numerical evaluation would require numerical quadrature formulas with many nodes. This would make discrete wavelet transformation very unattractive from the numerical point of view. Fortunately, there are possibilities to compute the coefficients of the scale m recursively from the coefficients on scale $m-1$. This is the core of Mallat's algorithm.

Let $\{V_m\}$ be an MRA of $L_2(\mathbb{R})$ and let us consider a signal $f \in V_0$. Then this signal has the following representation with respect to the scaling function φ:

$$f(x) = \sum_{k\in\mathbb{Z}} c_k^{(0)} \varphi(x-k). \tag{2.93}$$

Let us further introduce the following abbreviations

$$d_k^{(m)} = \langle f, \psi_{m,k} \rangle, \quad c_k^{(m)} = \langle f, \varphi_{m,k} \rangle. \tag{2.94}$$

Then with the help of the scaling equation (2.67) and the definition equation of the wavelet (2.79) one can conclude:

$$\begin{aligned}
d_k^{(m)} &= \langle f, \psi_{m,k} \rangle \\
&= \langle f, 2^{-m/2} \psi(2^{-m}x - k) \rangle \\
&= \Big\langle f, 2^{-m/2} \sqrt{2} \sum_{l \in \mathbb{Z}} g_l \varphi(2(2^{-m}x - k) - l) \Big\rangle \\
&= \sum_{l \in \mathbb{Z}} g_l \langle f, 2^{-\frac{m-1}{2}} \varphi(2^{-(m-1)}x - (2k + l)) \rangle \\
&= \sum_{l \in \mathbb{Z}} g_l \langle f, \varphi_{m-1,2k+l} \rangle \\
&= \sum_{l \in \mathbb{Z}} g_{l-2k} \langle f, \varphi_{m-1,l} \rangle \\
&= \sum_{l \in \mathbb{Z}} g_{l-2k} c_l^{(m-1)}.
\end{aligned}$$

In a completely analog way the corresponding result

$$c_k^{(m)} = \sum_{l \in \mathbb{Z}} h_{l-2k} c_l^{(m-1)}$$

can be proved. Together, we have the two recursions

$$c_k^{(m)} = \sum_{l \in \mathbb{Z}} h_{l-2k} c_l^{(m-1)}, \tag{2.95}$$

$$d_k^{(m)} = \sum_{l \in \mathbb{Z}} g_{l-2k} c_l^{(m-1)}. \tag{2.96}$$

The two equations (2.95) and (2.96) constitute already the Mallat algorithm for the fast computation of the wavelet coefficients of a given signal $f \in V_0$. If the representation (2.93) of the signal f with respect to the Riesz base of V_0 is already known, formula (2.95) computes the coefficients $c_k^{(1)}$ of the projection $P_1 f$ of f in the space V_1. Formula (2.96) computes the coefficients $d_k^{(1)}$ of the representation of the projection $Q_1 f$ of f in the space W_1. The coefficients $d_k^{(1)}$ are already a final result. They are the coefficients of the wavelet spectrum of f on the scale 1. In order to obtain the wavelet spectrum on the coarser scales 2, 3, ... the procedure will be repeated for $c_k^{(1)}$. The application of (2.95) will give the coefficients $c_k^{(2)}$ and formula (2.96) computes the wavelet spectrum $d_k^{(2)}$ on the scale 2. In this way the complete wavelet spectrum $d_k^{(m)}$ can be computed recursively. No integral has to be evaluated.

To be a bit more precise, the procedure will be described using decomposition operators H and G. Let us define

$$\begin{aligned} H : l_2(\mathbb{Z}) &\to l_2(\mathbb{Z}) \\ c \mapsto Hc &= \Big\{(Hc)_k = \sum_{l\in\mathbb{Z}} h_{l-2k} c_l\Big\} \end{aligned} \tag{2.97}$$

and

$$\begin{aligned} G : l_2(\mathbb{Z}) &\to l_2(\mathbb{Z}) \\ c \mapsto Gc &= \Big\{(Gc)_k = \sum_{l\in\mathbb{Z}} g_{l-2k} c_l\Big\}. \end{aligned} \tag{2.98}$$

This leads to the following scheme (Figure 2.23) for the computation of the wavelet spectrum.

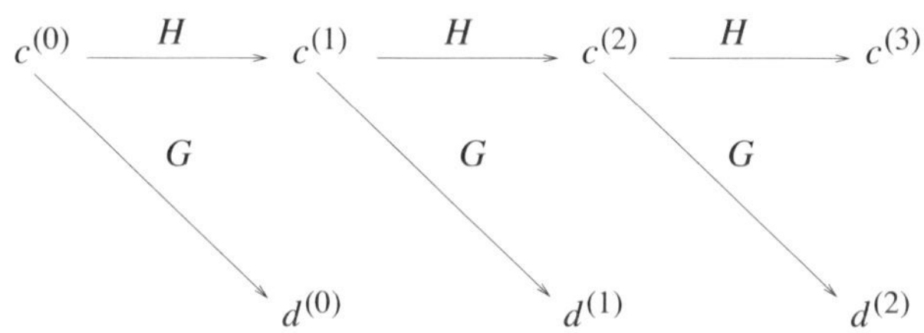

Figure 2.23. Scheme of the Mallat algorithm

Now the inverse problem has to be discussed. Suppose that the wavelet spectrum $\{d^{(1)}, d^{(2)}, d^{(3)}, \ldots\}$ is given. How can the signal f be reconstructed?

To answer this question, we consider the decomposition $V_0 = V_1 \oplus W_1$. For an element of V_0 holds

$$\begin{aligned} \sum_{k\in\mathbb{Z}} c_k^{(0)} \varphi_{0,k} &= \sum_{j\in\mathbb{Z}} c_j^{(1)} \varphi_{1,j} + \sum_{j\in\mathbb{Z}} d_j^{(1)} \psi_{1,j} \\ &= \sum_{j\in\mathbb{Z}} c_j^{(1)} \sum_{l\in\mathbb{Z}} h_l \varphi_{0,2j+l} + \sum_{j\in\mathbb{Z}} d_j^{(1)} \sum_{l\in\mathbb{Z}} g_l \varphi_{0,2j+l} \\ &= \sum_{j\in\mathbb{Z}} c_j^{(1)} \sum_{l\in\mathbb{Z}} h_{2j-l} \varphi_{0,l} + \sum_{j\in\mathbb{Z}} d_j^{(1)} \sum_{l\in\mathbb{Z}} g_{2j-l} \varphi_{0,l} \\ &= \sum_{l\in\mathbb{Z}} \Big(\sum_{j\in\mathbb{Z}} c_j^{(1)} h_{2j-l} + d_j^{(1)} g_{2j-l} \Big) \varphi_{0,l}. \end{aligned}$$

A comparison of coefficients yields

$$c_k^{(0)} = \Big(\sum_{j\in\mathbb{Z}} c_j^{(1)} h_{2j-k} + d_j^{(1)} g_{2j-k}\Big). \tag{2.99}$$

This means, given the coefficients $c^{(1)}$ of $P_1 f \in V_1$ and the coefficients $d^{(1)}$ of $Q_1 f \in W_1$ the coefficients of the original signal $f \in V_0$ can be reconstructed by means of (2.99).

Again, this reconstruction can be expressed by operators H^*, G^*.

Lemma 2.5.1. *Let H, G be the decomposition operators* (2.97), (2.98). *Then their adjoint operators H^*, G^* are given by*

$$\begin{aligned} H^* : l_2(\mathbb{Z}) &\to l_2(\mathbb{Z}) \\ c \mapsto H^*c &= \Big\{(H^*c)_k = \sum_{l\in\mathbb{Z}} h_{k-2l} c_l\Big\} \end{aligned} \tag{2.100}$$

and

$$\begin{aligned} G^* : l_2(\mathbb{Z}) &\to l_2(\mathbb{Z}) \\ c \mapsto G^*c &= \Big\{(G^*c)_k = \sum_{l\in\mathbb{Z}} g_{k-2l} c_l\Big\}. \end{aligned} \tag{2.101}$$

Proof.

$$\begin{aligned} \langle Hc, b\rangle_{l_2} &= \sum_k (Hc)_k b_k \\ &= \sum_k \Big(\sum_l h_{l-2k} c_l\Big) b_k \\ &= \sum_l c_l \Big(\sum_k h_{l-2k} b_k\Big) \\ &= \sum_l c_l (H^*b)_l \\ &= \langle c, H^*b\rangle_{l_2}. \end{aligned}$$

Hence H^* is the adjoint operator of H. In the same way it can be shown that G^* is the adjoint operator of G. □

With the help of the adjoint operators a single reconstruction step can be written as

$$c^{(m-1)} = H^*c^{(m)} + G^*d^{(m)}. \tag{2.102}$$

A schematic view of the inverse Mallat algorithm is shown in Figure 2.24.

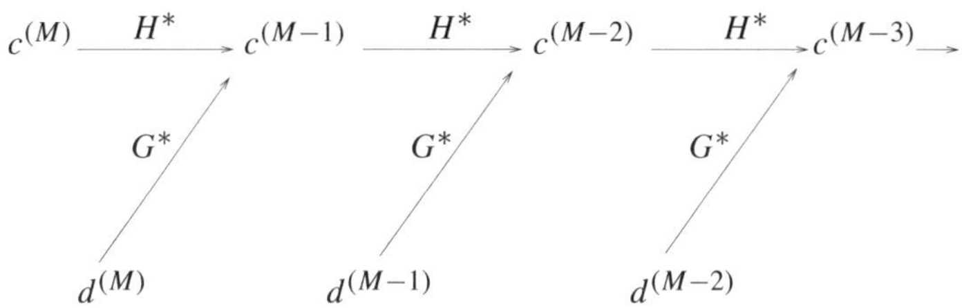

Figure 2.24. Scheme of inverse Mallat algorithm

So far, we have only considered signals from the space $L_2(\mathbb{R})$ having an infinite coefficient sequence $c^{(0)} \in l_2$. In practical applications the number of sampled signal values and therefore the number of coefficients $c_k^{(0)}$ will be finite. For a finite number of coefficients and for scaling equations of finite length the Mallat algorithm can be represented as a sequence of matrix multiplications.

Let us assume that the signal has a finite number of initial coefficients $c_0^{(0)}, \ldots, c_{15}^{(0)}$ and let us further assume that the scaling equation (2.67) has only four non-vanishing coefficients $h_0 \ldots, h_3$. The smoothing step (2.95) is

$$\begin{aligned}
c_0^{(1)} &= h_0 \cdot c_0^{(0)} + h_1 \cdot c_1^{(0)} + \cdots + h_4 \cdot c_4^{(0)} \\
c_1^{(1)} &= h_0 \cdot c_2^{(0)} + h_1 \cdot c_3^{(0)} + \cdots + h_4 \cdot c_6^{(0)} \\
&\vdots \\
c_6^{(1)} &= h_0 \cdot c_{12}^{(0)} + h_1 \cdot c_{13}^{(0)} + \cdots + h_4 \cdot c_{15}^{(0)} \\
c_7^{(1)} &= h_2 \cdot c_0^{(0)} + h_3 \cdot c_1^{(0)} + h_0 \cdot c_{14}^{(0)} + h_1 \cdot c_{15}^{(0)}.
\end{aligned}$$

The last equation of this group is generated by periodic continuation of the input stream $c_0^{(0)}, \ldots, c_{15}^{(0)}$. Basically, the last equation would be

$$c_7^{(1)} = h_0 \cdot c_{14}^{(0)} + h_1 \cdot c_{15}^{(0)} + h_2 \cdot c_{16}^{(0)} + h_3 \cdot c_{17}^{(0)}.$$

But since the coefficients $c_{16}^{(0)}, c_{17}^{(0)}$ do not exist, they are replaced by the periodic continuation $c_0^{(0)}, c_1^{(0)}$. In matrix notation this can be written as

$$\begin{pmatrix} c_0^{(1)} \\ c_1^{(1)} \\ \vdots \\ c_7^{(1)} \end{pmatrix} = \begin{pmatrix} h_0 & h_1 & h_2 & h_3 & & & & & & \\ & & h_0 & h_1 & h_2 & h_3 & & & & \\ \vdots & \vdots & & & & & h_0 & h_1 & h_2 & h_3 \\ h_2 & h_3 & \ldots & & & & & & h_0 & h_1 \end{pmatrix} \cdot \begin{pmatrix} c_0^{(0)} \\ c_1^{(1)} \\ \vdots \\ c_{15}^{(0)} \end{pmatrix}. \tag{2.103}$$

In a completely analog way the differencing step (2.96) can be written as

$$\begin{aligned}
d_0^{(1)} &= g_0 \cdot c_0^{(0)} + g_1 \cdot c_1^{(0)} + \cdots + g_4 \cdot c_4^{(0)} \\
d_1^{(1)} &= g_0 \cdot c_2^{(0)} + g_1 \cdot c_3^{(0)} + \cdots + g_4 \cdot c_6^{(0)} \\
&\vdots \\
d_6^{(1)} &= g_0 \cdot c_{12}^{(0)} + g_1 \cdot c_{13}^{(0)} + \cdots + g_4 \cdot c_{15}^{(0)} \\
d_7^{(1)} &= g_2 \cdot c_0^{(0)} + g_3 \cdot c_1^{(0)} + g_0 \cdot c_{14}^{(0)} + g_1 \cdot c_{15}^{(0)},
\end{aligned}$$

or in matrix form

$$\begin{pmatrix} d_0^{(1)} \\ d_1^{(1)} \\ \vdots \\ d_7^{(1)} \end{pmatrix} = \begin{pmatrix} g_0 & g_1 & g_2 & g_3 & & & & & & \\ & & g_0 & g_1 & g_2 & g_3 & & & & \\ \vdots & \vdots & & & & & & & & \\ & & & & & & h_0 & g_1 & g_2 & g_3 \\ g_2 & g_3 & \cdots & & & & & & g_0 & g_1 \end{pmatrix} \cdot \begin{pmatrix} c_0^{(0)} \\ c_1^{(1)} \\ \vdots \\ c_{15}^{(0)} \end{pmatrix}. \tag{2.104}$$

Therefore, for finite signals the operators H, G are the matrices

$$H = \begin{pmatrix} h_0 & h_1 & h_2 & h_3 & & & & & & \\ & & h_0 & h_1 & h_2 & h_3 & & & & \\ \vdots & \vdots & & & & & & & & \\ & & & & & & h_0 & h_1 & h_2 & h_3 \\ h_2 & h_3 & \cdots & & & & & & h_0 & h_1 \end{pmatrix} \tag{2.105}$$

and

$$G = \begin{pmatrix} g_0 & g_1 & g_2 & g_3 & & & & & & \\ & & g_0 & g_1 & g_2 & g_3 & & & & \\ \vdots & \vdots & & & & & & & & \\ & & & & & & h_0 & g_1 & g_2 & g_3 \\ g_2 & g_3 & \cdots & & & & & & g_0 & g_1 \end{pmatrix}. \tag{2.106}$$

For a finite signal the reconstruction (2.99) of $c^{(0)}$ from $c^{(1)}$ and $d^{(1)}$ can be written as

$$\begin{aligned}
c_0^{(0)} &= h_0 \cdot c_0^{(1)} + h_2 \cdot c_{15}^{(1)} + g_0 \cdot d_0^{(1)} + g_2 \cdot d_{15}^{(1)} \\
c_1^{(0)} &= h_1 \cdot c_0^{(1)} + h_3 \cdot c_{15}^{(1)} + g_1 \cdot d_0^{(1)} + g_3 \cdot d_{15}^{(1)} \\
c_2^{(0)} &= h_0 \cdot c_1^{(1)} + h_2 \cdot c_0^{(1)} + g_0 \cdot d_1^{(1)} + g_2 \cdot d_0^{(1)} \\
c_3^{(0)} &= h_3 \cdot c_0^{(1)} + h_1 \cdot c_1^{(1)} + h_3 \cdot c_0^{(1)} + h_1 \cdot c_1^{(1)} \\
&\vdots \\
c_{14}^{(0)} &= h_0 \cdot c_7^{(1)} + h_2 \cdot c_6^{(1)} + g_0 \cdot d_7^{(1)} + g_2 \cdot d_6^{(1)} \\
c_{15}^{(0)} &= h_1 \cdot c_7^{(1)} + h_3 \cdot c_6^{(1)} + g_1 \cdot d_7^{(1)} + g_3 \cdot d_6^{(1)},
\end{aligned}$$

or in matrix form

$$\begin{pmatrix} c_0^{(0)} \\ c_1^{(0)} \\ \vdots \\ c_{14}^{(0)} \\ c_{15}^{(0)} \end{pmatrix} = \begin{pmatrix} h_0 & & & h_2 \\ h_1 & & & h_3 \\ & \ddots & & \\ & & h_2 & h_0 \\ & & h_3 & h_1 \end{pmatrix} \cdot \begin{pmatrix} c_0^{(1)} \\ c_1^{(1)} \\ \vdots \\ c_6^{(1)} \\ c_7^{(1)} \end{pmatrix} + \begin{pmatrix} g_0 & & & g_2 \\ g_1 & & & g_3 \\ & \ddots & & \\ & & g_2 & g_0 \\ & & g_3 & g_1 \end{pmatrix} \cdot \begin{pmatrix} c_0^{(1)} \\ c_1^{(1)} \\ \vdots \\ c_6^{(1)} \\ c_7^{(1)} \end{pmatrix}. \tag{2.107}$$

Evidently, the reconstruction operators H^*, G^* are the matrices

$$H^* = \begin{pmatrix} h_0 & & & h_2 \\ h_1 & & & h_3 \\ & \ddots & & \\ & & h_2 & h_0 \\ & & h_3 & h_1 \end{pmatrix}, \tag{2.108}$$

$$G^* = \begin{pmatrix} g_0 & & & g_2 \\ g_1 & & & g_3 \\ & \ddots & & \\ & & g_2 & g_0 \\ & & g_3 & g_1 \end{pmatrix}. \tag{2.109}$$

Obviously $H^* = H^\top$, $G^* = G^\top$ holds, which is quite natural since for matrices the adjoint operator coincides with the transposed matrix.

Example 2.5.1. As an example the Mallat decomposition and reconstruction is to be considered for a simple 16-elements $c^{(0)}$ sequence:

$$c^{(0)} = \{0, 0, 1, 1, 1, 1, 1, 2, 3, 4, 5, 5, 5, 5, 5, 5\}.$$

The initial sequence is displayed in Figure 2.25.

The smoothing and the differencing filters are assumed to have only two non-vanishing elements

$$h_0 = 0.707107, \quad h_1 = 0.707107, \quad g_0 = 0.707107, \quad g_1 = -0.707107. \tag{2.110}$$

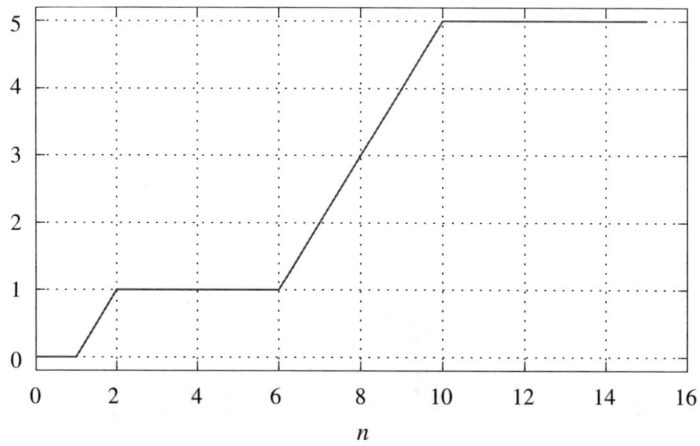

Figure 2.25. Initial sequence

Consequently, after the first decomposition step the following sequences are obtained:

$$c^{(1)} = \{0, 1.414214, 1.414214, 3.535535, 6.363963, 7.07107, 7.07107, 7.07107\}, \tag{2.111}$$

$$d^{(1)} = \{0, 0, 0, -0.707107, -0.707107, 0, 0, 0\}. \tag{2.112}$$

The smoothed and differenced sequences are shown in the Figure 2.26.

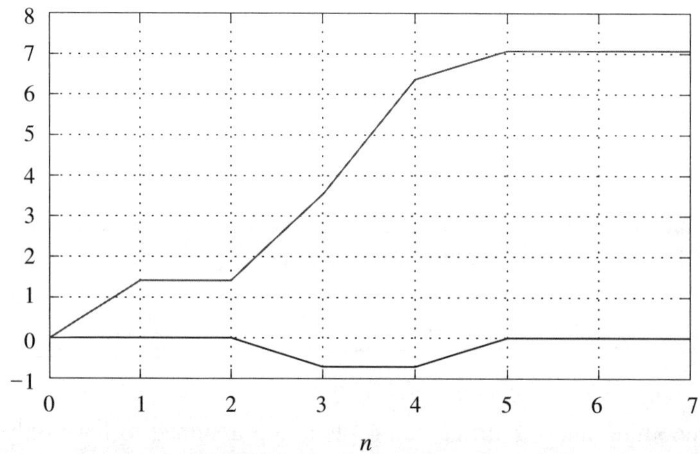

Figure 2.26. Outcome of the first decomposition step

This procedure can be repeated leading to the sequences

$$c^{(2)} = \{1, 3.5, 9.5, 10\}, \tag{2.113}$$

$$d^{(2)} = \{-1, -3.5, -0.707107, 0\}, \tag{2.114}$$

which are plotted in the Figure 2.27.

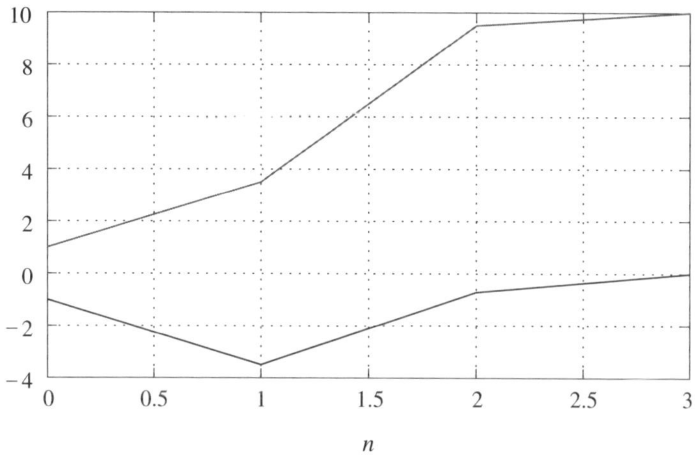

Figure 2.27. Outcome of the second decomposition step

Conversely, the reconstruction step applied to $c^{(2)}$, $d^{(2)}$ recovers $c^{(1)}$ and the reconstruction step applied to $c^{(1)}$, $d^{(1)}$ gives back $c^{(0)}$.

2.6 Wavelet packages

Frequently, in a frame $\lambda_0 = 2$ is chosen as scale spacing. Therefore the frame coefficients for different scales indicate the signal behavior in different octaves. For some applications it is necessary to have a finer frequency resolution. The simplest way to achieve this goal would be to decrease the frequency spacing parameter. But this would increase the computational load to an unacceptable value. To avoid this disadvantage, wavelet packages can be applied. Wavelet packages are generalizations of wavelets in such a way that each octave is split into finer frequency bands. This splitting trick is due to I. Daubechies.

Theorem 2.6.1. *Let $\{f(\bullet - k) \mid k \in \mathbb{Z}\}$ be an orthonormal base of $L_2(\mathbb{R})$, fulfilling the scaling equation*

$$f(x) = \sum_{k \in \mathbb{Z}} h_k f(2x - k). \tag{2.115}$$

With the scaling coefficients h_h *and the corresponding coefficients*

$$g_k := (-1)^k h_{1-k}, \tag{2.116}$$

two new functions can be defined:

$$F_1(x) := \sum_{k\in\mathbb{Z}} h_k f(x-k), \tag{2.117}$$
$$F_2(x) := \sum_{k\in\mathbb{Z}} g_k f(x-k). \tag{2.118}$$

Then $\{F_1(\bullet - 2k), F_2(\bullet - 2k)\}$ *is an orthonormal base for* $\text{span}\{f(\bullet - k)\}$.

Proof. [15]. □

The equations (2.117), (2.118) can be considered as two filters applied to the signal f. Their filter characteristics are:

$$H(\omega) = \sqrt{2}\sum_{k\in\mathbb{Z}} h_k e^{-\imath k\omega}, \tag{2.119}$$
$$G(\omega) = \sqrt{2}\sum_{k\in\mathbb{Z}} g_k e^{-\imath k\omega}. \tag{2.120}$$

As Figure 2.28 shows, H is a low-pass and G is a high-pass filter.

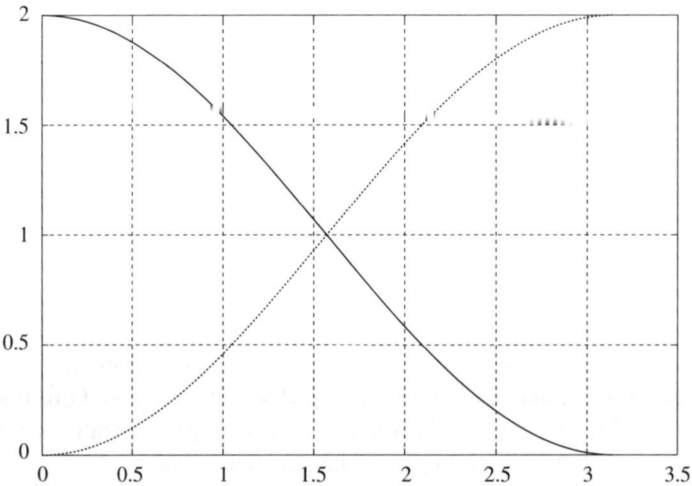

Figure 2.28. Filter characteristics H and G

Hence the functions F_1 and F_2 are the lower and the higher frequency part of f as it is shown in Figure 2.29.

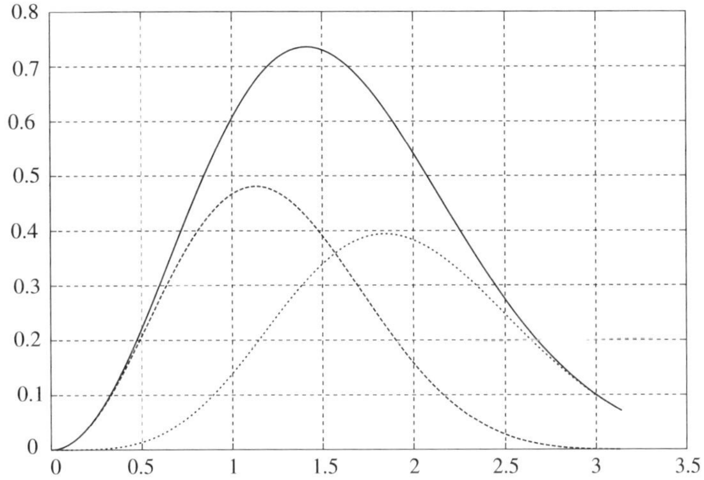

Figure 2.29. Spectra of f, F_1 and F_2

The functions F_1 and F_2 obtained by the splitting of f have a better frequency localization but of course at the price of having larger supports.

Now the idea of wavelet packages is to use a wavelet ψ as the function f to be split. This will give two new wavelets

$$\psi^{(1)}(x) = \sum_{k \in \mathbb{Z}} h_k \psi(x-k), \tag{2.121}$$

$$\psi^{(2)}(x) = \sum_{k \in \mathbb{Z}} g_k \psi(x-k), \tag{2.122}$$

which together span the space W_1:

$$\begin{aligned} W_1 &= \operatorname{span}\{\psi(\bullet - k) \mid k \in \mathbb{Z}\} \\ &= \operatorname{span}\{\psi^{(1)}(\bullet - 2k) \mid k \in \mathbb{Z}\} \oplus \{\psi^{(2)}(\bullet - 2k) \mid k \in \mathbb{Z}\} \\ &= W_1^1 \oplus W_1^2. \end{aligned}$$

The space W_1^1 contains the lower frequency part and the space W_1^2 contains the higher frequency part of the original space W_1. Using in the next step $f(x) = \frac{1}{\sqrt{2}}\psi(2^{-1}x)$ as function to be split, the space W_2 can be split into two bands:

$$W_2 = W_2^1 \oplus W_2^2. \tag{2.123}$$

This procedure can be repeated again and again leading to the MRA scheme shown in Figure 2.30. In contrast to the standard MRA scheme, for every differentiation step an additional differentiation and smoothing step has to be carried out.

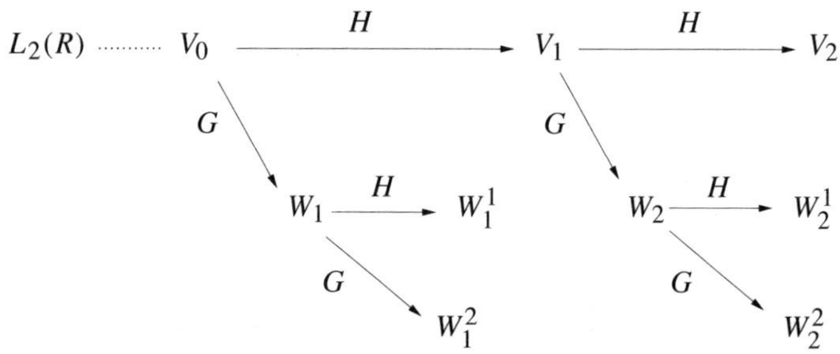

Figure 2.30. MRA scheme for a wavelet package

Example 2.6.1. For the simple Haar wavelet

$$\psi(x) = \chi_{[0,\frac{1}{2}]} - \chi_{[\frac{1}{2},1]},$$

having the scaling coefficients

$$h_0 = h_1 = \frac{1}{\sqrt{2}}, \quad g_0 = \frac{1}{\sqrt{2}}, \quad g_1 = -\frac{1}{\sqrt{2}}$$

the corresponding wavelet packages are

$$\psi^{(1)}(x) = h_0\psi(x) + h_1\psi(x-1) = \frac{1}{\sqrt{2}}\big(\chi_{[0,\frac{1}{2}]} - \chi_{[\frac{1}{2},1]} + \chi_{[1,\frac{3}{2}]} - \chi_{[\frac{3}{2},2]}\big)$$

and

$$\psi^{(1)}(x) = g_0\psi(x) + g_1\psi(x-1) = \frac{1}{\sqrt{2}}\big(\chi_{[0,\frac{1}{2}]} - \chi_{[\frac{1}{2},\frac{3}{2}]} + \chi_{[\frac{3}{2},2]}\big).$$

The graph of the wavelet packages is shown in Figure 2.31.

Now a modified form of the Mallat algorithm for wavelet packages has to be developed. Let

$$c_{m,k} = \langle f, \varphi_{m,k}\rangle, \tag{2.124}$$
$$d_{m,k} = \langle f, \psi_{m,k}\rangle, \tag{2.125}$$
$$d^{(1)}_{m,k} = \langle f, \psi^{(1)}_{m,k}\rangle, \tag{2.126}$$
$$d^{(2)}_{m,k} = \langle f, \psi^{(2)}_{m,k}\rangle. \tag{2.127}$$

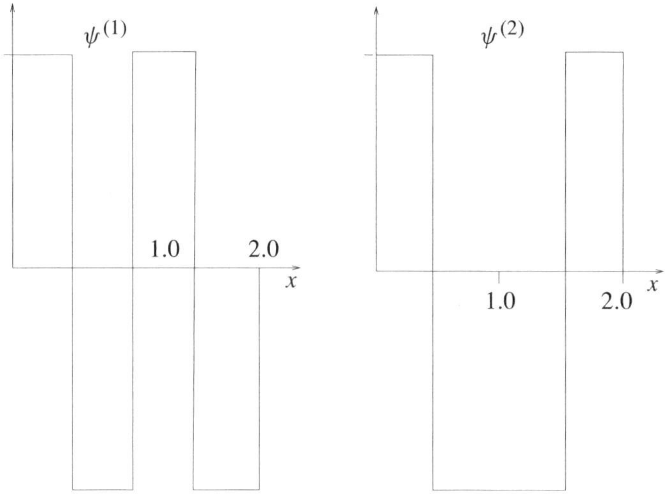

Figure 2.31. Haar wavelet packages

Then, using the scaling equations, one obtains

$$\begin{aligned} c_{m,k} &= \Big\langle f, \sum_{l\in\mathbb{Z}} h_l \varphi_{m-1,2k+l} \Big\rangle \\ &= \sum_{l\in\mathbb{Z}} h_l c_{m-1,2k+l} \\ &= \sum_{l\in\mathbb{Z}} h_{l-2k} c_{m-1,l}, \end{aligned} \tag{2.128}$$

$$\begin{aligned} d_{m,k} &= \Big\langle f, \sum_{l\in\mathbb{Z}} g_l \varphi_{m-1,2k+l} \Big\rangle \\ &= \sum_{l\in\mathbb{Z}} g_l c_{m-1,2k+l} \\ &= \sum_{l\in\mathbb{Z}} g_{l-2k} c_{m-1,l}, \end{aligned} \tag{2.129}$$

$$\begin{aligned} d^{(1)}_{m,k} &= \langle f, 2^{-m/2} \psi^{(1)}(2^{-m}\bullet - 2k) \rangle \\ &= \Big\langle f, 2^{-m/2} \sum_{n\in\mathbb{Z}} h_n \psi(2^{-m}\bullet - 2k - n) \Big\rangle \\ &= \sum_{n\in\mathbb{Z}} h_n \langle f, \psi_{m,2k+n} \rangle \\ &= \sum_{n\in\mathbb{Z}} h_n d_{m,2k+n} = \sum_{n\in\mathbb{Z}} h_{n-2k} d_{m,n}, \end{aligned} \tag{2.130}$$

$$
\begin{aligned}
d_{m,k}^{(2)} &= \langle f, 2^{-m/2}\psi^{(2)}(2^{-m}\bullet - 2k)\rangle \\
&= \Big\langle f, 2^{-m/2}\sum_{n\in\mathbb{Z}} g_n\psi(2^{-m}\bullet - 2k - n)\Big\rangle \\
&= \sum_{n\in\mathbb{Z}} g_n\langle f, \psi_{m,2k+n}\rangle \\
&= \sum_{n\in\mathbb{Z}} g_n d_{m,2k+n} \\
&= \sum_{n\in\mathbb{Z}} g_{n-2k} d_{m,n}.
\end{aligned}
\tag{2.131}
$$

Using again the operators H and G, defined in (2.97) and (2.98), the Mallat algorithm for wavelet packages can be written in a more compact form:

$$c_m = Hc_{m-1}, \tag{2.132}$$

$$d_m = Gc_{m-1}, \tag{2.133}$$

$$d_m^{(1)} = Hd_m, \tag{2.134}$$

$$d_m^{(2)} = Gd_m. \tag{2.135}$$

In the same way, with thc help of the adjoint operators H^* and G^* from formulas (2.100) and (2.101) the reconstruction scheme can be written as

$$d_m = H^*d_m^{(1)} + G^*d_m^{(2)}, \tag{2.136}$$

$$c_{m-1} = H^*c_m + G^*d_m. \tag{2.137}$$

Of course, the splitting trick can be also applied to $\psi^{(1)}$ and $\psi^{(2)}$ leading to a further increased frequency resolution on the price of a reduced time resolution.

Example 2.6.2. In order to convey an impression on the effect of wavelet package analysis, a piecewise constant signal is analyzed once by the Haar wavelet and once by the Haar wavelet package. The analysis is visualized in Figure 2.32.

In both spectra the signal is entirely located in at the scale 4, i.e. the only scales which occur in the signal are between 8 and 16. The wavelet spectrum cannot discriminate scales between 8 and 12 from scales between 12 and 16. It is wildly oscillating in this range. In contrast to the wavelet spectrum the wavelet package spectrum clearly shows that the signal consists of a superposition of a signal with a scale of 8 and a signal of scale 12. The first having twice the amplitude of the second. This proves the increased scale resolution of a wavelet package compared to the scale resolution of the simple wavelet.

2.7 Biorthogonal wavelets

Orthogonal wavelets have the advantage that for the decomposition and for the reconstruction the same filter coefficients can be used. On the other hand, orthogonal

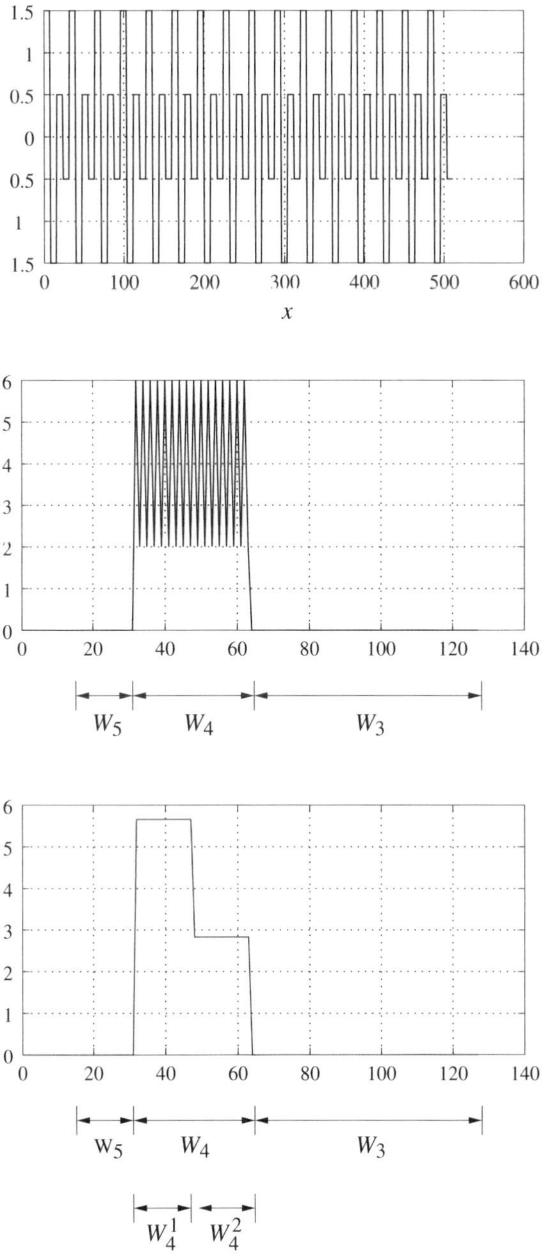

Figure 2.32. Piecewise constant signal (top), its wavelet spectrum (middle) and its wavelet package spectrum (bottom)

wavelets are always asymmetric and have comparatively large supports.

If one is interested in wavelets with smaller supports or in symmetric wavelets the only way out is to use different filters for decomposition and for reconstruction:

$$f = \sum_{m\in\mathbb{Z}} \sum_{k\in\mathbb{Z}} \langle f, \tilde{\psi}_{m,k}\rangle \psi_{m,k} = \sum_{m\in\mathbb{Z}} \sum_{k\in\mathbb{Z}} \langle f, \psi_{m,k}\rangle \tilde{\psi}_{m,k}. \tag{2.138}$$

This is only possible if scaling functions and wavelets of the decomposition and of the reconstruction filters are mutually orthogonal.

Definition 2.7.1. The wavelets $\{\psi, \tilde{\psi}\}$ are called *biorthogonal*, if for their scaled translates

$$\psi_{m,k}(t) := 2^{-m/2}\psi(2^{-m}t - n), \quad \tilde{\psi}_{m,k}(t) := 2^{-m/2}\tilde{\psi}(2^{-m}t - n) \tag{2.139}$$

the following relation holds:

$$\langle \psi_{m,k}, \tilde{\psi}_{\mu,\nu}\rangle = \delta_{m,\mu}\delta_{n,\nu}. \tag{2.140}$$

As in the case of orthogonal wavelets also in the case of biorthogonal wavelets the wavelet ψ and the dual wavelet $\tilde{\psi}$ are related to a scaling function φ and a dual scaling function $\tilde{\varphi}$ by a pair of scaling equations:

$$\varphi(t) = \sqrt{2}\sum_{k\in\mathbb{Z}} h_k \varphi(2t - k), \tag{2.141}$$

$$\tilde{\varphi}(t) = \sqrt{2}\sum_{k\in\mathbb{Z}} \tilde{h}_k \tilde{\varphi}(2t - k). \tag{2.142}$$

The equations (2.140), (2.141) and (2.142) express that in the case of biorthogonality we have two multi-resolution analyses of $L_2(\mathbb{R})$:

$$\{0\} \subset \cdots \subset V_{-2} \subset V_{-1} \subset V_0 \subset V_1 \cdots \subset L_2(\mathbb{R}), \tag{2.143}$$

$$\{0\} \subset \cdots \subset \tilde{V}_{-2} \subset \tilde{V}_{-1} \subset \tilde{V}_0 \subset \tilde{V}_1 \cdots \subset L_2(\mathbb{R}). \tag{2.144}$$

The scaling spaces V_j and the dual scaling spaces $\tilde{V}_j$ are the spans of the integer translates of the scaling function φ and the dual scaling function $\tilde{\varphi}$:

$$V_j = \mathrm{span}_{k\in\mathbb{Z}}\{2^{-j/2}\varphi(2^{-j}\bullet - k)\}, \quad \tilde{V}_j = \mathrm{span}_{k\in\mathbb{Z}}\{2^{-j/2}\tilde{\varphi}(2^{-j}\bullet - k)\}. \tag{2.145}$$

Quite similar, the wavelet spaces are the spans of the scaled and translated wavelets:

$$W_j = \mathrm{span}_{k\in\mathbb{Z}}\{2^{-j/2}\psi(2^{-j}\bullet - k)\}, \quad \tilde{W}_j = \mathrm{span}_{k\in\mathbb{Z}}\{2^{-j/2}\tilde{\psi}(2^{-j}\bullet - k)\}. \tag{2.146}$$

The important difference to orthogonal wavelets is that the wavelet space W_j is *not* the orthogonal complement of V_j in V_{j+1} ! This role is played by $\tilde{W}_j$:

$$V_j \perp \tilde{W}_j, \quad \tilde{V}_j \perp W_j. \tag{2.147}$$

The decomposition and reconstruction algorithms for biorthogonal wavelets are only a slight modification of the Mallat algorithm for the decomposition and reconstruction with orthogonal wavelets.

First the operators $H, G, \tilde{H}, \tilde{G}$ have to be defined:

$$\begin{aligned} H\colon l_2(\mathbb{Z}) &\to l_2(\mathbb{Z}) \\ c &\mapsto \Big\{(Hc)_k = \sum_{l\in\mathbb{Z}} h_{l-2k} c_l\Big\} \end{aligned} \tag{2.148}$$

$$\begin{aligned} G\colon l_2(\mathbb{Z}) &\to l_2(\mathbb{Z}) \\ c &\mapsto \Big\{(Gc)_k = \sum_{l\in\mathbb{Z}} g_{l-2k} c_l\Big\}, \quad g_k = (-1)^k \tilde{h}_{1-k}, \end{aligned} \tag{2.149}$$

$$\begin{aligned} \tilde{H}\colon l_2(\mathbb{Z}) &\to l_2(\mathbb{Z}) \\ c &\mapsto \Big\{(\tilde{H}c)_k = \sum_{l\in\mathbb{Z}} \tilde{h}_{l-2k} c_l\Big\} \end{aligned} \tag{2.150}$$

$$\begin{aligned} \tilde{G}\colon l_2(\mathbb{Z}) &\to l_2(\mathbb{Z}) \\ c &\mapsto \Big\{(\tilde{G}c)_k = \sum_{l\in\mathbb{Z}} \tilde{g}_{l-2k} c_l\Big\}, \quad \tilde{g}_k = (-1)^k h_{1-k}. \end{aligned} \tag{2.151}$$

Let us now assume that for an element $f \in V_0$ the representation with respect to the base $\{\varphi(\bullet - k)\}$ is already given:

$$f(t) = \sum_{k\in\mathbb{Z}} c_k^{(0)} \varphi(t-k). \tag{2.152}$$

As for orthogonal wavelets, the scaling and the wavelet coefficients on scale m are defined by

$$c_k^{(m)} := \langle f, \varphi_{m,k}\rangle, \quad d_k^{(m)} := \langle f, \psi_{m,k}\rangle. \tag{2.153}$$

Then the following relations are valid:

$$\begin{aligned} d_k^{(m)} &= \langle f, \psi_{m,k}\rangle \\ &= \sum_{l\in\mathbb{Z}} g_l \langle f, \varphi_{m-1,2k+l}\rangle \\ &= \sum_{l\in\mathbb{Z}} g_l c_{2k+l}^{(m-1)} \\ &= \sum_{l\in\mathbb{Z}} g_{l-2k} c_l^{(m-1)} \\ &= (Gc^{(m-1)})_k, \end{aligned} \tag{2.154}$$

$$
\begin{aligned}
c_k^{(m)} &= \langle f, \varphi_{m,k} \rangle \\
&= \sum_{l \in \mathbb{Z}} h_l \langle f, \varphi_{m-1,2k+l} \rangle \\
&= \sum_{l \in \mathbb{Z}} h_l c_{2k+l}^{(m-1)} \\
&= \sum_{l \in \mathbb{Z}} h_{l-2k} c_l^{(m-1)} \\
&= (Hc^{(m-1)})_k.
\end{aligned} \tag{2.155}
$$

This means that also in the case of biorthogonal wavelets the decomposition follows the scheme displayed in Figure 2.33.

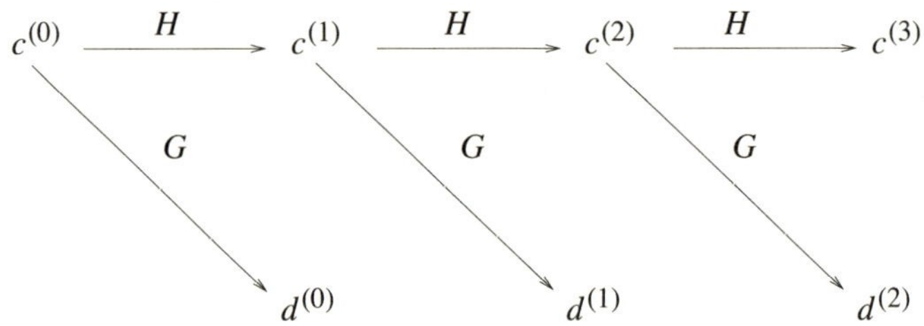

Figure 2.33. Decomposition for biorthogonal wavelets

Now let us consider the reconstruction. In contrast to the orthogonal wavelets case instead of the adjoint operators H^*, G^* the corresponding operators $\tilde{H}^*$ and $\tilde{G}^*$ will be used. These operators are defined in the following way:

$$
\begin{aligned}
\tilde{H}^* \colon l_2(\mathbb{Z}) &\to l_2(\mathbb{Z}) \\
c &\mapsto \Big\{ (\tilde{H}^* c)_k = \sum_{l \in \mathbb{Z}} \tilde{h}_{k-2l} c_l \Big\}
\end{aligned} \tag{2.156}
$$

$$
\begin{aligned}
\tilde{G}^* \colon l_2(\mathbb{Z}) &\to l_2(\mathbb{Z}) \\
c &\mapsto \Big\{ (\tilde{G}^* c)_k = \sum_{l \in \mathbb{Z}} \tilde{g}_{k-2l} c_l \Big\}, \quad \tilde{g}_k = (-1)^k h_{1-k}.
\end{aligned} \tag{2.157}
$$

Lemma 2.7.1. *It holds*

$$
\tilde{H}^* H + \tilde{G}^* G = I. \tag{2.158}
$$

Proof. Since φ, $\tilde{\varphi}$ and ψ, $\tilde{\psi}$ are scaling functions and wavelet respectively, which are constructed according to Definition 2.7.1, the following relation holds:

$$\begin{aligned}
\delta_{0,k} &= \int_{\mathbb{R}} \varphi(x)\tilde{\varphi}(x-k)\,dx \\
&= \int_{\mathbb{R}} \sqrt{2}\sum_{l\in\mathbb{Z}} h_l\varphi(2x-l)\cdot\sqrt{2}\sum_{m\in\mathbb{Z}} \tilde{h}_m\tilde{\varphi}(2x-2k-m)\,dx \\
&= 2\sum_{l,m\in\mathbb{Z}} h_l\tilde{h}_m \int_{\mathbb{R}} \varphi(2x-l)\cdot\tilde{\varphi}(2x-2k-m)\,dx \\
&= \sum_{l,m\in\mathbb{Z}} h_l\tilde{h}_m \int_{\mathbb{R}} \varphi(\xi)\cdot\tilde{\varphi}(\xi+l-2k-m)\,d\xi \\
&= \sum_{l,m\in\mathbb{Z}} h_l\tilde{h}_m\cdot\delta_{l,2k+m} \\
&= \sum_{l\in\mathbb{Z}} h_l\cdot\tilde{h}_{l-2k}.
\end{aligned}$$

In the same way the validity of

$$\delta_{0,k} = \sum_{l\in\mathbb{Z}} g_l\cdot\tilde{g}_{l-2k}$$

can be shown. Finally, it can be concluded that

$$\begin{aligned}
([\tilde{H}^*H+\tilde{G}^*G]c)_k &= (\tilde{H}^*Hc)_k + (\tilde{G}^*Gc)_k \\
&= \sum_{l\in\mathbb{Z}} \tilde{h}_{k-2l}(Hc)_l + \sum_{l\in\mathbb{Z}} \tilde{g}_{k-2l}(Gc)_l \\
&= \sum_{l\in\mathbb{Z}} \tilde{h}_{k-2l}\sum_{m\in\mathbb{Z}} h_{m-2l}c_m + \sum_{l\in\mathbb{Z}} \tilde{g}_{k-2l}\sum_{m\in\mathbb{Z}} g_{m-2l}c_m \\
&= \sum_{m\in\mathbb{Z}} c_m\Big(\sum_{l\in\mathbb{Z}} \tilde{h}_{k-2l}h_{m-2l} + \sum_{l\in\mathbb{Z}} \tilde{g}_{k-2l}g_{m-2l}\Big) \\
&= \sum_{m\in\mathbb{Z}} c_m\Big(\sum_{l\in\mathbb{Z}} \tilde{h}_{k+2l}h_{m+2l} + \sum_{l\in\mathbb{Z}} \tilde{g}_{k-2l}g_{m-2l}\Big) \\
&= \sum_{m\in\mathbb{Z}} c_m\Big(\sum_{l\in\mathbb{Z}} \tilde{h}_{k+2l}h_{m+2l} + (-1)^{m+k}\sum_{l\in\mathbb{Z}} h_{2l+1-k}\tilde{h}_{2l+1-m}\Big).
\end{aligned}$$

Now four cases have to be distinguished:

1. Both m and k are even.

2. Both m and k are odd.

3. m is even and k is odd.

4. m is odd and k is even.

Case 1:

$$
\begin{aligned}
&c_m\Big(\sum_{l\in\mathbb{Z}}\tilde{h}_{k+2l}h_{m+2l} + (-1)^{m+k}\sum_{l\in\mathbb{Z}}h_{2l+1-k}\tilde{h}_{2l+1-m}\Big)\\
&= c_m\Big(\sum_{l\in\mathbb{Z}}\tilde{h}_{2l+(k-m)}h_{2l} + \sum_{l\in\mathbb{Z}}h_{2l+1}\tilde{h}_{2l+1+(k-m)}\Big)\\
&= c_m\sum_{l\in\mathbb{Z}}\tilde{h}_{l+(k-m)}h_l\\
&= c_m\cdot\delta_{k,m}.
\end{aligned}
$$

Case 2:

$$
\begin{aligned}
&c_m\Big(\sum_{l\in\mathbb{Z}}\tilde{h}_{k+2l}h_{m+2l} + (-1)^{m+k}\sum_{l\in\mathbb{Z}}h_{2l+1-k}\tilde{h}_{2l+1-m}\Big)\\
&= c_m\Big(\sum_{l\in\mathbb{Z}}\tilde{h}_{2l+1+(k-m)}h_{2l+1} + \sum_{l\in\mathbb{Z}}h_{2l}\tilde{h}_{2l+(k-m)}\Big)\\
&= c_m\sum_{l\in\mathbb{Z}}\tilde{h}_{l+(k-m)}h_l\\
&= c_m\cdot\delta_{k,m}.
\end{aligned}
$$

Case 3:

$$
\begin{aligned}
&c_m\Big(\sum_{l\in\mathbb{Z}}\tilde{h}_{k+2l}h_{m+2l} + (-1)^{m+k}\sum_{l\in\mathbb{Z}}h_{2l+1-k}\tilde{h}_{2l+1-m}\Big)\\
&= c_m\Big(\sum_{l\in\mathbb{Z}}\tilde{h}_{2l+(k-m)}h_{2l} - \sum_{l\in\mathbb{Z}}h_{2l}\tilde{h}_{2l+(k-m)}\Big)\\
&= 0.
\end{aligned}
$$

Case 4:

$$
\begin{aligned}
&c_m\Big(\sum_{l\in\mathbb{Z}}\tilde{h}_{k+2l}h_{m+2l} + (-1)^{m+k}\sum_{l\in\mathbb{Z}}h_{2l+1-k}\tilde{h}_{2l+1-m}\Big)\\
&= c_m\Big(\sum_{l\in\mathbb{Z}}\tilde{h}_{2l}h_{2l+(m-k)} - \sum_{l\in\mathbb{Z}}h_{2l+(m-k)}\tilde{h}_{2l}\Big)\\
&= 0.
\end{aligned}
$$

Consequently, for the computation of $([\tilde{H}^*H + \tilde{G}^*G]c)_k$ again two cases have to be considered.

Case 1: *k even*

$$
\begin{aligned}
&([\tilde{H}^*H + \tilde{G}^*G]c)_k \\
&= \sum_{m\in\mathbb{Z}} c_m\Big(\sum_{l\in\mathbb{Z}} \tilde{h}_{k+2l}h_{m+2l} + (-1)^{m+k}\sum_{l\in\mathbb{Z}} h_{2l+1-k}\tilde{h}_{2l+1-m}\Big)\\
&= \sum_{m\in\mathbb{Z}} c_{2m}\Big(\sum_{l\in\mathbb{Z}} \tilde{h}_{k+2l}h_{2m+2l} + (-1)^{2m+k}\sum_{l\in\mathbb{Z}} h_{2l+1-k}\tilde{h}_{2l+1-2m}\Big)\\
&\quad + \sum_{m\in\mathbb{Z}} c_{2m+1}\Big(\sum_{l\in\mathbb{Z}} \tilde{h}_{k+2l}h_{2m+1+2l} + (-1)^{2m+1+k}\sum_{l\in\mathbb{Z}} h_{2l+1-k}\tilde{h}_{2l+1-2m-1}\Big)\\
&= \sum_{m\in\mathbb{Z}} c_{2m}\delta_{k,2m}\\
&= c_k.
\end{aligned}
$$

Case 2: *k odd*

$$
\begin{aligned}
&([\tilde{H}^*H + \tilde{G}^*G]c)_k \\
&= \sum_{m\in\mathbb{Z}} c_m\Big(\sum_{l\in\mathbb{Z}} \tilde{h}_{k+2l}h_{m+2l} + (-1)^{m+k}\sum_{l\in\mathbb{Z}} h_{2l+1-k}\tilde{h}_{2l+1-m}\Big)\\
&= \sum_{m\in\mathbb{Z}} c_{2m}\Big(\sum_{l\in\mathbb{Z}} \tilde{h}_{k+2l}h_{2m+2l} + (-1)^{2m+k}\sum_{l\in\mathbb{Z}} h_{2l+1-k}\tilde{h}_{2l+1-2m}\Big)\\
&\quad + \sum_{m\in\mathbb{Z}} c_{2m+1}\Big(\sum_{l\in\mathbb{Z}} \tilde{h}_{k+2l}h_{2m+1+2l} + (-1)^{2m+1+k}\sum_{l\in\mathbb{Z}} h_{2l+1-k}\tilde{h}_{2l+1-2m-1}\Big)\\
&= \sum_{m\in\mathbb{Z}} c_{2m+1}\delta_{k,2m+1}\\
&= c_k.
\end{aligned}
$$

□

Schematically, the reconstruction step for biorthogonal wavelets is displayed in Figure 2.34.

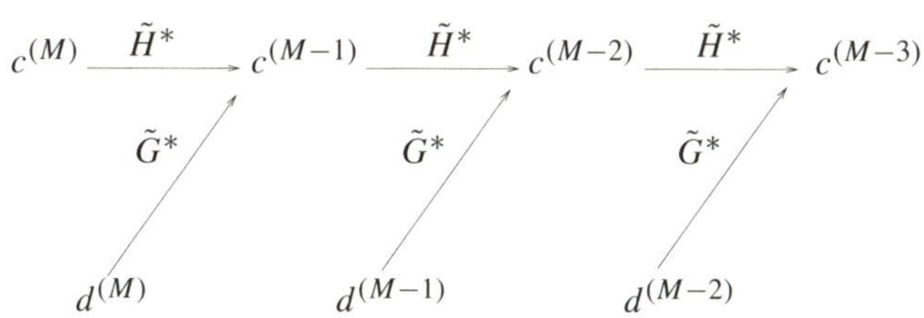

Figure 2.34. Reconstruction for biorthogonal wavelets

As in the case of orthogonal wavelets, also for biorthogonal wavelets their construction starts with scaling functions $\{\varphi, \tilde{\varphi}\}$ and their scaling equations (2.141) and

(2.142). If the Fourier series of the scaling coefficients $\{h_k, \tilde{h}_k\}$ are denoted by $H(\omega)$ and $\tilde{H}(\omega)$, i.e.

$$H(\omega) = \frac{1}{2}\sqrt{2}\sum_{k\in\mathbb{Z}} h_k e^{-\imath k\omega}, \tag{2.159}$$

$$\tilde{H}(\omega) = \frac{1}{2}\sqrt{2}\sum_{k\in\mathbb{Z}} \tilde{h}_k e^{-\imath k\omega}, \tag{2.160}$$

the procedure for the construction of biorthogonal wavelets is described in the following theorem.

Theorem 2.7.1. *Let $H, \tilde{H}$ be trigonometric polynomials which fulfill*

$$\overline{H(\omega)}\tilde{H}(\omega) + \overline{H(\omega+\pi)}\tilde{H}(\omega+\pi) = 1, \tag{2.161}$$

$$H(0) = \tilde{H}(0) = 1. \tag{2.162}$$

Additionally, suppose

$$H(\omega) = \left(\frac{1+e^{\imath\omega}}{2}\right)^N p(\omega), \tag{2.163}$$

$$\tilde{H}(\omega) = \left(\frac{1+e^{\imath\omega}}{2}\right)^{\tilde{N}} \tilde{p}(\omega), \tag{2.164}$$

with trigonometric polynomials p, $\tilde{p}$ having the properties

$$B_j := \max_{\omega\in\mathbb{R}} |\prod_{k=1}^{j} p(2^{-k}\omega)|^{1/j}, \quad \operatorname{supp}_{j\in\mathbb{N}} B_j < 2^{N-1/2}, \tag{2.165}$$

$$\tilde{B}_j := \max_{\omega\in\mathbb{R}} |\prod_{k=1}^{j} \tilde{p}(2^{-k}\omega)|^{1/j}, \quad \operatorname{supp}_{j\in\mathbb{N}} \tilde{B}_j < 2^{\tilde{N}-1/2}. \tag{2.166}$$

Then the following results are true:

1. $$\varphi(t) = \prod_{j\geq 1} H(2^{-j}\omega), \quad \tilde{\varphi}(t) = \prod_{j\geq 1} \tilde{H}(2^{-j}\omega), \tag{2.167}$$

2. $$\int_{\mathbb{R}} \varphi(t)\tilde{\varphi}(t-k)\,dt = 0, \quad k \in \mathbb{Z}\setminus\{0\}, \tag{2.168}$$

3. *the functions*

$$\hat{\psi}(\omega) := e^{-\imath\omega/2}\overline{\tilde{H}(\omega/2+\pi)} \cdot \hat{\varphi}(\omega/2), \tag{2.169}$$

$$\hat{\tilde{\psi}}(\omega) := e^{-\imath\omega/2}\overline{H(\omega/2+\pi)} \cdot \hat{\tilde{\varphi}}(\omega/2) \tag{2.170}$$

fulfill

- $\psi(t) = \sqrt{2}\sum_{k\in\mathbb{Z}}(-1)^k \tilde{h}_{1-k}\varphi(2t-k)$,
- $\tilde{\psi}(t) = \sqrt{2}\sum_{k\in\mathbb{Z}}(-1)^k h_{1-k}\tilde{\varphi}(2t-k)$,
- $\int_{\mathbb{R}} \psi_{m,n}(t)\tilde{\psi}_{\mu,\nu}(t)\,dt = \delta_{m,\mu}\delta_{n,\nu}$.

Proof. [15], [12]. □

In order to illustrate the concept of biorthogonal wavelets in the following example a pair of biorthogonal wavelets is constructed.

Example 2.7.1. The simplest possible orthogonal wavelet is the Haar wavelet with the box-function

$$\varphi(x) = \begin{cases} 1, & 0 \le x < 1 \\ 0, & \text{else} \end{cases}$$

as scaling function. The next better choice for a scaling function is the hat-function, i.e. the convolution of the box-function $B_0(x) := \chi_{[-1,1]}(2x)$ with itself:

$$\varphi(x) := \int_{\mathbb{R}} B_0(t)B_0(x-t)\,dt = \int_0^1 \varphi(x-t)\,dt = \begin{cases} x+1, & -1 \le x < 0 \\ 1 \quad x, & 0 \le x < 1 \\ 0, & \text{else.} \end{cases}$$

The hat-function is displayed in Figure 2.35.

The Fourier spectrum of the hat function φ can be computed with the help of the convolution theorem:

$$\hat{\varphi}(\omega) = \sqrt{2\pi}\,\hat{B_0}^2(\omega) = \frac{1}{\sqrt{2\pi}}\left(\frac{1}{\omega/2}\sin(\omega/2)\right)^2 = \frac{1}{\sqrt{2\pi}}\frac{1}{(\omega/2)^2}\sin^2(\omega/2).$$

A short computation yields

$$\hat{\varphi}(\omega) = \frac{1}{\sqrt{2\pi}}\frac{1}{4(\omega/4)^2}4\sin^2(\omega/4)\cos^2(\omega/4).$$

Hence, defining H by

$$H(\omega) = \frac{1}{2} + \frac{1}{2}\cos(\omega),$$

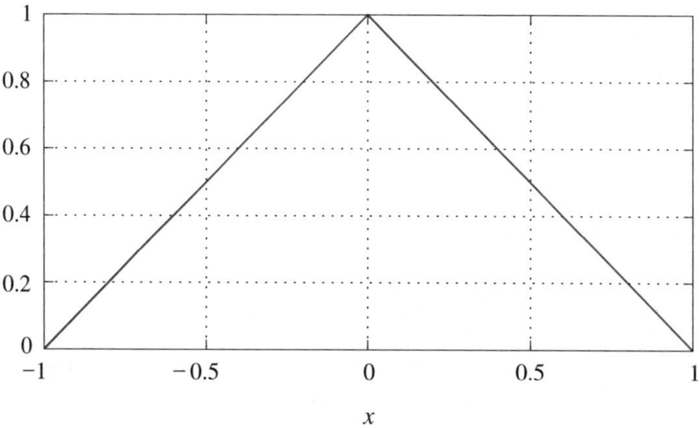

Figure 2.35. Hat-function

the following scaling equation in the frequency domain holds:

$$\hat{\varphi}(\omega) = \cos^2(\omega/4) \cdot \left(\frac{1}{\sqrt{2\pi}} \frac{1}{4(\omega/4)^2} 4 \sin^2(\omega/4) \right) = H(\omega/2) \cdot \hat{\varphi}(\omega/2).$$

This means that the coefficients h_k of the scaling equations are the Fourier coefficients of the filter function H,

$$H(\omega) = \frac{1}{4} e^{-\imath\omega} + \frac{1}{2} + \frac{1}{4} e^{\imath\omega},$$

i.e. the decomposition low-pass filter H has the filter coefficients

$$h_{-1} = 1/4\sqrt{2}, \quad h_0 = 1/2\sqrt{2}, \quad h_1 = 1/4\sqrt{2}.$$

The coefficients g_k of the corresponding reconstruction high-pass filter easily can be derived from the relation $\tilde{g}_k = (-1)^k h_{1-k}$, i.e.

$$\tilde{g}_0 = 1/4\sqrt{2}, \quad \tilde{g}_1 = -1/2\sqrt{2}, \quad \tilde{g}_2 = 1/4\sqrt{2}.$$

The filter characteristics of the decomposition high-pass filter and the decomposition low-pass filter are shown in Figure 2.36.

The determination of the coefficients $\tilde{h}_k$ of the reconstruction low-pass filter is governed by the equation

$$\sum_{l \in \mathbb{Z}} h_l \tilde{h}_{l-2k} = \delta_{0,k}.$$

Due to the condition (2.161) the unknown coefficients $\tilde{h}_k$ have to fulfill the normalization condition

$$\sum_{k \in \mathbb{Z}} \tilde{h}_k = \sqrt{2}. \tag{2.171}$$

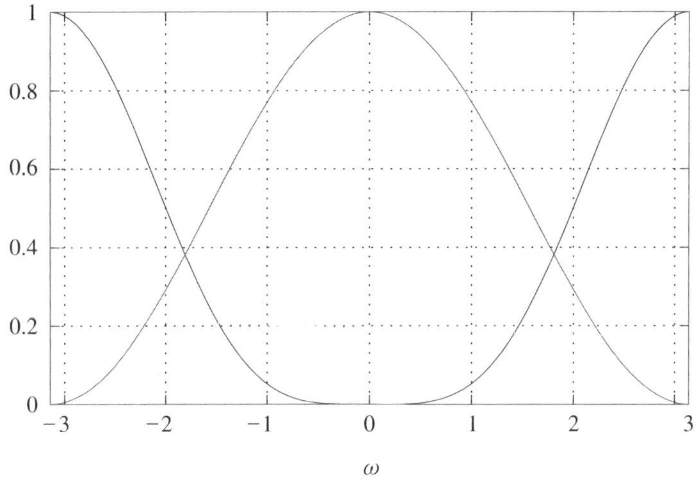

Figure 2.36. Filter characteristic of decomposition high-pass filter and decomposition low-pass filter

Looking for a reconstruction filter with 5 coefficients, this condition yields 3 equations for the 5 unknown parameters:

$$\begin{aligned} \frac{1}{2}\tilde{h}_{-2} + \frac{1}{4}\tilde{h}_{-1} &= 0, \\ \frac{1}{4}\tilde{h}_{-1} + \frac{1}{2}\tilde{h}_0 + \frac{1}{4}\tilde{h}_1 &= 1/\sqrt{2}, \\ \frac{1}{2}h_2 + \frac{1}{4}h_1 &= 0. \end{aligned} \tag{2.172}$$

In order to have a unique solution of these linear equations, two additional conditions can be imposed:

- a normalization condition (2.171) and
- a symmetry condition $\tilde{h}_{-2} = \tilde{h}_2$, $\tilde{h}_{-1} = \tilde{h}_1$.

Hence the remaining system of linear equations is

$$\begin{aligned} \frac{1}{2}\tilde{h}_2 + \frac{1}{4}\tilde{h}_1 &= 0, \\ \frac{1}{2}\tilde{h}_0 + \frac{1}{2}\tilde{h}_1 &= \frac{1}{\sqrt{2}}, \\ 2\tilde{h}_2 + 2\tilde{h}_1 + \tilde{h}_0 &= \sqrt{2}. \end{aligned} \tag{2.173}$$

One of the solutions of these equations is

$$\tilde{h} = \left\{-\frac{1}{8}\sqrt{2}, \frac{1}{4}\sqrt{2}, \frac{3}{4}\sqrt{2}, \frac{1}{4}\sqrt{2}, -\frac{1}{8}\sqrt{2}\right\}$$

and to the corresponding reconstruction high-pass filter has the coefficients

$$g = \left\{\frac{1}{8}\sqrt{2}, \frac{1}{4}\sqrt{2}, -\frac{3}{4}\sqrt{2}, \frac{1}{4}\sqrt{2}, \frac{1}{8}\sqrt{2}\right\}.$$

The filter characteristics of these reconstruction filters are displayed in Figure 2.37.

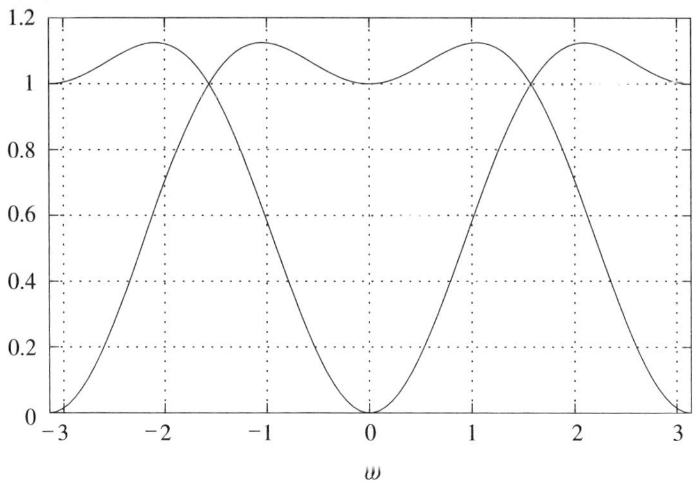

Figure 2.37. Filter characteristics of reconstruction low- and high-pass filter

Using the iterative technique for the solution of the scaling equation a graphical representation of the reconstruction scaling function $\tilde{\varphi}$ can be computed. With the known scaling functions φ, $\tilde{\varphi}$ and the known wavelet coefficients g_k, $\tilde{g}_k$ the recursions of Theorem 2.7.1 can be used for the computation of the decomposition and the reconstruction wavelet ψ and $\tilde{\psi}$. The graphical representations of these four function is given in Figure 2.38.

In the example given above, the decomposition and the reconstruction filter differ considerably in length. This is a typical feature of biorthogonal wavelets. Several studies, e.g. in [12] were carried out to construct biorthogonal wavelet bases, where the reconstruction and the decomposition filters have similar lengths. Some of the filters which are taken from [72] are given below:

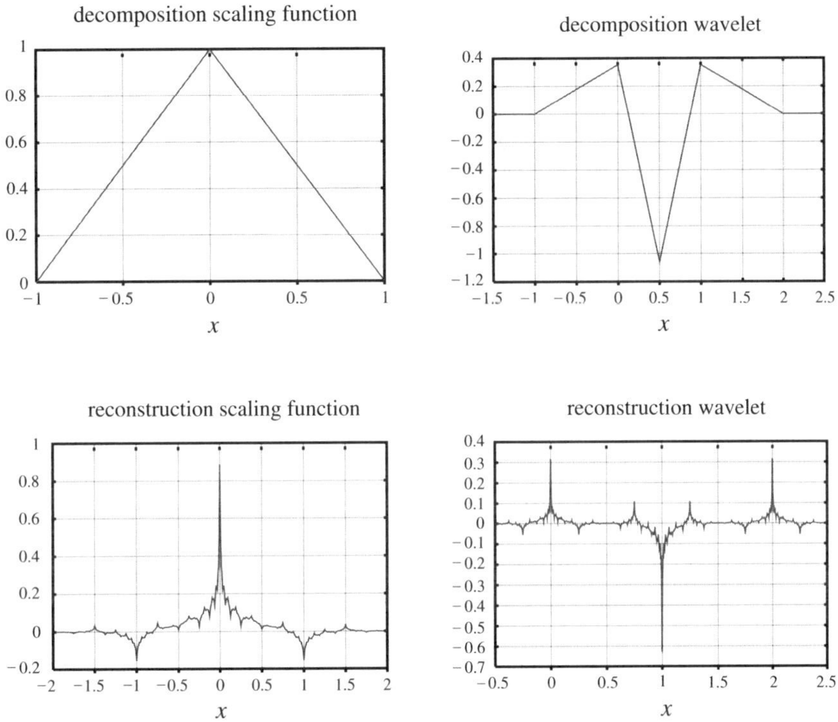

Figure 2.38. Biorthogonal pair of scaling functions and wavelets

n	h_n	$\tilde{h}_n$
−4	0.0	0.037828445554969
−3	−0.06453888262876	−0.02384946501956
−2	−0.04068941760920	−0.11062440441844
−1	0.41809227322204	0.37740285561283
0	0.78848561640637	0.85269867900889
1	0.41809227322204	0.37740285561283
2	−0.04068941760920	−0.11062440441844
3	−0.06453888262876	−0.02384946501956
4	0.0	0.037828445554969
−5	0.00992177208685	0.0
−4	0.02183057133337	0.05382683783789
−3	0.00027989343090	−0.11843354319764
−2	0.06115645341349	−0.36378609009851

n	h_n	$\tilde{h}_n$
−1	0.34335173921766	0.47198693379091
0	0.541132373169141	1.32702528570780
1	0.34335173921766	0.47198693379091
2	0.06115645341349	−0.36378609009851
3	0.00027989343090	−0.11843354319764
4	0.02183057133337	0.05382683783789
5	0.00992177208685	0.0

The first biorthogonal wavelet filter given in the above table is of particular importance. It belongs to the class of Cohen–Daubechies–Feauveau [12] biorthogonal wavelets, which were used in the FBI Fingerprint Compression Standard.

Orthogonal and biorthogonal wavelet bases differ in some important items:

- Orthogonal wavelet filters and scaling filters must be of the same length, and the length must be even. This restriction can be dropped for biorthogonal wavelet bases.

- Symmetric wavelets and symmetric scaling functions are possible in the framework of biorthogonal wavelets. In the framework of orthogonal wavelet bases this is impossible.

- For biorthogonal wavelet bases Parseval's identity does no longer hold. This is probably the main disadvantage of biorthogonal systems.

From the practical point of view, in many cases it will be more suitable to use biorthogonal bases instead of orthogonal wavelet bases for the signal processing. Biorthogonal bases give more freedom in their construction. So they can be better adapted to the problem under consideration.

2.8 Compactly supported orthogonal wavelets

Mallat's algorithm is only useful if the length of the filter sequences $\{h_k\}$ and $\{g_k\}$ is finite. Combined with their orthogonality this will lead to wavelets with a compact support. The way how to construct such wavelets was found by I. Daubechies. Here, we will follow the basic ideas of the construction of the so-called Daubechies wavelets. The complete theory is rather complicated so that some steps have to be skipped. The reasoning starts with a consideration of the filter properties of the wavelet transformation.

2.8.1 Daubechies wavelets

The idea of the Daubechies wavelets is to construct orthogonal wavelets with a compact support. Here we will follow this construction in its main features. Some steps have to be left out since their proofs are a bit tedious and of technical nature.

Let us construct an orthogonal wavelet with $2N$ non-vanishing scaling coefficients $h_0, \dots, h_{2N-1}$. Since the wavelet is to be orthogonal, the scaling coefficients have to fulfill the following two equations:

$$\sqrt{2} = \sum_{n=0}^{2N-1} h_n, \tag{2.174}$$

$$\delta_{0m} = \sum_{k=0}^{2N-1-2m} h_k h_{k+2m}, \qquad m = 0, \dots, N-1. \tag{2.175}$$

These are $N+1$ quadratic equations for $2N$ unknown coefficients. In order to get a unique solution, $N-1$ additional equations have to be supplemented. Usually, these additional equations express that the wavelet to be constructed has an sufficiently high order.

The question of how to solve this system of quadratic equations is open in most cases. Therefore an alternative technique for the construction of compactly supported orthogonal wavelets will be sketched here. It is based on the following idea.

The filter characteristic $H(\omega)$ of an orthogonal scaling function φ has to fulfill the following conditions:

- orthogonality $\quad |H(\omega)|^2 + |H(\omega+\pi)|^2 = 1, \qquad (2.176)$
- regularity $\quad H(\omega) = \left(\frac{1+e^{-\imath\omega}}{2}\right)^N L(\omega),\ L(\omega)$ a trigonometric polynomial.

Theorem 2.8.1. *There exists a polynomial P fulfilling*

$$(1-y)^N P(y) + y^N P(1-y) = 1, \quad y \in [0,1], \tag{2.177}$$

such that

$$|H(\omega)|^2 = \left(\cos^2\frac{\omega}{2}\right)^N |L(\omega)|^2,$$

$$|H(\omega)|^2 = \left(\cos^2\frac{\omega}{2}\right)^N P\left(\sin^2\frac{\omega}{2}\right)$$

holds.

Proof. Obviously,

$$\left|\left(\frac{1+e^{-\imath\omega}}{2}\right)^N\right|^2 = \left(\left|\frac{1+e^{-\imath\omega}}{2}\right|^2\right)^N = \left(\frac{1+\cos\omega}{2}\right)^N = \left(\cos^2\frac{\omega}{2}\right)^N$$

holds, which leads to

$$|H(\omega)|^2 = \left|\left(\cos^2\frac{\omega}{2}\right)^N\right|^2 |L(\omega)|^2.$$

Since L is a trigonometric polynomial its square $|L(\omega)|^2$ is a polynomial p in $\cos\omega$. Because of

$$\sin^2\frac{\omega}{2} = \frac{1-\cos\omega}{2}$$

it is convenient to write $|L(\omega)|^2$ as a polynomial P in $\sin^2\omega/2$:

$$|H(\omega)|^2 = \left(\cos^2\frac{\omega}{2}\right)^N P\left(\sin^2\frac{\omega}{2}\right).$$

Since

$$\cos^2\frac{\omega+\pi}{2} = \left(\cos\frac{\omega}{2}\cos\frac{\pi}{2} - \sin\frac{\omega}{2}\sin\frac{\pi}{2}\right)^2 = \sin^2\frac{\omega}{2}$$

is valid, it can be concluded that

$$\begin{aligned} 1 &= |H(\omega)|^2 + |H(\omega+\pi)|^2 \\ &= \left(\cos^2\frac{\omega}{2}\right)^N P\left(\sin^2\frac{\omega}{2}\right) + \left(\cos^2\frac{\omega+\pi}{2}\right)^N P\left(\sin^2\frac{\omega+\pi}{2}\right) \\ &= \left(\cos^2\frac{\omega}{2}\right)^N P\left(\sin^2\frac{\omega}{2}\right) + \left(\sin^2\frac{\omega}{2}\right)^N P\left(\cos^2\frac{\omega}{2}\right). \end{aligned}$$

Setting $y = \sin^2(\omega/2)$,

$$1 = (1-y)^N P(y) + y^N P(1-y)$$

follows. □

Remark. The theorem allows the construction of H^2 and L^2 from the solution of (2.177). Hence there are two problems left:

1. The solution of (2.177) and
2. the extraction of the square roots of H^2 and L^2.

The first problem will be tackled by Bezouz's theorem.

Theorem 2.8.2 (Bezouz's theorem). *If p_1, p_2 are two polynomials of degree n_1, n_2 with no common zeros, then there exist unique polynomials q_1, q_2 of degree $n_2 - 1$, $n_1 - 1$ with*

$$p_1(x)q_1(x) + p_2(x)q_2(x) = 1.$$

Proof. [15]. □

Bezouz's theorem guarantees the existence of unique polynomials q_1, q_2 of degree $\leq N-1$ with

$$(1-y)^N q_1(y) + y^N q_2(y) = 1.$$

Substituting $u = 1-y$ yields

$$(1-u)^N q_2(u) + u^N q_1(u) = 1.$$

Due to the uniqueness of q_1, q_2 the relation $q_1 = q_2 = P$ follows. This leads to

$$\begin{aligned} q_1(y) &= (1-y)^{-N}(1-y^N q_1(y)) \\ &= (1-y)^{-N} - y^N \sum_{k=0}^{N-1} q_k (1-y)^{k-N} \\ &= \sum_{k=0}^{\infty} \binom{-N}{k} (-1)^k y^k - \sum_{l=0}^{N-1} q_l \sum_{k=0}^{\infty} \binom{l-N}{k} (-1)^k y^{k+N} \\ &= \sum_{k=0}^{N-1} \binom{-N}{k} (-1)^k y^k + O(y^N) \\ &= \sum_{k=0}^{N-1} \binom{N+k+1}{k} y^k + O(y^N). \end{aligned}$$

Since q_1 is a polynomial of degree $\leq N-1$, the equation

$$P_N(y) = q_1(y) = \sum_{k=0}^{N-1} \binom{N+k-1}{k} y^k$$

follows. This is an explicit solution of (2.177).

Once having the solution $P(y)$ at hand one could compute $|H(\omega)|^2$ and by extracting the square root, the coefficients h_k of the scaling equation would be available. But $P =: P_N$ is only the solution of the lowest possible degree $N-1$. Besides this solution there are many other solutions of higher degree. For a solution P of degree higher than $N-1$ we have

$$(1-y)^N [P(y) - P_N(y)] + y^N [P(1-y) - P_N(1-y)] = 0. \tag{2.178}$$

The polynomial y^N has 0 as a zero of the multiplicity N. Hence 0 has to be a zero of multiplicity N of $P(y) - P_N(y)$ as well. This implies that $P - P_N$ is divisible by y^N:

$$P(y) = P_N(y) + y^N \tilde{P}(y), \tag{2.179}$$

with $\tilde{P}$ being a polynomial of degree smaller than the degree of P. Obviously,

$$\begin{aligned}
\tilde{P}(y) + \tilde{P}(1-y) &= \frac{P(y) - P_N(y)}{y^N} + \frac{P(1-y) - P_N(1-y)}{(1-y)^N} \\
&= \frac{(1-y)^N (P(y) - P_N(y)) + y^N (P(1-y) - P_N(1-y))}{y^N (1-y)^N} \\
&= 0. \qquad (2.180)
\end{aligned}$$

Consequently, $\tilde{P}$ is symmetric with respect to $1/2$. The results obtained so far are to be summarized in the following lemma.

Lemma 2.8.1. *A trigonometric polynomial H of the form*

$$H(\omega) = \left(\frac{1 + e^{-\imath\omega}}{2}\right)^N L(\omega) \qquad (2.181)$$

satisfies (2.176) *if and only if*

$$|L(\omega)|^2 = P\left(\sin^2 \frac{\omega}{2}\right) \qquad (2.182)$$

with

$$P(y) = P_N(y) + y^N R\left(\frac{1}{2} - y\right), \qquad (2.183)$$

where

$$P_N(y) = \sum_{k=0}^{N-1} \binom{N-1+k}{k} y^k \qquad (2.184)$$

and R is an odd polynomial chosen such that $P(y) \geq 0$ for $y \in [0, 1]$.

The lemma provides a tool to characterize $|H(\omega)|^2$. For the determination of the scaling equation coefficients $H(\omega)$ is needed instead. The proof of the following lemma is constructive and describes how – so to say – the square root of $|H(\omega)|^2$ can be drawn.

Lemma 2.8.2. *Let*

$$A = \sum_{m=0}^{M} a_m \cos m\omega, \quad a_m \in \mathbb{R} \qquad (2.185)$$

be a positive trigonometric polynomial. Then there exist a polynomial

$$B(\omega) = \sum_{m=0}^{M} b_m e^{\imath m\omega}, \quad b_m \in \mathbb{R}, \qquad (2.186)$$

such that $|B(\omega)|^2 = A(\omega)$.

Proof. 1. Since

$$\cos m\omega = \cos^m \omega - \binom{m}{2} \cos^{m-2} \omega \sin^2 \omega + \binom{m}{4} \cos^{m-4} \omega \sin^4 \omega - \cdots,$$

it can be written

$$A(\omega) = p_A(\cos \omega),$$

with p_A being a polynomial of degree M. This polynomial has the product representation

$$p_A(c) = \alpha \prod_{j=1}^{M} (c - c_j),$$

where the zeros c_j are either real or conjugate complex. On the other hand

$$\begin{aligned} A(\omega) &= \sum_{m=0}^{M} a_m \cos m\omega \\ &= \sum_{m=0}^{M} a_m \frac{e^{-\imath m\omega} + e^{\imath m\omega}}{2} \\ &= e^{\imath M\omega} \sum_{m=0}^{M} \frac{e^{-\imath (m+M)\omega} + e^{\imath (m-M)\omega}}{2} \\ &= e^{\imath M\omega} P_A(e^{-\imath\omega}), \end{aligned}$$

where P_A is a polynomial of degree $2M$. Hence for $|z| = |e^{-\imath\omega}| = 1$ we have

$$p_A(\cos \omega) = A(\omega) = e^{\imath M\omega} P_A(e^{-\imath\omega})$$

or

$$P_A(e^{-\imath\omega}) = e^{-\imath M\omega} p_A(\cos\omega) = e^{-\imath\omega} p_A \left(\frac{e^{\imath\omega} + e^{-\imath\omega}}{2} \right).$$

For $z = e^{-\imath\omega}$ it follows that

$$P_A(z) = z^M \alpha \prod_{j=1}^{M} \left(\frac{z + z^{-1}}{2} - c_j \right) = \alpha \prod_{j=1}^{M} \left(\frac{1}{2} - c_j z + \frac{1}{2} z^2 \right).$$

2. If c_j is real, then the zeros of $q_j(z) := \frac{1}{2} - c_j z + \frac{1}{2} z^2$ are $c_j \pm \sqrt{c_j^2 - 1}$. For $|c_j| \geq 1$ the two zeros are real. Since

$$\left(c_j + \sqrt{c_j^2 - 1}\right) \left(c_j - \sqrt{c_j^2 - 1}\right) = c_j^2 - (c_j^2 - 1) = 1$$

holds, these real zeros are of the form r_j, r_j^{-1}. For $|c_j| < 1$ these two zeros are conjugate complex and of the absolute value 1, i.e. they are of the form $e^{\iota\gamma_j}, e^{-\iota\gamma_j}$. In this case

$$A(\gamma_j) = e^{\iota M\gamma_j}\alpha\prod_{j=1}^{M}\left(\frac{1}{2} - c_j e^{-\iota\gamma_j} + \frac{1}{2}(e^{-\iota\gamma_j})^2\right) = 0$$

holds, i.e., the γ_j are zeros of A.

3. If c_j is not real the conjugate complex pair $c_j, \bar{c}_j$ is considered. The polynomial

$$q_j(z) = \left(\frac{1}{2} - c_j z + \frac{1}{2}z^2\right)\left(\frac{1}{2} - \bar{c}_j z + \frac{1}{2}z^2\right)$$

has four zeros: $c_j \pm \sqrt{c_j^2 - 1}$ and $\bar{c}_j \pm \sqrt{\bar{c}_j^2 - 1}$. Since

$$\left(c_j + \sqrt{c_j^2 - 1}\right)\left(c_j - \sqrt{c_j^2 - 1}\right) = c_j^2 - (c_j^2 - 1) = 1,$$

$$\left(\bar{c}_j + \sqrt{\bar{c}_j^2 - 1}\right)\left(\bar{c}_j - \sqrt{\bar{c}_j^2 - 1}\right) = \bar{c}_j^2 - (\bar{c}_j^2 - 1) = 1$$

holds, the zeros form a quadruple $z_j, z_j^{-1}, \bar{z}_j, \bar{z}_j^{-1}$.

4. As a preliminary summary, the following factorization of P_A is obtained:

$$P_A(z) = \frac{1}{2}\alpha_M \prod_{j=1}^{J}(z - z_j)(z - \bar{z}_j)(z - z_j^{-1})(z - \bar{z}_j^{-1}) \prod_{l=1}^{L}(z - r_l)(z - r_l^{-1}).$$

5. For $z = e^{-\iota\omega}$ holds

$$\begin{aligned}
|(e^{-\iota\omega} - z_0)(e^{-\iota\omega} - \bar{z}_0^{-1})| &= \sqrt{(e^{-\iota\omega} - z_0)(e^{\iota\omega} - \bar{z}_0)(e^{-\iota\omega} - \bar{z}_0^{-1})(e^{\iota\omega} - z_0^{-1})} \\
&= \sqrt{(2 - e^{-\iota\omega}z_0^{-1} - e^{\iota\omega}z_0)(2 - e^{-\iota\omega}\bar{z}_0 - e\iota\omega\bar{z}_0^{-1})} \\
&= \sqrt{\frac{-e^{\iota\omega}}{z_0}(-2z_0 e^{-\iota\omega} + e^{-2\iota\omega} + z_0^2)} \\
&\quad \times \sqrt{\frac{-e^{-\iota\omega}}{\bar{z}_0}(-2\bar{z}_0 e^{\iota\omega} + \bar{z}_0^2 + e^{2\iota\omega})} \\
&= \frac{1}{|z_0|}\sqrt{(e^{-\iota\omega} - z_0)^2(e^{\iota\omega} - \bar{z}_0)^2} \\
&= \frac{1}{|z_0|}|e^{-\iota\omega} - z_0|^2.
\end{aligned}$$

This relation can be used for a factorization of the polynomial $A(\omega)$:

$$
\begin{aligned}
A(\omega) &= |A(\omega)| \\
&= |P_A(e^{-\imath\omega})| \\
&= \left|\frac{1}{2}\alpha_M \prod_{j=1}^{J}(e^{-\imath\omega} - z_j)(e^{-\imath\omega} - \bar{z}_j)(e^{-\imath\omega} - z_j^{-1})(e^{-\imath\omega} - \bar{z}_j^{-1})\right| \\
&\quad \times \left|\prod_{l=1}^{L}(e^{-\imath\omega} - r_l)(e^{-\imath\omega} - r_l^{-1})\right| \\
&\quad \times \left|\prod_{k=1}^{K}(e^{-\imath\omega} - e^{\imath\gamma_k})(e^{-\imath\omega} - e^{-\imath\gamma_k})\right|^2 \\
&= \left|\frac{1}{2}\alpha_M \prod_{j=1}^{J}\left(\frac{1}{|z_j|}|e^{-\imath\omega} - z_j|^2 \frac{1}{|\bar{z}_j|}|e^{-\imath\omega} - \bar{z}_j|^2\right)\right| \\
&\quad \times \left|\prod_{l=1}^{L}\frac{1}{|r_l|}(e^{-\imath\omega} - r_l)^2\right|\left|\prod_{k=1}^{K}(e^{-\imath\omega} - e^{\imath\gamma_k})(e^{-\imath\omega} - e^{-\imath\gamma_k})\right| \\
&= \left[\frac{1}{2}|\alpha_M|\prod_{j=1}^{J}\frac{1}{|z_j|^2}\prod_{l=1}^{L}\frac{1}{|r_l|}\right]\left|\prod_{j=1}^{J}(e^{-\imath\omega} - z_j)(e^{-\imath\omega} - \bar{z}_j)\right|^2 \\
&\quad \times \left|\prod_{k=1}^{K}(e^{-\imath\omega} - e^{\imath\gamma_k})(e^{-\imath\omega} - e^{-\imath\gamma_k})\right|^2\left|\prod_{l=1}^{L}\frac{1}{|r_l|}(e^{-\imath\omega} - r_l)\right|^2 \\
&= \left(\left[\frac{1}{2}|\alpha_M|\prod_{j=1}^{J}\frac{1}{|z_j|^2}\prod_{l=1}^{L}\frac{1}{|r_l|}\right]^{1/2}\right)^2\left|\prod_{j=1}^{J}(e^{-\imath\omega} - z_j)(e^{-\imath\omega} - \bar{z}_j)\right|^2 \\
&\quad \times \left|\prod_{k=1}^{K}(e^{-\imath\omega} - e^{\imath\gamma_k})(e^{-\imath\omega} - e^{-\imath\gamma_k})\right|^2\left|\prod_{l=1}^{L}\frac{1}{|r_l|}(e^{-\imath\omega} - r_l)\right|^2.
\end{aligned}
$$

This means with

$$
\begin{aligned}
B(\omega) = \left(\left[\frac{1}{2}|\alpha_M|\prod_{j=1}^{J}\frac{1}{|z_j|^2}\prod_{l=1}^{L}\frac{1}{|r_l|}\right]^{1/2}\right)&\left|\prod_{j=1}^{J}(e^{-\imath\omega} - z_j)(e^{-\imath\omega} - \bar{z}_j)\right| \\
\times \left|\prod_{k=1}^{K}(e^{-\imath\omega} - e^{\imath\gamma_k})(e^{-\imath\omega} - e^{-\imath\gamma_k})\right|&\left|\prod_{l=1}^{L}\frac{1}{|r_l|}(e^{-\imath\omega} - r_l)\right|
\end{aligned}
$$

we derive $B^2(\omega) = A(\omega)$. □

The proof of the theorem is constructive and provides an algorithm for the determination of $H(\omega)$:

1. Choose a degree N of regularity.

2. Construct the polynomial

$$P_N(y) = \sum_{k=0}^{N-1} \binom{N-1+k}{k} y^k.$$

3. Find an odd polynomial R such that

$$P(y) := P_N(y) + y^N R\left(\frac{1}{2} - y\right) \geq 0, \quad y \in [0, 1].$$

4. Compute

$$A(\omega) := P\left(\sin^2 \frac{\omega}{2}\right).$$

5. Find the polynomial p_A with $A(\omega) = p_A(\cos \omega)$.

6. Find the zeros c_j of $p_A(c)$.

7. For real c_j with $|c_j| \geq 1$ denote the zeros of $\frac{1}{2} - c_j z + \frac{1}{2} z^2$ by r_j, r_j^{-1}.

8. For real c_j with $|c_j| < 1$ denote the zeros of $\frac{1}{2} - c_j z + \frac{1}{2} z^2$ by $e^{\imath \gamma_j}, e^{-\imath \gamma_j}$.

9. For conjugate complex zeros $c_j, \bar{c}_j$ denote the zeros of

$$\left(\tfrac{1}{2} - c_j z + \tfrac{1}{2} z^2\right)\left(\tfrac{1}{2} - \bar{c}_j z + \tfrac{1}{2} z^2\right)$$

by $z_j, z_j^{-1}, \bar{z}_j, \bar{z}_j^{-1}$.

10. Set

$$\begin{aligned} H(\omega) = &\left(\frac{1 + e^{-\imath\omega}}{2}\right)^N \left(\left[\frac{1}{2}|\alpha_M| \prod_{j=1}^{J} \frac{1}{|z_j|^2} \prod_{l=1}^{L} \frac{1}{|r_l|}\right]^{1/2}\right) \\ &\times |\prod_{j=1}^{J} (e^{-\imath\omega} - z_j)(e^{-\imath\omega} - \bar{z}_j)| \times \\ &\times \left|\prod_{k=1}^{K} (e^{-\imath\omega} - e^{\imath\gamma_k})(e^{-\imath\omega} - e^{-\imath\gamma_k})\right| \left|\prod_{l=1}^{L} (e^{-\imath\omega} - r_l)\right|. \end{aligned}$$

Example 2.8.1. In order to illustrate this procedure, the simplest example $N = 1$ will be discussed here. The polynomial $P_1(y)$ is given by

$$P_1(y) = 1.$$

Since this polynomial is already non-negative, no other solution is needed:

$$P(y) = P_1(y) = 1.$$

This polynomial P is so simple that it is not necessary to apply the factorization algorithm. The function H can be found directly. Following Theorem 2.8.1,

$$H(\omega) = \left(\frac{1 + e^{-\imath\omega}}{2}\right) |P(\sin^2 \omega)|^2 = \left(\frac{1 + e^{-\imath\omega}}{2}\right)$$

defines the scaling equation of an orthogonal scaling function of order $N = 1$. The coefficients h_k of this scaling equation are given by the Fourier series of H:

$$H(\omega) = \frac{1}{\sqrt{2}}\left(\frac{1}{\sqrt{2}} + \frac{1}{\sqrt{2}} e^{-\imath\omega}\right).$$

In other words, the scaling coefficients are

$$h_0 = h_1 = \frac{1}{\sqrt{2}}.$$

Therefore the simplest orthogonal wavelet with compact support is the well-known Haar wavelet.

This algorithm only generates the scaling coefficients h_k and g_k of the scaling function and the wavelet. How to find these functions themselves will be discussed later. First of all we will illustrate the above mentioned algorithm for another examples of Daubechies wavelets. The greatest disadvantage of the Haar wavelet is its irregularity. Since the regularity increases with in creasing length N of the filters, the next possible candidate will be $N = 2$.

Example 2.8.2 (Daubechies 4 wavelet). For $N = 2$ one obtains

$$P_2(y) = \binom{1}{0} + \binom{2}{1} y = 1 + 2y.$$

Since P_2 is already non-negative on $[0, 1]$, the relation

$$P(y) = P_2(y) = 1 + 2y$$

follows. Consequently,

$$\begin{aligned}
A(\omega) &= \left(1 + 2\sin^2\frac{\omega}{2}\right) \\
&= (2 - \cos\omega) \\
&= \left(2 - \frac{1}{2}(e^{\imath\omega} + e^{-\imath\omega})\right) \\
&= e^{\imath\omega}e^{-\imath\omega}\left(2 - \frac{1}{2}(e^{\imath\omega} + e^{-\imath\omega})\right) \\
&= e^{\imath\omega}\left(2e^{-\imath\omega} - \frac{1}{2} - \frac{1}{2}e^{-2\imath\omega}\right) \\
&= -\frac{e^{\imath\omega}}{2}(e^{-2\imath\omega} - 4e^{-\imath\omega} + 1) \\
&= -\frac{e^{\imath\omega}}{2}(e^{-\imath\omega} - (2+\sqrt{3}))(e^{-\imath\omega} - (2-\sqrt{3})) \\
&= -\frac{e^{\imath\omega}}{2}\frac{1}{|2+\sqrt{3}|}|e^{-\imath\omega} - (2+\sqrt{3})|^2.
\end{aligned}$$

Since

$$H(\omega) \sim \left(\frac{1+e^{-\imath\omega}}{2}\right)^2 |A(\omega)|^{1/2}$$

is true, it follows that

$$\begin{aligned}
H(\omega) &\sim \left(\frac{1+e^{-\imath\omega}}{2}\right)^2 \frac{1}{\sqrt{2}}\frac{1}{\sqrt{2+\sqrt{3}}}(e^{-\imath\omega} - (2+\sqrt{3})) \\
&= \left(\frac{1+e^{-\imath\omega}}{2}\right)^2 \frac{1}{\sqrt{2}}\frac{1}{(\sqrt{2+\sqrt{3}})(1-\sqrt{3})}(e^{-\imath\omega}(1-\sqrt{3}) + (1+\sqrt{3})) \\
&= \frac{1}{\sqrt{2}}\frac{1}{(\sqrt{2+\sqrt{3}})(1-\sqrt{3})}\frac{1+2e^{-\imath\omega}+e^{-2\imath\omega}}{4}(e^{-\imath\omega}(1-\sqrt{3}) + (1+\sqrt{3})) \\
&= \frac{1}{\sqrt{2}}\frac{1}{(\sqrt{2+\sqrt{3}})(1-\sqrt{3})}\left[\frac{1+\sqrt{3}}{4} + \frac{3+\sqrt{3}}{4}e^{-\imath\omega}\right. \\
&\qquad \left. + \frac{3-\sqrt{3}}{4}e^{-2\imath\omega} + \frac{1-\sqrt{3}}{4}e^{-3\imath\omega}\right].
\end{aligned}$$

The normalization condition

$$|H(\omega)|^2 + |H(\omega+\pi)|^2 = 1$$

yields

$$H(\omega) = \frac{1}{\sqrt{2}}\left[\frac{1+\sqrt{3}}{4\sqrt{2}} + \frac{3+\sqrt{3}}{4\sqrt{2}}e^{-\imath\omega} + \frac{3-\sqrt{3}}{4\sqrt{2}}e^{-2\imath\omega} + \frac{1-\sqrt{3}}{4\sqrt{2}}e^{-3\imath\omega}\right].$$

A comparison of coefficients gives

$$h_0 = \frac{1}{4\sqrt{2}}(1+\sqrt{3}),\ h_1 = \frac{1}{4\sqrt{2}}(3+\sqrt{3}),\ h_2 = \frac{1}{4\sqrt{2}}(3-\sqrt{3}),\ h_3 = \frac{1}{4\sqrt{2}}(1-\sqrt{3}). \tag{2.187}$$

The following table contains the scaling coefficients for Daubechies wavelets of different orders.

order N	index n	scaling coefficient h_n
$N=1$	0	0.70710678119
	1	0.70710678119
$N=2$	0	0.48296291314
	1	0.83651630374
	2	0.22414386804
	3	−0.12940952255
$N=3$	0	0.33267055295
	1	0.80689150931
	2	0.45987750212
	3	−0.13501102001
	4	−0.08544127388
	5	0.03522269189
$N=4$	0	0.23037781331
	1	0.71484657055
	2	0.63088076794
	3	−0.02798376942
	4	−0.18703481172
	5	0.03084138184
	6	0.03288301167
	7	−0.01059740179

So far we have a procedure to construct the scaling coefficients h_k, g_k of an orthogonal compactly supported wavelet ψ and its scaling function φ. Now the questions arises how do these function look like? One possible answer comes with the Mallat algorithm. The inverse Mallat algorithm allows the reconstruction of a signal with given wavelet coefficients.

Let $\psi \in W_1$. Since $W_1 \subset V_{-N}$ the mother wavelet ψ has the representation

$$\psi(x) = \sum_{k\in\mathbb{Z}} c_k^{(-N)} \varphi_{-N,k}(x).$$

On the other hand, like any other element of V_{-N} it also has the representation

$$\psi(x) = \sum_{m=-N}^{1} \sum_{k\in\mathbb{Z}} d_k^{(m)} \psi_{m,k}(x) + \sum_{k\in\mathbb{Z}} c_k^{(1)} \varphi_{1,k}(x)$$

with

$$d_0^{(1)} = 1, \quad d_k^{(m)} = 0 \text{ else}, \quad c_k^{(1)} = 0.$$

With the help of the inverse Mallat algorithm the coefficients $c_k^{(-N)}$ can be computed recursively from the given coefficients $d_k^{(m)}$, $c_k^{(1)}$. Under certain circumstances, which will be discussed later, the coefficients $c_k^{(-N)}$ are close to the sampled values

$$\psi(k2^{-N}) \approx c_k^{(-N)}$$

and in this way the mother wavelet and the scaling function respectively can be computed pointwise. The scaling function φ and the wavelet ψ for the Daubechies wavelet with four non-vanishing coefficients, computed with the above mentioned procedure, are displayed in Figure 2.39 and Figure 2.40.

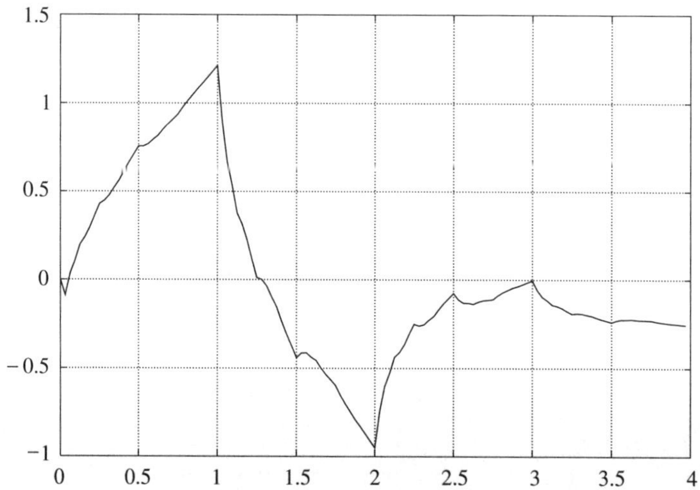

Figure 2.39. Scaling function of Daubechies 4 wavelet

Obviously, the Daubechies 4 wavelet is rather irregular and the question arises, if for an increasing number of coefficients the regularity increases. The following theorem gives a positive answer to this question.

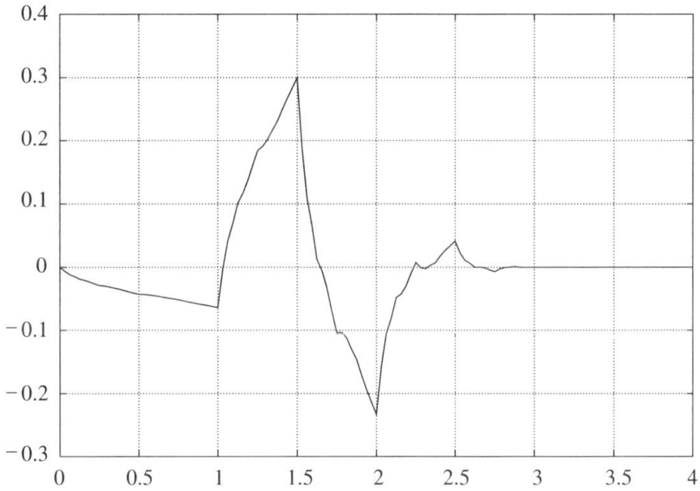

Figure 2.40. Daubechies 4 wavelet

Theorem 2.8.3. *Let φ_N and ψ_N be the scaling function and the wavelet with $2N$ scaling coefficients. Then*

$$\psi_n \in H^s, \quad s < (\ln N)/4\ln 2 \tag{2.188}$$

holds.

Proof. [68]. □

The set H^s is a so-called Sobolev space. Without going deeply into details, a rough imagination can be given by the statement that H^s contains all functions which are s-times differentiable. Here the differentiation is meant in the generalized sense. This means that the regularity of the wavelet increases indeed, but very slowly.

In some applications the regularity of the wavelet is not so important. Frequently it is more important that the wavelet is orthogonal to all polynomials of maximum degree N.

$$\int_{\mathbb{R}} x^n \psi(x)\,dx = 0, \quad n = 0, \dots, N. \tag{2.189}$$

Lemma 2.8.3. *Let ψ_N be the Daubechies wavelet with $2N$ scaling coefficients. Then*

$$\int_{\mathbb{R}} x^n \psi_N(x)\,dx = 0, \quad n = 0, \dots, N-1. \tag{2.190}$$

Proof. [68]. □

2.8.2 Solution of scaling equations

In many cases the only knowledge about a scaling function φ is the sequence $\{h_k\}$ of its scaling coefficients:

$$\varphi(x) = \sum_{k\in\mathbb{Z}} h_k\varphi(2x-k). \tag{2.191}$$

It remains the question of how the scaling equation (2.191) can be solved for the unknown scaling function φ. Of course there is a larger number of possible solution techniques, each of them with specific advantages and disadvantages. Here, only one methods will be discussed. This method is

- particularly simple,
- well-suited to generate a graphical representation of the scaling function φ.

The basic idea of this method is, to represent $\varphi \in V_0$ with respect to a base of $V_{-m} \supset V_0$:

$$\varphi(x) = \sum_{k\in\mathbb{Z}} c_k^{(-m)} 2^{m/2}\varphi(2^m x - k). \tag{2.192}$$

For increasing numbers of m the scaled and shifted version $2^{m/2}\varphi(2^m x - k)$ approximates the delta distribution $\delta(x - 2^{-m}k)$ better and better. Hence

$$\begin{aligned}
\varphi(2^{-m}k) &= (\delta(x-2^{-m}k), \varphi(x)) \approx \langle 2^{-m/2}\varphi(2^m x - k), \varphi(x)\rangle \\
&= \langle 2^{m/2}\varphi(2^m x - k), \sum_{l\in\mathbb{Z}} c_l^{(-m)} 2^{m/2}\varphi(2^m x - l)\rangle \\
&= \sum_{l\in\mathbb{Z}} c_l^{(-m)} \langle 2^{m/2}\varphi(2^m x - k), 2^{m/2}\varphi(2^m x - l)\rangle \\
&= \sum_{l\in\mathbb{Z}} c_l^{(-m)} \delta_{l,k} \\
&= c_k^{(-m)}.
\end{aligned}$$

This means, for m large enough the unknown development coefficients $c_k^{(-m)}$ can be approximated by the sample values of the scaling function φ at the places $x = 2^{-m}k$.

Under certain circumstances it can be expected that the step-function

$$\varphi_m(x) := \sum_{k\in\mathbb{Z}} c_k^{(-m)} \chi_{[-1/2,1/2]}(2^m x - k) \tag{2.193}$$

with $\chi_{[a,b]}$ being the indicator function of the interval $[a, b]$

$$\chi_{[a,b]}(x) := \begin{cases} 1, & x \in [a,b] \\ 0, & \text{else} \end{cases} \tag{2.194}$$

converges to the scaling function φ. The only question remaining is: How to compute the development coefficients $c_k^{(-m)}$? For this purpose the inverse Mallat algorithm can be used. Obviously

$$\varphi(x) = \sum_{k<0} 0 \cdot \varphi(x-k) + 1 \cdot \varphi(x) + \sum_{k>0} 0 \cdot \varphi(x-k)$$

holds. Hence, setting

$$c_k^{(0)} = \delta_{0,k}, \quad d_k^{(0)} = 0$$

and

$$c^{(-1)} = H^* c^{(0)} + G^* d^{(0)} = H^* c^{(0)},$$

the scaling function can be represented with respect to a base of V_{-1}:

$$\varphi(x) = \sum_{k \in \mathbb{Z}} c_k^{(-1)} 2^{-1/2} \varphi(2^{-1}x - k) + \sum_{k \in \mathbb{Z}} 0 \cdot 2^{-1/2} \psi(2^{-1}x - k). \tag{2.195}$$

Iterating again,

$$c^{(-2)} = H^* c^{(-1)} + G^* d^{(-1)} = H^* c^{(-1)}, \tag{2.196}$$

the coefficients $c_k^{(-2)}$ of the representation of φ with respect to a base of V_{-2} are obtained. Consequently, after m steps of inverse Mallat algorithm the coefficients $c_k^{(-m)}$ are obtained:

$$c_k^{(0)} = \delta_{0,k}, \tag{2.197}$$

$$c^{(-m)} = (H^*)^m c^{(0)}. \tag{2.198}$$

The precise conditions which a sequence $\{h_k\}$ of scaling coefficients has to fulfill in order to guarantee convergence are rather complicated. They will not be discussed here. An intensive study of these coefficients can be found in [68].

Example 2.8.3. As an example for the convergence of this iteration process to the scaling function and to the wavelets respectively, the scaling coefficients of the Daubechies wavelet of order 6 will be used. The scaling coefficients are given in the following table:

n	h_n
0	0.3326705529500825
1	0.8068915093110924
2	0.4598775021184914
3	−0.1350110200102546
4	−0.0854412738820267
5	0.0352262918857095

With the help of these scaling coefficients the iteration was carried out up to $m = 8$. In Figure 2.41 the elements $\varphi_{-4}, \varphi_{-5}, \varphi_{-8}$ and $\psi_{-4}, \psi_{-5}, \psi_{-8}$ respectively are shown.

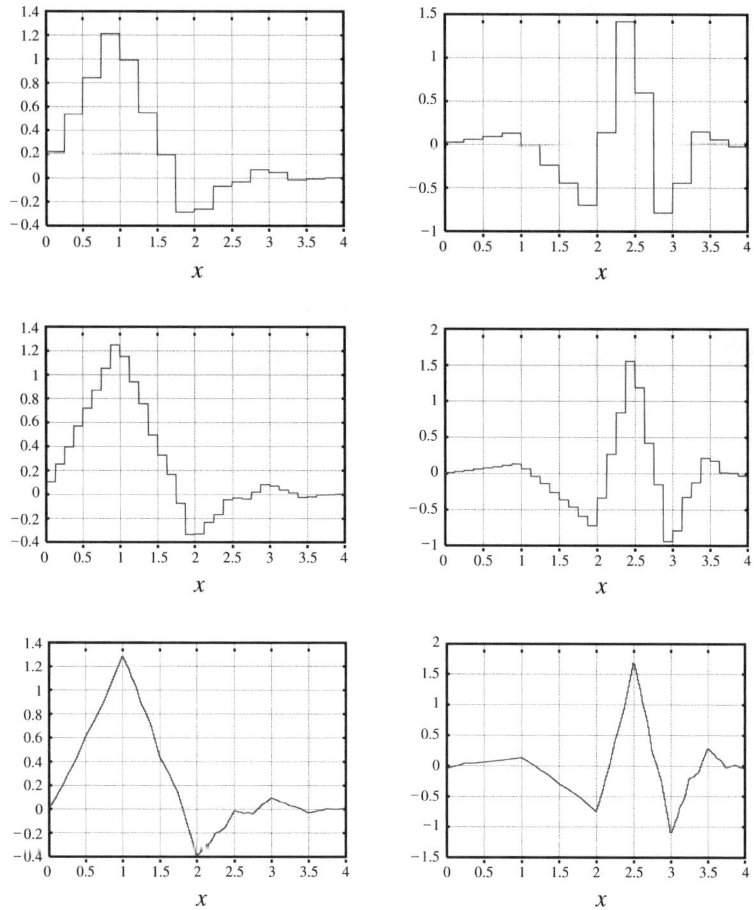

Figure 2.41. Iteration steps 4, 5 and 8 for scaling function φ (left) and wavelet ψ (right)

2.9 Wavelet bases on an interval

So far we have only considered MRAs on $L_2(\mathbb{R})$, i.e. we only dealt with the representation of signals which are defined on the whole real axis. In practical applications there is always only a finite registration length of a signal, meaning that only a signal defined on a finite interval is available. For this reason, an MRA of $L_2([0, 1])$ is needed. One very simple method is to take the scaling function φ and the wavelet ψ of an MRA of $L_2(\mathbb{R})$ and make them periodic:

$$\varphi_{m,n}^{\text{period}}(x) := \sum_{l \in \mathbb{Z}} \varphi_{m,n}(x + l), \qquad \psi_{m,n}^{\text{period}}(x) := \sum_{l \in \mathbb{Z}} \psi_{m,n}(x + l). \tag{2.199}$$

Of course, this method is only applicable if both φ and ψ have a finite support.

Lemma 2.9.1. *For fixed m the scaling functions $\varphi_{m,n}^{\text{period}}$ and the wavelets $\psi_{m,n}^{\text{period}}$ are orthogonal to each other:*

$$\langle \varphi_{m,n}^{\text{period}}, \psi_{m,k}^{\text{period}} \rangle = 0. \tag{2.200}$$

Proof.

$$\begin{aligned}
\langle \varphi_{m,n}^{\text{period}}, \psi_{m,k}^{\text{period}} \rangle &= \Big\langle \sum_{p\in\mathbb{Z}} \varphi_{m,n}(x+p), \sum_{q\in\mathbb{Z}} \psi_{m,k}(x+q) \Big\rangle \\
&= \sum_{p\in\mathbb{Z}} \sum_{q\in\mathbb{Z}} \langle \varphi_{m,n}(x+p), \psi_{m,k}(x+q) \rangle \\
&= 0.
\end{aligned}$$

□

The scaling spaces V_m^{period} and the wavelet spaces W_m^{period} are now defined as the linear span of the scaling functions $\varphi_{m,n}^{\text{period}}$ and the wavelets $\psi_{m,k}^{\text{period}}$, respectively:

$$V_m^{\text{period}} := \operatorname{span}\{\varphi_{m,n}^{\text{period}} \mid 0 \le n \le 2^{|m|-1}\}, \tag{2.201}$$

$$W_m^{\text{period}} := \operatorname{span}\{\psi_{m,n}^{\text{period}} \mid 0 \le n \le 2^{|m|-1}\}. \tag{2.202}$$

Lemma 2.9.2. *The spaces V_m^{period} form an MRA of $L_2([0,1])$.*

Proof. Let $f \in V_m^{\text{period}}$. Then with certain coefficients α_k the following representation is valid:

$$\begin{aligned}
f(x) &= \sum_{k=0}^{2^{|m|}-1} \alpha_k \cdot \varphi_{m,k}^{\text{period}}(x) \\
&= \sum_{k=0}^{2^{|m|}-1} \alpha_k \cdot \Big(\sum_l \varphi_{m,k}(x+l) \Big) \\
&= \sum_{k=0}^{2^{|m|}-1} \alpha_k \cdot \Big(\sum_l 2^{-m/2} \varphi(2^{-m}(x+l) - k) \Big) \\
&= \sum_{k=0}^{2^{|m|}-1} \alpha_k \cdot \Big(\sum_l \sqrt{2} \sum_{p\in\mathbb{Z}} h_p 2^{-m/2} \varphi(2^{-m+1}x + (2^{-m+1}l - 2k - p)) \Big) \\
&= \sum_{k=0}^{2^{|m|}-1} \alpha_k \cdot \Big(\sum_l \sum_{p\in\mathbb{Z}} h_p 2^{-(m-1)/2} \varphi(2^{-(m-1)}x + (2^{-m+1}l - 2k - p)) \Big) =
\end{aligned}$$

$$
\begin{aligned}
&= \sum_{k=0}^{2^{|m|}-1} \alpha_k \cdot \sum_l \sum_{p\in\mathbb{Z}} h_p \varphi_{m-1,2k+p}(x+l) \\
&= \sum_{k=0}^{2^{|m|}-1} \alpha_k \sum_{p\in\mathbb{Z}} h_p \varphi_{m-1,2k+p}^{\text{period}}(x) \\
&= \sum_{i=0}^{2^{|m-1|}-1} \beta_i \varphi_{m-1,i}^{\text{period}}(x).
\end{aligned}
$$

Consequently, $f \in V_{m-1}^{\text{period}}$ holds, which leads to the hierarchical decomposition

$$
\{0\} \subset \cdots \subset V_1^{\text{period}} \subset V_0^{\text{period}} \subset V_{-1}^{\text{period}} \subset \cdots \subset L_2([0,1]).
$$

Additionally,

$$
\begin{aligned}
f(2x) &= \sum_{k=0}^{2^{|m|}-1} \alpha_k \cdot \varphi_{m,k}^{\text{period}}(2x) \\
&= \sum_{k=0}^{2^{|m|}-1} \alpha_k \cdot \frac{1}{\sqrt{2}} \varphi_{m-1,k}^{\text{period}}(x)
\end{aligned}
$$

is valid. This means

$$
f(x) \subset V_m^{\text{period}} \Leftrightarrow f(2x) \in V_{m-1}^{\text{period}}.
$$

These two conditions are the basic properties of an MRA. The proof of the remaining, more technical conditions is left out. □

The scaling spaces V_m^{period} and the wavelet spaces W_m^{period} inherit not only the MRA property but also the orthogonality between scaling and wavelet space.

Lemma 2.9.3. *For the scaling and wavelet spaces V_m^{period}, W_m^{period} and for $j \le 0$ holds*

$$
V_j^{\text{period}} \perp W_j^{\text{period}}, \quad V_{j-1}^{\text{period}} = V_j^{\text{period}} \oplus W_j^{\text{period}}. \tag{2.203}
$$

Proof. The relation $V_j^{\text{period}} \perp W_j^{\text{period}}$ follows immediately from (2.200).

Let $f \in L_2[0,1]$. By extending this function with zeros outside $[0,1]$ a function $\bar{f} \in L_2(\mathbb{R})$ is generated:

$$
\bar{f}(x) := \begin{cases} f(x), & x \in [0,1] \\ 0, & x \in \mathbb{R} \setminus [0,1]. \end{cases}
$$

On the other hand, a function $g \in L_2(\mathbb{R})$ can be transformed into a function $g^{\text{period}} \in L_2([0,1])$ by setting

$$g^{\text{period}}(x) = \sum_{l \in \mathbb{Z}} g(x+l).$$

Obviously, for an arbitrary $f \in L_2([0,1])$

$$\bar{f}^{\text{period}} = f$$

holds. Now let $f \in V_j$. For its periodic equivalent holds

$$\begin{aligned} f^{\text{period}}(x) &= \sum_{l \in \mathbb{Z}} f(x+l) \\ &= \sum_{l \in \mathbb{Z}} \sum_{k \in \mathbb{Z}} f_k \varphi_{j,k}(x+l) \\ &= \sum_{k \in \mathbb{Z}} f_k \sum_{l \in \mathbb{Z}} \varphi_{j,k}(x+l) \\ &= \sum_{k \in \mathbb{Z}} f_k \varphi_{j,k}^{\text{period}}(x) \\ &\in V_j^{\text{period}}. \end{aligned}$$

In the same way $f \in W_j \Rightarrow f^{\text{period}} \in W_j^{\text{period}}$ can be shown. For an arbitrary $f \in V_{j-1}^{\text{period}}$ its extension $\bar{f}$ belongs to V_{j-1}. Due to the MRA properties of the sequence $\{V_j\}$, $\bar{f}$ can be decomposed in

$$\bar{f} = P_j \bar{f} + Q_j \bar{f}, \quad P_j \bar{f} \in V_j, \; Q_j \bar{f} \in W_j.$$

Finally, for $f \in V_{j-1}^{\text{period}}$ follows

$$\begin{aligned} f(x) &= \bar{f}^{\text{period}}(x) \\ &= (P_j \bar{f})^{\text{period}} + (Q_j \bar{f})^{\text{period}} \\ &\in V_j^{\text{period}} \oplus W_j^{\text{period}}. \end{aligned}$$ □

With the MRA properties also the fast wavelet decomposition and reconstruction is carried over to the finite-interval case.

Lemma 2.9.4. *Suppose*

$$f = \sum_k c_k^{(j-1)} \varphi_{j-1,k}^{\text{period}} \in V_{j-1}^{\text{period}}. \tag{2.204}$$

The function f has a unique decomposition

$$f = \sum_k c_k^{(j)} \varphi_{j,k}^{\text{period}} + \sum_k d_k^{(j)} \psi_{j,k}^{\text{period}}, \tag{2.205}$$

with the coefficients $c_k^{(j)}, d_k^{(j)}$ related to $c_k^{(j-1)}$ by

$$c_k^{(j)} = \sum_l h_l c_{l+2k}^{(j-1)}, \quad d_k^{(j)} = \sum_l g_l c_{l+2k}^{(j-1)}, \tag{2.206}$$

where the coefficients fulfill $c_{l+2^{-j}}^{(j)} = c_l^{(j)}$.

2.10 Two-dimensional wavelets

2.10.1 Continuous two-dimensional wavelets

As in the one-dimensional case, also in the two-dimensional case there are two kinds of wavelet transforms: the continuous and the discrete wavelet transform. Following the outline in the one-dimensional case, first the definition of a two-dimensional wavelet will be given. Then the two-dimensional continuous wavelet transform will be discussed before going to the two-dimensional discrete wavelet transform.

Definition 2.10.1. A function $\psi \in L_2(\mathbb{R}^2)$ is called a *two-dimensional wavelet* if it fulfills the admissibility condition

$$0 < c_\psi := 4\pi^2 \int_{\mathbb{R}^2} \frac{|\hat{\psi}(\omega)|^2}{\|\omega\|^2} \, d\omega < \infty. \tag{2.207}$$

Similar to the one-dimensional case, also here some important conclusions about the nature of a two-dimensional wavelet can be drawn from this definition. Obviously,

$$0 = \hat{\psi}(0) = \frac{1}{2\pi} \int_{\mathbb{R}^2} \psi(x) e^{-\imath \mathbf{0}^\top \boldsymbol{x}} \, d\boldsymbol{x} = \frac{1}{2\pi} \int_{\mathbb{R}^2} \psi(x) \, d\boldsymbol{x} \tag{2.208}$$

must hold. This means that the function ψ has to oscillate in some way.

In the one-dimensional case the signal was compared to scaled and shifted versions of the mother wavelet. In the two-dimensional case besides the translation the rotation is an additional movement of the Euclidean plane. Therefore in the two-dimensional case there are four wavelet parameters:

- the scale parameter a,
- the translation vector $\boldsymbol{b} = (b_1, b_2)^\top$,
- the rotation angle ϑ.

Therefore, the scaled, shifted and rotated version of the two-dimensional mother-wavelet ψ is

$$\frac{1}{a}\psi\left(\frac{1}{a}\boldsymbol{R}^\top(\vartheta)(\boldsymbol{x}-\boldsymbol{b})\right),$$

with the rotation matrix $\boldsymbol{R}$ defined by

$$\boldsymbol{R}(\vartheta) = \begin{bmatrix} \cos\vartheta & -\sin\vartheta \\ \sin\vartheta & \cos\vartheta \end{bmatrix}. \tag{2.209}$$

Naturally, also in the two-dimensional case the wavelet transform is a comparison of the signal with scaled, rotated and shifted versions of the mother-wavelet.

Definition 2.10.2. Let ψ be a two-dimensional wavelet and let $f \in L_2(\mathbb{R}^2)$. Then

$$\mathcal{W}^{eu}\{f\}(a,\vartheta,\boldsymbol{b}) := \frac{1}{\sqrt{c_\psi}} \int_{\mathbb{R}^2} \frac{1}{a}\psi\left(\frac{1}{a}\boldsymbol{R}^\top(\vartheta)(\boldsymbol{x}-\boldsymbol{b})\right) \cdot f(\boldsymbol{x})\, d\boldsymbol{x} \tag{2.210}$$

is called the *two-dimensional continuous wavelet transform* of f.

The two-dimensional wavelet transform measures the intensity of the signal components at scales of the size a close to the location $\boldsymbol{b}$, which have direction preference of ϑ. If the two-dimensional continuous wavelet transform is used for example for image analysis, not only the location of edges in the image can be detected but also their orientations.

Surprisingly, the inverse two-dimensional wavelet transform requires only a single integration over all scales a and all rotations ϑ. An integration over the translation can be omitted.

Lemma 2.10.1. *Let ψ be a two-dimensional wavelet and let $f \in L_2(\mathbb{R}^2)$. Let their Fourier transforms $\hat{\psi}$ and $\hat{f}$ be integrable. Then*

$$f(\boldsymbol{x}) = \frac{2\pi}{\sqrt{c_\psi}} \int_0^{2\pi} \int_0^\infty \mathcal{W}^{eu}\{f\}(a,\vartheta,\boldsymbol{b}) \frac{da}{a^2}\, d\vartheta. \tag{2.211}$$

Proof. [68]. □

Example 2.10.1. The simplest way to construct a two-dimensional wavelet is the tensor product of two one-dimensional wavelets ψ:

$$\psi(x_1, x_2) = \psi(x_1) \cdot \psi(x_2). \tag{2.212}$$

Since

$$\hat{\psi}(\omega_1, \omega_2) = \hat{\psi}(\omega_1)\hat{\psi}(\omega_2)$$

and the inequality

$$\|\omega\|^2 \geq 2|\omega_1||\omega_2|$$

is valid, for the admissibility condition it can be concluded that

$$\begin{aligned}\int_{\mathbb{R}^2} \frac{|\hat{\psi}(\omega_1, \omega_2)|^2}{\|\omega^2\|^2}\,d\omega &\le \frac{1}{2}\int_{\mathbb{R}} \frac{|\hat{\psi}(\omega_1)|^2}{|\omega_1|}\,d\omega_1 \cdot \int_{\mathbb{R}} \frac{|\hat{\psi}(\omega_2)|^2}{|\omega_2|}\,d\omega_2 \\ &- \frac{c_\psi^2}{8\pi^2} \\ &< \infty.\end{aligned} \tag{2.213}$$

If the rotation parameter ϑ and the scale parameter a are arbitrary but fixed, for the numerical implementation of the two-dimensional wavelet transformation the FFT method can be used. Let

$$\psi_\vartheta(\boldsymbol{x}) := \psi(-\boldsymbol{R}^\top(\vartheta)\boldsymbol{x}). \tag{2.214}$$

Then the wavelet transformation can be expressed as a convolution:

$$\begin{aligned}\mathcal{W}_\psi^{\mathrm{eu}}(a, \vartheta, \boldsymbol{b}) &= \frac{1}{\sqrt{c_\psi}}\int_{\mathbb{R}^2} \frac{1}{a}\psi\left(\frac{1}{a}\boldsymbol{R}^\top(\vartheta)(\boldsymbol{x}-\boldsymbol{b})\right) \cdot f(\boldsymbol{x})\,d\boldsymbol{x} \\ &= \frac{1}{\sqrt{c_\psi}}\int_{\mathbb{R}^2} \frac{1}{a}\psi_\vartheta\left(-\frac{1}{a}(\boldsymbol{x}-\boldsymbol{b})\right) f(\boldsymbol{x})\,d\boldsymbol{x} \\ &= \frac{1}{\sqrt{c_\psi}}\left(f * \psi_\vartheta\left(\frac{1}{a}\bullet\right)\right)(\boldsymbol{b}) \\ &= \frac{2\pi a^{2\,2}}{\sqrt{c_\psi}}\mathcal{F}^{-1}\{\hat{f}(\omega)\hat{\psi_\vartheta}(a\omega)\}.\end{aligned} \tag{2.215}$$

For the computation of the spectra $\hat{f}$ and $\hat{\psi_\vartheta}$ FFT can be efficiently used.

2.10.2 Discrete two-dimensional wavelets

One of the main application field of wavelets is image processing. Since images are two-dimensional data, they can only be analyzed by two-dimensional wavelets. The construction of two-dimensional wavelets is the straightforward generalization of the construction of one-dimensional wavelets.

It starts with a two-dimensional MRA.

Definition 2.10.3. A nested sequence

$$\{0\} \subset \cdots \subset V_2 \subset V_1 \subset V_0 \subset V_{-1} \subset \cdots \subset L_2(\mathbb{R}^2) \tag{2.216}$$

of closed subsets $V_m \subset L_2(\mathbb{R}^2)$ is called a *multi-resolution analysis* (MRA) of $L_2(\mathbb{R}^2)$, if the following four statements hold:

(i)

$$\overline{\bigcup_{m\in\mathbb{Z}} V_m} = L_2(\mathbb{R}^2). \tag{2.217}$$

(ii)

$$\bigcap_{m\in\mathbb{Z}} V_m = \{0\}. \tag{2.218}$$

(iii)

$$f(\bullet) \in V_m \Leftrightarrow f(\boldsymbol{A}^m \bullet) \in V_0, \quad \boldsymbol{A} \text{ a regular } 2\times 2 \text{ matrix.} \tag{2.219}$$

(iv) There is a function $\varphi \in L_2(\mathbb{R}^2)$with

$$V_0 = \overline{\operatorname{span}\{\varphi(\bullet - \boldsymbol{k}) \mid \boldsymbol{k} \in \mathbb{Z}^2\}} \tag{2.220}$$

and

$$A \sum_{\boldsymbol{k}\in\mathbb{Z}^2} c_{\boldsymbol{k}}^2 \le \Big\| \sum_{\boldsymbol{k}\in\mathbb{Z}^2} c_{\boldsymbol{k}} \varphi(\bullet - \boldsymbol{k}) \Big\|_{L_2}^2 \le B \sum_{\boldsymbol{k}\in\mathbb{Z}^2} c_{\boldsymbol{k}}^2, \quad 0 < A, B \tag{2.221}$$

for all $\{c_{\boldsymbol{k}}\}_{\boldsymbol{k}\in\mathbb{Z}^2} \in l_2(\mathbb{Z}^2)$.

The matrix $\boldsymbol{A}$ is called *dilatation matrix*. As in the one-dimensional case, the space V_0 can be decomposed into its smoothed version V_1 and the orthogonal complement W_1 of V_1 in V_0:

$$V_0 = V_1 \oplus W_1.$$

In contrast to the one-dimensional case the orthogonal complement is not spanned any longer by a single function ψ but by $|\det(\boldsymbol{A})| - 1$ independent functions.

Theorem 2.10.1. *Let $\{V_m\}_{m\in\mathbb{Z}}$ be an MRA with the dilatation matrix $\boldsymbol{A}$. Then there are $|\det(\boldsymbol{A}) - 1|$ wavelets*

$$\psi^{(1)}, \psi^{(2)}, \ldots, \psi^{(|\det(\boldsymbol{A})|-1)} \in V_{-1} \tag{2.222}$$

which span a CONS [2] of the orthogonal complement of V_0 in V_{-1}.

In other words

$$\{\psi_{m,\boldsymbol{k}}^{(j)}(\boldsymbol{x}) = |\det(\boldsymbol{A})|^{-m/2}\psi^{(j)}(\boldsymbol{A}^{-m}\boldsymbol{x} - \boldsymbol{k}) \mid j = 1, \ldots, |\det(\boldsymbol{A})|-1, m \in \mathbb{Z}, \boldsymbol{k} \in \mathbb{Z}^2\} \tag{2.223}$$

is a CONS of $L_2(\mathbb{R}^2)$.

Proof. [73]. □

A natural choice for the dilatation matrix is

$$\boldsymbol{A} = \begin{pmatrix} 2 & 0 \\ 0 & 2 \end{pmatrix}, \tag{2.224}$$

[2]Complete orthonormal system, see Definition A.2.5 on p. 224

i.e. a dilatation in each coordinate direction by the factor 2. Since $|\det(\boldsymbol{A})| = 4$, there are three independent wavelets $\psi^{(1)}, \psi^{(2)}, \psi^{(3)}$. The simplest way to find these two-dimensional wavelets is to form the tensor products from their one-dimensional counterparts.

Lemma 2.10.2. *Let φ be an orthogonal scaling function and let ψ be the corresponding wavelet. Then*

$$\varphi(\boldsymbol{x}) := \varphi(x_1) \cdot \varphi(x_2) \tag{2.225}$$

is the two-dimensional scaling function, and the functions

$$\psi^{(1)}(\boldsymbol{x}) = \psi(x_1)\varphi(x_2), \quad \psi^{(2)}(\boldsymbol{x}) = \varphi(x_1)\psi(x_2), \quad \psi^{(3)}(\boldsymbol{x}) = \psi(x_1)\psi(x_2) \tag{2.226}$$

are the two-dimensional wavelets belonging to the dilatation matrix (2.224).

Proof. [68]. □

Let us now consider the Mallat algorithm for two-dimensional tensor wavelets. The scaling equations for the two-dimensional wavelets can be derived from their one-dimensional counterparts.

$$\begin{aligned} \varphi(\boldsymbol{x}) &:= \varphi(x_1)\varphi(x_2) \\ &= \sqrt{2}\sum_{k_1\in\mathbb{Z}} h_{k_1}\varphi(2x_1 - k_1) \times \sqrt{2}\sum_{k_2\in\mathbb{Z}} h_{k_2}\varphi(2x_2 - k_2) \\ &= 2\sum_{(k_1,k_2)\in\mathbb{Z}^2} h_{k_1}h_{k_2}\varphi(\boldsymbol{A}\boldsymbol{x} - (k_1, k_2)). \end{aligned} \tag{2.227}$$

$$\begin{aligned} \psi^{(1)}(\boldsymbol{x}) &:= \psi(x_1)\varphi(x_2) \\ &= \sqrt{2}\sum_{k_1\in\mathbb{Z}} g_{k_1}\varphi(2x_1 - k_1) \times \sqrt{2}\sum_{k_2\in\mathbb{Z}} h_{k_2}\varphi(2x_2 - k_2) \\ &= 2\sum_{(k_1,k_2)\in\mathbb{Z}^2} g_{k_1}h_{k_2}\varphi(\boldsymbol{A}\boldsymbol{x} - (k_1, k_2)). \end{aligned} \tag{2.228}$$

$$\begin{aligned} \psi^{(2)}(\boldsymbol{x}) &:= \varphi(x_1)\psi(x_2) \\ &= \sqrt{2}\sum_{k_1\in\mathbb{Z}} h_{k_1}\varphi(2x_1 - k_1) \times \sqrt{2}\sum_{k_2\in\mathbb{Z}} g_{k_2}\varphi(2x_2 - k_2) \\ &= 2\sum_{(k_1,k_2)\in\mathbb{Z}^2} h_{k_1}g_{k_2}\varphi(\boldsymbol{A}\boldsymbol{x} - (k_1, k_2)). \end{aligned} \tag{2.229}$$

$$\begin{aligned}\psi^{(3)}(\boldsymbol{x}) &:= \psi(x_1)\psi(x_2)\\ &= \sqrt{2}\sum_{k_1\in\mathbb{Z}} g_{k_1}\varphi(2x_1-k_1)\times\sqrt{2}\sum_{k_2\in\mathbb{Z}} g_{k_2}\varphi(2x_2-k_2) \qquad (2.230)\\ &= 2\sum_{(k_1,k_2)\in\mathbb{Z}^2} g_{k_1}g_{k_2}\varphi(\boldsymbol{A}\boldsymbol{x}-(k_1,k_2)).\end{aligned}$$

With the help of these scaling equations the decomposition of V_0 into V_1 and its orthogonal complement can be described. Let $f \in V_0$ and $\boldsymbol{k} = (k_1, k_2)$, $\mathbf{l} = (l_1, l_2)$. Then

$$f(\boldsymbol{x}) = \sum_{k\in\mathbb{Z}^2} c_{0,\boldsymbol{k}}\varphi(\boldsymbol{x}-\boldsymbol{k}) \qquad (2.231)$$

holds. The coefficients on the next coarser scale can be computed by

$$\begin{aligned}c_{0,\boldsymbol{k}} &= \langle f, \varphi_{0,\boldsymbol{k}}\rangle\\ &= \sum_{l\in\mathbb{Z}^2} 2h_{l_1}h_{l_2}\langle f, \varphi(\boldsymbol{A}\boldsymbol{x}-\boldsymbol{A}\boldsymbol{k}-\mathbf{l})\rangle\\ &= \sum_{l\in\mathbb{Z}^2} h_{l_1}h_{l_2}\langle f, \varphi_{-1,2k_1+l_1}\varphi_{-1,2k_2+l_2}\rangle \qquad (2.232)\\ &= \sum_{l_1\in\mathbb{Z}} h_{l_1-2k_1}\sum_{l_2\in\mathbb{Z}} h_{l_2-2k_2}\langle f, \varphi_{-1,l_1}\varphi_{-1,l_2}\rangle\\ &= \sum_{l_1\in\mathbb{Z}} h_{l_1-2k_1}\sum_{l_2\in\mathbb{Z}} h_{l_2-2k_2}c_{-1,\mathbf{l}}.\end{aligned}$$

The same argumentation can be applied to $\psi^{(i)}$, leading to the final results

$$c_{0,\boldsymbol{k}} = \sum_{l_1\in\mathbb{Z}} h_{l_1-2k_1}\sum_{l_2\in\mathbb{Z}} h_{l_2-2k_2}c_{-1,\mathbf{l}}, \qquad (2.233)$$

$$d^{(1)}_{0,\boldsymbol{k}} = \sum_{l_1\in\mathbb{Z}} g_{l_1-2k_1}\sum_{l_2\in\mathbb{Z}} h_{l_2-2k_2}c_{-1,\mathbf{l}}, \qquad (2.234)$$

$$d^{(2)}_{0,\boldsymbol{k}} = \sum_{l_1\in\mathbb{Z}} h_{l_1-2k_1}\sum_{l_2\in\mathbb{Z}} g_{l_2-2k_2}c_{-1,\mathbf{l}}, \qquad (2.235)$$

$$d^{(3)}_{0,\boldsymbol{k}} = \sum_{l_1\in\mathbb{Z}} g_{l_1-2k_1}\sum_{l_2\in\mathbb{Z}} g_{l_2-2k_2}c_{-1,\mathbf{l}}. \qquad (2.236)$$

If the coefficients $c_{0,\boldsymbol{k}}$ are considered as a two-dimensional array of real numbers, the coefficients $c_{1,\boldsymbol{k}}, d^{(1)}_{1,\boldsymbol{k}}, d^{(2)}_{1,\boldsymbol{k}}, d^{(3)}_{1,\boldsymbol{k}}$ on the next coarser scale are the result of the column- and row-wise application of the one-dimensional operators H and G. If we define

$$(H_R c)_{k_1,k_2} := \sum_{l\in\mathbb{Z}} h_{l-2k_2} c_{k_1,l}, \tag{2.237}$$

$$(H_C c)_{k_1,k_2} := \sum_{l\in\mathbb{Z}} h_{l-2k_1} c_{l,k_2}, \tag{2.238}$$

$$(G_R c)_{k_1,k_2} := \sum_{l\in\mathbb{Z}} g_{l-2k_2} c_{k_1,l}, \tag{2.239}$$

$$(G_C c)_{k_1,k_2} := \sum_{l\in\mathbb{Z}} g_{l-2k_1} c_{l,k_2}, \tag{2.240}$$

the two-dimensional Mallat algorithm can be written as

$$c_0 = H_R H_C c_{-1}, \quad d_0^{(1)} = G_R H_C c_{-1} \tag{2.241}$$

$$d_0^{(2)} = H_R G_C c_{-1}, \quad d_0^{(3)} = G_R G_C c_{-1}. \tag{2.242}$$

For a finite set of coefficients c_{-1} the new coefficients $c_0, d_0^{(1)}, d_0^{(2)}, d_0^{(3)}$ can be stored in the place of the old coefficients. This is shown in Figure 2.42.

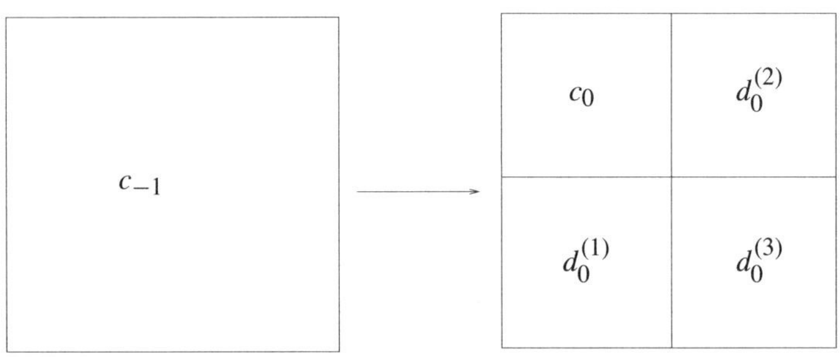

Figure 2.42. One step of two-dimensional Mallat algorithm

In a second step c_0 can be decomposed into $c_1, d_1^{(1)}, d_1^{(2)}, d_1^{(3)}$ and these coefficients can be stored in the place of c_0 as it is shown in Figure 2.43.

Example 2.10.2. In order to demonstrate the effect of the two dimensional Mallat algorithm, the gray-values of the Lena-image 2.44 are considered as the $c^{(0)}$ coefficients of an input signal.

The results of the first two steps of Mallat's algorithm are displayed in Figure 2.45.

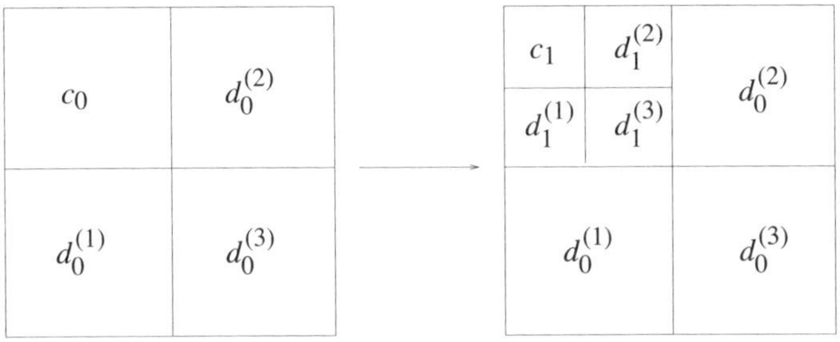

Figure 2.43. Second step of two-dimensional Mallat algorithm

Figure 2.44. Test image

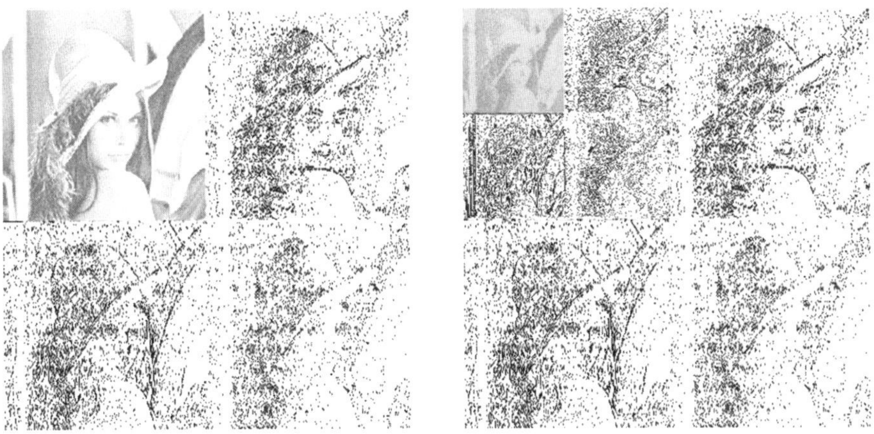

Figure 2.45. Results of the first (left) and the second (right) step of Mallat's algorithm

It is visible that the c_0 coefficients give a smoothed version of the original picture. The $d_0^{(1)}$ and $d_0^{(2)}$ coefficients contain the diagonal elements of the picture and finally the $d_0^{(3)}$ coefficients give the remaining smaller-scale content of the picture.

2.11 Wavelets on a sphere

2.11.1 Harmonic wavelets

Spherical Harmonics. Here, first some compact notations related to spherical harmonics on and outside the unit sphere σ will be introduced. Recalling the definition of a fully normalized surface spherical harmonics given in (A.60), we have

$$\bar{Y}_{nm}(\vartheta, \lambda) := \sqrt{\frac{1}{\pi(1+\delta_{0n})}} \begin{cases} \bar{P}_n^{(|m|)}(\cos\vartheta)\cos(m\lambda), & m \geq 0 \\ \bar{P}_n^{(|m|)}(\cos\vartheta)\sin(|m|\lambda), & m < 0. \end{cases} \tag{2.243}$$

The first index n is called the *degree* and the second index m is called the *order* of the fully normalized surface spherical harmonic $\bar{Y}_{nm}$.

The fully normalized surface spherical harmonics are closely related to the harmonic functions outside the unit sphere. This relationship will be discussed later. For the time being this postulated relationship will only give a motivation for the denotation of the span of all surface spherical harmonics of the same degree n as Harm_n.

Definition 2.11.1. The span of all surface spherical harmonics $\bar{Y}_{nm}$ of the same degree n is called Harm_n:

$$\mathrm{Harm}_n := \overline{\mathrm{span}\{\bar{Y}_{nm} \mid m = -n, \ldots, n\}}. \tag{2.244}$$

The direct sum of all spaces Harm_k for $k = p, \ldots, q$ will be denoted by $\mathrm{Harm}_{p,\ldots,q}$.

$$\mathrm{Harm}_{p,\ldots,q} = \bigoplus_{k=p}^{q} \mathrm{Harm}_k . \tag{2.245}$$

Legendre polynomials are special cases of the associated Legendre functions defined in (A.51):

$$P_n(t) := P_n^{(0)}(t). \tag{2.246}$$

The fully normalized surface spherical harmonics are closely related to the Legendre polynomials by the addition theorem:

Theorem 2.11.1 (Addition theorem of spherical harmonics). *Let ξ and η be two points on the unit sphere σ. Then the following relation is true:*

$$\sum_{m=-n}^{n} \bar{Y}_{mn}(\xi)\bar{Y}_{mn}(\eta) = \frac{2n+1}{4\pi} P_n(\xi^\top \eta). \tag{2.247}$$

Proof. [27]. □

Now the already mentioned relationship between the surface spherical harmonics and the harmonic functions will be addressed.

Definition 2.11.2. Let $x \in \mathbb{R}^3 \setminus \sigma$ be a point in the exterior of the unit sphere. The functions

$$H_{n,m}(x) = |x|^{-n-1} \bar{Y}_{n,m}\left(\frac{x}{|x|}\right) \tag{2.248}$$

are called *spherical harmonics* of degree n and order m.

Lemma 2.11.1. *The spherical harmonics are harmonic outside the unit sphere.*

Proof. [27]. □

Obviously, the spherical harmonics are the harmonic continuations of the corresponding surface spherical harmonics using the Poisson integral. Let $\mathrm{Pot}(\sigma)$ be the set of all harmonic functions f, which are regular at infinity, i.e. which fulfill

$$\lim_{|x|\to\infty} f(x) = 0. \tag{2.249}$$

Then

$$\mathrm{Pot}(\sigma) = \overline{\operatorname{span}_{\substack{n=0,1,\ldots \\ m=-n,\ldots,n}} H_{n,m}}. \tag{2.250}$$

In plain words (2.250) means that every harmonic function has a unique series expansion

$$f(x) = \sum_{n=0}^{\infty} \sum_{m=-n}^{n} f_{n,m} H_{n,m}(x). \tag{2.251}$$

Correspondingly, let us denote the set of all infinitely often differentiable harmonic functions, which are regular at infinity by $\mathrm{Pot}^{(\infty)}(\sigma)$.

Sobolev spaces. In many cases only such harmonic functions are important, which have special spectral properties. Since the relationship between harmonic functions and their traces on the surface of the unit sphere are unique, the spectral behavior of a harmonic function is completely defined by the spectral behavior of its trace. Sobolev spaces collect all functions with traces of a certain spectral decay.

Let $\{A_n\}$ be a sequence of positive real numbers and let $f, g \in \mathrm{Pot}^{(\infty)}(\sigma)$. Then f, g have series expansions according to (2.251). Suppose now that the coefficients $f_{n,m}, g_{n,m}$ of these expansions fulfill

$$\sum_{n=0}^{\infty} \sum_{m=-n}^{n} f_{n,m}^2 A_n^2 < \infty, \quad \sum_{n=0}^{\infty} \sum_{m=-n}^{n} g_{n,m}^2 A_n^2 < \infty. \tag{2.252}$$

Then

$$\langle f, g\rangle := \sum_{n=0}^{\infty} \sum_{m=-n}^{n} A_n^2 f_{n,m} g_{n,m} \tag{2.253}$$

defines a scalar product in $\mathrm{Pot}^{(\infty)}(\sigma)$. The corresponding norm is

$$\|f\| := \Big(\sum_{n=0}^{\infty} \sum_{m=-n}^{n} A_n^2 f_{n,m}^2\Big)^{1/2}. \tag{2.254}$$

Let $\mathcal{E}(\{A_n\})$ be the subset of all f which have a finite norm:

$$\mathcal{E}(\{A_n\}) := \{f \in \mathrm{Pot}^{(\infty)}(\sigma) \mid \|f\| < \infty\}. \tag{2.255}$$

Definition 2.11.3. The Sobolev space $\mathcal{H}(\{A_n\}, \sigma)$ is the completion of $\mathcal{E}(\{A_n\})$ under the norm (2.254).

An intuitive but very unprecise interpretation of a Sobolev space is that this space consists of all harmonic functions with series development coefficients decaying faster than A_n^{-1} for $n \to \infty$. This means the membership to a Sobolev space is restricted to functions which have a certain decay of their short wavelength spectrum.

Definition 2.11.4. Let $\{A_n\}$ be a sequence of positive real numbers fulfilling

$$\sum_{n=0}^{\infty} A_n^2 \frac{2n+1}{4\pi} < \infty, \tag{2.256}$$

then

$$K_{\mathcal{H},A_n}(\boldsymbol{x}, \boldsymbol{y}) = \sum_{n=0}^{\infty} A_n^{-2} \sum_{m=-n}^{n} H_{n,m}(\boldsymbol{x}) H_{n,m}(\boldsymbol{y}) \tag{2.257}$$

is called a *reproducing kernel* of $\mathcal{H}(\{A_n\}, \sigma)$.

The name reproducing kernel is motivated by the property

$$\begin{aligned}\langle K_{\mathcal{H},A_n}(\boldsymbol{x}, \bullet), f\rangle &= \sum_{n=0}^{\infty} \sum_{m=-n}^{n} A_n^2 (A_n^{-2} H_{n,m}(\boldsymbol{x}) f_{n,m}) \\ &= f(\boldsymbol{x}).\end{aligned}$$

Example 2.11.1. Here, some examples for reproducing kernels are given.

Shannon kernel. The Shannon kernel is characterized by the following setting of the weight coefficients A_n:

$$A_n := \begin{cases} 1, & 0 \le n < M \\ 0, & M \le n < \infty. \end{cases} \tag{2.258}$$

Due to this particular definition of the weight coefficients, a closed expression for the Shannon kernel can be found:

$$\begin{aligned} K_{\mathcal{H},A_n}(\boldsymbol{x},\boldsymbol{y}) &= \sum_{n=0}^{M-1} \sum_{m=-n}^{n} H_{n,m}(\boldsymbol{x}) H_{n,m}(\boldsymbol{y}) \\ &= \sum_{n=0}^{M-1} (|\boldsymbol{x}||\boldsymbol{y}|)^{-n-1} \sum_{m=-n}^{n} \bar{Y}_{n,m}\left(\frac{\boldsymbol{x}}{|\boldsymbol{x}|}\right) \bar{Y}_{n,m}\left(\frac{\boldsymbol{y}}{|\boldsymbol{y}|}\right) \\ &= \sum_{n=0}^{M-1} (|\boldsymbol{x}||\boldsymbol{y}|)^{-n-1} \frac{2n+1}{4\pi} P_n\left(\frac{\boldsymbol{x}^\top}{|\boldsymbol{x}|} \frac{\boldsymbol{y}}{|\boldsymbol{y}|}\right). \end{aligned} \tag{2.259}$$

Using the well-known recurrence relation for Legendre polynomials,

$$(n+1)[P_{n+1}(t) - P_n(t)] - n[P_n(t) - P_{n-1}(t)] = (2n+1)(t-1)P_n(t), \tag{2.260}$$

for $|\boldsymbol{x}| = |\boldsymbol{y}| = 1$ the closed expression

$$K_{\mathcal{H},A_n}(\boldsymbol{x},\boldsymbol{y}) = \left(\frac{\boldsymbol{x}^\top}{|\boldsymbol{x}|} \frac{\boldsymbol{y}}{|\boldsymbol{y}|} - 1\right)^{-1} \frac{M}{4\pi} \left[P_M\left(\frac{\boldsymbol{x}^\top}{|\boldsymbol{x}|} \frac{\boldsymbol{y}}{|\boldsymbol{y}|}\right) - P_{M-1}\left(\frac{\boldsymbol{x}^\top}{|\boldsymbol{x}|} \frac{\boldsymbol{y}}{|\boldsymbol{y}|}\right)\right] \tag{2.261}$$

is found. Figure 2.46 shows the Shannon kernel for $M = 16$ and $M = 32$.

Abel–Poisson kernel. For the Abel–Poisson kernel the weights are equal to the elements of a converging geometrical sequence

$$A_n = \rho^n, \quad 0 < \rho < 1.$$

The series expansion of the Abel–Poisson kernel is

$$K_{\mathcal{H},A_n}(\boldsymbol{x},\boldsymbol{y}) = \sum_{n=0}^{\infty} \frac{\rho^n}{(|\boldsymbol{x}||\boldsymbol{y}|)^{n+1}} \frac{2n+1}{4\pi} P_n\left(\frac{\boldsymbol{x}^\top}{|\boldsymbol{x}|} \frac{\boldsymbol{y}}{|\boldsymbol{y}|}\right). \tag{2.262}$$

For $|\boldsymbol{x}| = |\boldsymbol{y}| = 1$ also this series has a closed expression, the kernel of the Abel–Poisson integral formula

$$K_{\mathcal{H},A_n}(\boldsymbol{x},\boldsymbol{y}) = \frac{1}{4\pi} \frac{1-\rho^2}{\left(1+\rho^2 - 2\rho\boldsymbol{x}^\top\boldsymbol{y}\right)^{3/2}}. \tag{2.263}$$

The Abel–Poisson kernel and the corresponding weight sequence for the values $\rho = 0.7$ and $\rho = 0.9$ are shown in Figure 2.47.

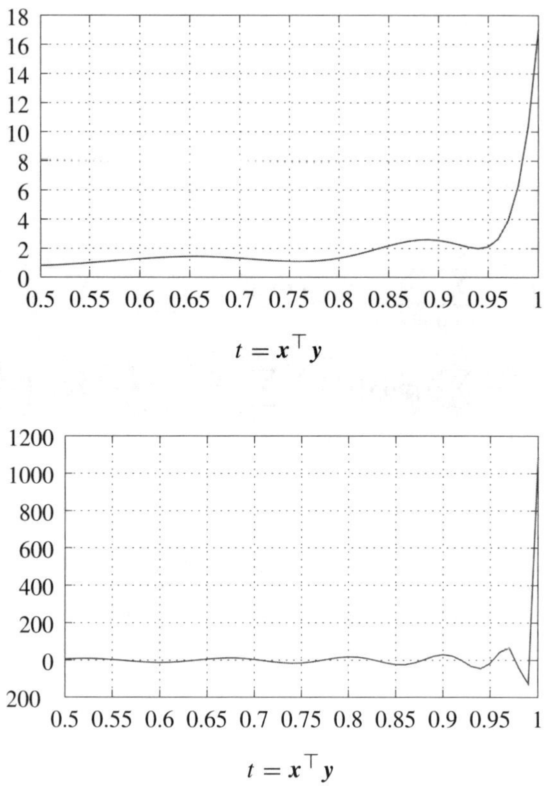

Figure 2.46. Shannon kernels for $M = 16$ and $M = 32$

Harmonic wavelets. Now all preparing steps have been taken to give the definition of a harmonic wavelet on the sphere. As in the Euclidean case the definition of a harmonic wavelet on the sphere starts with an MRA of $\mathcal{H}(\{A_n\}, \sigma)$.

Definition 2.11.5. The subsets $V_j \subset \mathcal{H}(\{A_n\}, \sigma)$ defined by

$$V_j = \bigoplus_{0 \le n \le 2^j} \mathrm{Harm}_n = \mathrm{Harm}_{0,\dots,2^j} \tag{2.264}$$

are called *scaling spaces*. The spaces W_j defined by

$$W_j = V_{j+1} \ominus V_j \tag{2.265}$$

are called *detail spaces*.

Obviously, the scaling spaces are nested

$$V_0 \subset V_1 \subset \cdots \subset V_j \subset \cdots \tag{2.266}$$

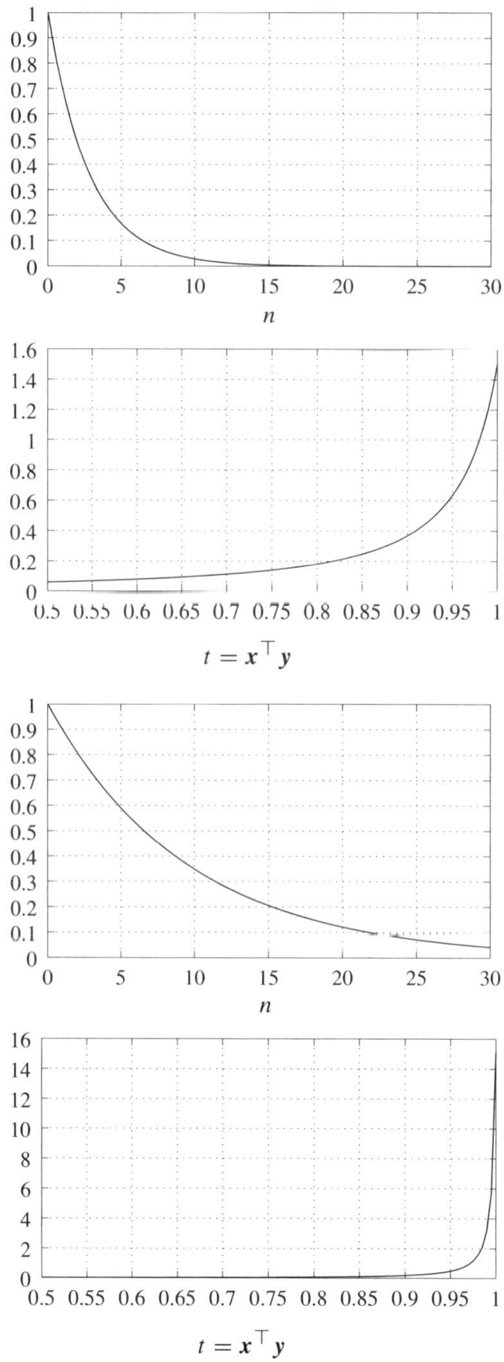

Figure 2.47. Weights A_n and Abel–Poisson kernel K for $\rho = 0.7$ and $\rho = 0.9$

and the detail space W_j is the complement of V_j in V_{j+1}:

$$V_{j+1} = V_j \oplus W_j. \tag{2.267}$$

In addition, the following equation holds:

$$\overline{\bigcup_{j\in\mathbb{N}} V_j} = \mathcal{H}(\{A_n\}, \sigma). \tag{2.268}$$

These properties are quite similar to the MRA in the Euclidean case. For this reason the nested sequence $\{V_j\}$ is also called an MRA of $\mathcal{H}(\{A_n\}, \sigma)$.

Some properties which are expected from the Euclidean case cannot be carried over directly, since the operations scaling and shifting have a different meaning in the spherical case. Therefore it has to be discussed now what is to be understood by *translation* and by *dilatation* in the spherical case.

For this purpose the following sequences of kernel functions will be considered:

$$\Phi_j(\boldsymbol{x}, \boldsymbol{y}) := \sum_{n=0}^{2^j} A_n^{-2} \sum_{m=-n}^{n} H_{n,m}(\boldsymbol{x}) H_{n,m}(\boldsymbol{y}), \tag{2.269}$$

$$\Psi_j(\boldsymbol{x}, \boldsymbol{y}) := \sum_{n=2^j}^{2^{j+1}} A_n^{-2} \sum_{m=-n}^{n} H_{n,m}(\boldsymbol{x}) H_{n,m}(\boldsymbol{y}). \tag{2.270}$$

For these kernel functions the dilatation operator $\mathcal{D}_J$ can be defined easily by

$$\mathcal{D}_J \Phi_j(\boldsymbol{x}, \boldsymbol{y}) := \Phi_{j+J}(\boldsymbol{x}, \boldsymbol{y}), \quad \mathcal{D}_J \Psi_j(\boldsymbol{x}, \boldsymbol{y}) := \Psi_{j+J}(\boldsymbol{x}, \boldsymbol{y}). \tag{2.271}$$

In order to be able to define a shifting operator, a relationship between the kernel functions $\Phi_j(\boldsymbol{x}, \boldsymbol{y})$, $\Psi_j(\boldsymbol{x}, \boldsymbol{y})$ and a point $\boldsymbol{y}/|\boldsymbol{y}|$ on the sphere has to be established. The neatest way to do this is to fix one argument of the kernel functions to $\boldsymbol{y}$:

$$\Phi_{j,y}(\bullet) := \Phi_j(\bullet, \boldsymbol{y}), \quad \Psi_{j,y}(\bullet) := \Psi_j(\bullet, \boldsymbol{y}). \tag{2.272}$$

Now the translation operator can be defined quite naturally by

$$\mathcal{S}_y \Phi_j(\bullet, \bullet) := \Phi_{j,y}(\bullet), \quad \mathcal{S}_y \Psi_j(\bullet, \bullet) := \Psi_{j,y}(\bullet). \tag{2.273}$$

The function $\Phi_0(\bullet, \bullet)$ is called the mother scaling function and the function $\Psi_0(\bullet, \bullet)$ the mother wavelet. By the combination of the translation and the dilatation operator scaled and shifted versions of the mother functions can be generated:

$$\mathcal{S}_y \mathcal{D}_j \Phi_0(\bullet, \bullet) = \Phi_{j,y}(\bullet, \boldsymbol{y}), \quad \mathcal{S}_y \mathcal{D}_j \Psi_0(\bullet, \bullet) = \Psi_{j,y}(\bullet, \boldsymbol{y}). \tag{2.274}$$

In this way quite a number of similarities between the MRA in the Euclidean space and on the sphere have been established. Only the scaling property

$$f(\bullet) \in V_j \Leftrightarrow f(2\bullet) \in V_{j+1} \tag{2.275}$$

is not valid in the spherical case. Since this scaling property is the key to the algorithms of fast wavelet transform, on the sphere an analog of Mallat's algorithm cannot be found.

Example 2.11.2. As an example the scaling function and the wavelet will be constructed for Sobolev spaces based on the Shannon and the Abel–Poisson kernel, respectively.

Shannon kernel. For the Shannon kernel the scaling function Φ_j is given by

$$\Phi_j(\boldsymbol{x}, \boldsymbol{y}) = \sum_{n=0}^{\min\{2^j, M-1\}} (|\boldsymbol{x}||\boldsymbol{y}|)^{-n-1} \frac{2n+1}{4\pi} P_n\left(\frac{\boldsymbol{x}^\top}{|\boldsymbol{x}|}\frac{\boldsymbol{y}}{|\boldsymbol{y}|}\right) \tag{2.276}$$

and the wavelet Ψ_j by

$$\Psi_j(\boldsymbol{x}, \boldsymbol{y}) = \sum_{\min\{2^j, M-1\}}^{\min\{2^{j+1}, M-1\}} (|\boldsymbol{x}||\boldsymbol{y}|)^{-n-1} \frac{2n+1}{4\pi} P_n\left(\frac{\boldsymbol{x}^\top}{|\boldsymbol{x}|}\frac{\boldsymbol{y}}{|\boldsymbol{y}|}\right). \tag{2.277}$$

For $j = 4$ both functions are shown in Figure 2.48.

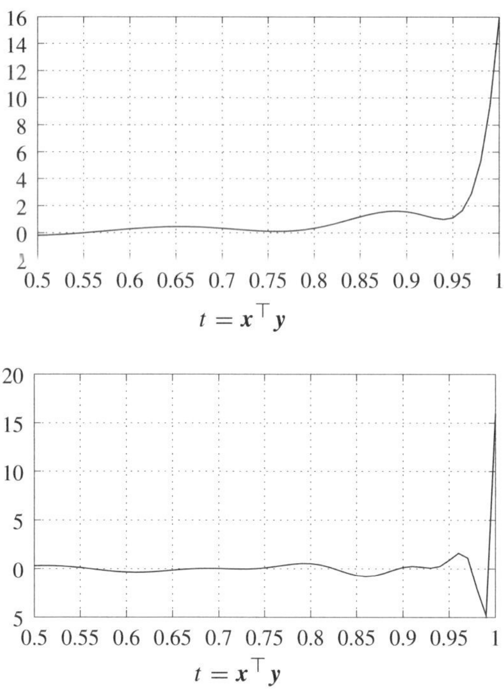

Figure 2.48. Shannon scaling function Φ_j and wavelet Ψ_j for the scale $j = 4$

Abel–Poisson kernel. For the Abel–Poisson kernel the scaling function Φ_j is given by

$$\Phi_j(\boldsymbol{x}, \boldsymbol{y}) = \sum_{n=0}^{2^j} \rho^n (|\boldsymbol{x}||\boldsymbol{y}|)^{-n-1} \frac{2n+1}{4\pi} P_n \left(\frac{\boldsymbol{x}^\top}{|\boldsymbol{x}|} \frac{\boldsymbol{y}}{|\boldsymbol{y}|} \right) \tag{2.278}$$

and the wavelet Ψ_j by

$$\Psi_j(\boldsymbol{x}, \boldsymbol{y}) = \sum_{2^j+1}^{2^{j+1}} \rho^n (|\boldsymbol{x}||\boldsymbol{y}|)^{-n-1} \frac{2n+1}{4\pi} P_n \left(\frac{\boldsymbol{x}^\top}{|\boldsymbol{x}|} \frac{\boldsymbol{y}}{|\boldsymbol{y}|} \right). \tag{2.279}$$

For the value $\rho = 0.7$ both the scaling function and the wavelet are shown in Figure 2.49.

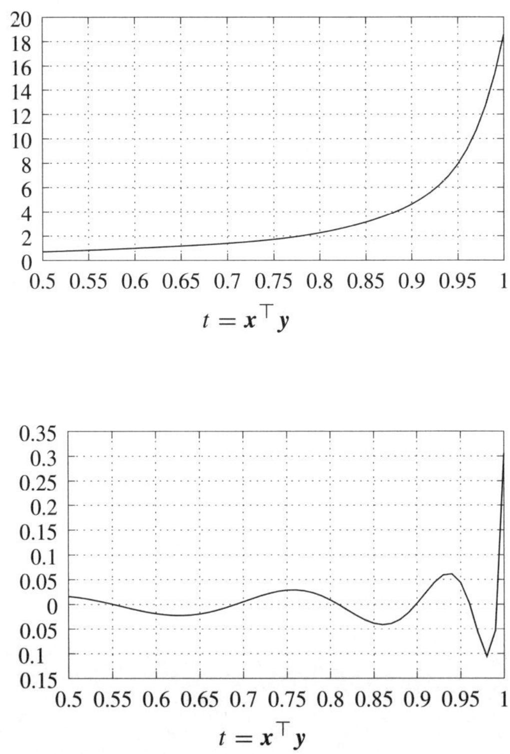

Figure 2.49. Abel–Poisson scaling function Φ_j and wavelet Ψ_j for the scale $j = 4$

Spherical wavelet transform. The spherical wavelet transform aims at a decomposition of a scalar function, defined in the exterior of the unit sphere, into parts, where every part represents a certain frequency band in the function. As in the Euclidean case the wavelet transformation is carried out by comparing the given signal $f \in \mathcal{H}(\{A_n\}, \sigma)$ with the scaled and shifted versions $\Psi_{j,y}$ of a particular wavelet.

Definition 2.11.6. Let $f \in \mathcal{H}(\{A_n\}, \sigma)$ and let Ψ be the wavelet belonging to an MRA of $\mathcal{H}(\{A_n\}, \sigma)$. The function

$$(WT)f(j, y) := \langle \Psi_{j,y}, f \rangle \tag{2.280}$$

is called *scale discrete spherical wavelet transform* of f.

The value $(WT)f(j, y)$ indicates the amount of energy, which is contained in the signal f in a neighborhood of the point $\boldsymbol{y} \in \sigma$ and on the scale j. The scale j contains all spherical harmonics of the order $2^j + 1$ to 2^{j+1}. This wavelet transform is invertible as it is shown in the following lemma.

Lemma 2.11.2. *Let $f \in \mathcal{H}(\{A_n\}, \sigma)$ and let $(WT)f(j, y)$ be its scale discrete spherical wavelet transform. Then the following reconstruction formula is valid:*

$$f(\boldsymbol{x}) = \sum_{j=0}^{\infty} \langle \Psi_j(\boldsymbol{x}, \bullet), (WT)f(j, \bullet) \rangle. \tag{2.281}$$

Proof. Let $f \in \mathcal{H}(\{A_n\}, \sigma)$. Then f has a spherical harmonics expansion

$$f(\boldsymbol{x}) = \sum_{n=0}^{\infty} \sum_{m=-n}^{n} f_{n,m} H_{n,m}(\boldsymbol{x})$$

with

$$\sum_{n=0}^{\infty} A_n^2 \sum_{m=-n}^{n} f_{n,m}^2 < \infty.$$

The wavelet $\Psi_{j,y}(\boldsymbol{x})$ has a spherical harmonics expansion as well:

$$\Psi_{j,y}(\boldsymbol{x}) = \sum_{n=2^j}^{2^{j+1}} \sum_{m=-n}^{n} \Psi_{n,m} H_{n,m}(\boldsymbol{x}),$$

with

$$\Psi_{n,m} = A_n^{-2} H_{n,m}(\boldsymbol{y}).$$

Consequently, the wavelet transform $(WT)f(j, y)$ has the series expansion

$$(WT)f(j, y) = \sum_{n=2^j}^{2^{j+1}} \sum_{m=-n}^{n} f_{n,m} H_{n,m}(\boldsymbol{y}).$$

Inserting this into the reconstruction formula (2.281) yields

$$\begin{aligned}
\sum_{j=0}^{\infty}\langle\Psi_j(\boldsymbol{x},\bullet),(WT)f(j,\bullet)\rangle &= \sum_{j=0}^{\infty}\Big\langle\sum_{n=2^j}^{2^{j+1}}\sum_{m=-n}^{n}\Psi_{n,m}H_{n,m},\sum_{\nu=2^j}^{2^{j+1}}\sum_{\mu=-\nu}^{\nu}f_{\nu,\mu}H_{\nu,\mu}\Big\rangle \\
&= \sum_{j=0}^{\infty}\Big\langle\sum_{n=2^j}^{2^{j+1}}\sum_{m=-n}^{n}A_n^{-2}H_{n,m}(\boldsymbol{x})H_{n,m}(\bullet), \\
&\qquad\qquad \sum_{\nu=2^j}^{2^{j+1}}\sum_{\mu=-\nu}^{\nu}f_{\nu,\mu}H_{\nu,\mu}\Big\rangle \\
&= \sum_{j=0}^{\infty}\sum_{n=2^j}^{2^{j+1}}\sum_{m=-n}^{n}f_{n,m}H_{n,m}(\boldsymbol{x}) \\
&= \sum_{n=0}^{\infty}\sum_{m=-n}^{n}f_{n,m}H_{n,m}(\boldsymbol{x}) \\
&= f(\boldsymbol{x}).
\end{aligned}$$

□

The scale discrete spherical wavelet transform is closely related to an MRA of $\mathcal{H}(\{A_n\},\sigma)$ since

$$(WT)f(j,\bullet)\in W_j. \tag{2.282}$$

In contrast to the Euclidean case the spherical wavelet decomposition follows the scheme displayed in Figure 2.50.

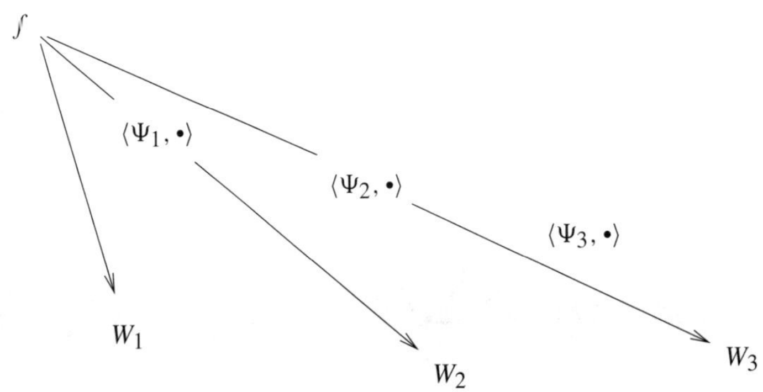

Figure 2.50. Schematic view of the spherical wavelet decomposition

Instead of being recursive, in the spherical case each decomposition step hast to start from the original signal f. The reason for this non-recursive character of the spherical wavelet decomposition is the absence of a scaling relation (2.275) between two consecutive scaling spaces.

Example 2.11.3. In order to give a certain feeling for the effect of the spherical wavelet transform, the so-called *disturbing potential* T of the Earth will be subjected to such a transformation. The potential of the gravity field of the Earth is usually denoted by W. Since the potential W is very irregular a model U is often used as an approximation for W. The function U is the potential of a homogeneous ellipsoid having the same mass as the Earth and being in an hydrostatic equilibrium. The model potential U is called *normal potential* and can be expressed by a simple formula. The disturbing potential T is defined as the difference between the actual potential W and the normal potential U:

$$T := W - U. \tag{2.283}$$

Since, both W and U contain the centrifugal potential the difference T is harmonic and can therefore be analyzed by harmonic wavelets.

As input data the EGM96 model for the disturbing potential T is used. The so-called geoid undulations, i.e. the values

$$N := \frac{T}{\|\nabla U\|} \tag{2.284}$$

are computed based on the EGM96 model and displayed in Figure 2.51, in the upper left corner. For the wavelet analysis of the geoid undulation field N the Shannon kernel for $M = 360$ is used. The field N is analyzed on the scales 3, 5, 7. The results of the wavelet transforms on these scales are displayed in Figure 2.51 from top right to bottom right.

As a rough estimate it can be said that the scale J corresponds to details in the signal with a size between $4 \cdot 10^4/2^{J+1}$ km to $4 \cdot 10^4/2^J$ km, as it is given in Table 2.1.

Table 2.1. Scale dependent detail sizes

Scale J	detail size [km]
1	20 000–40 000
2	10 000–20 000
3	5 000–10 000
4	2 500–5 000
5	1 250–2 500
6	625–1 250
7	312–625

Figure 2.51 shows clearly that the spherical wavelet analysis is a filter bank returning on the scale J only features of the original signal with sizes between $4 \cdot 10^4/2^{J+1}$ km and $4 \cdot 10^4/2^J$ km.

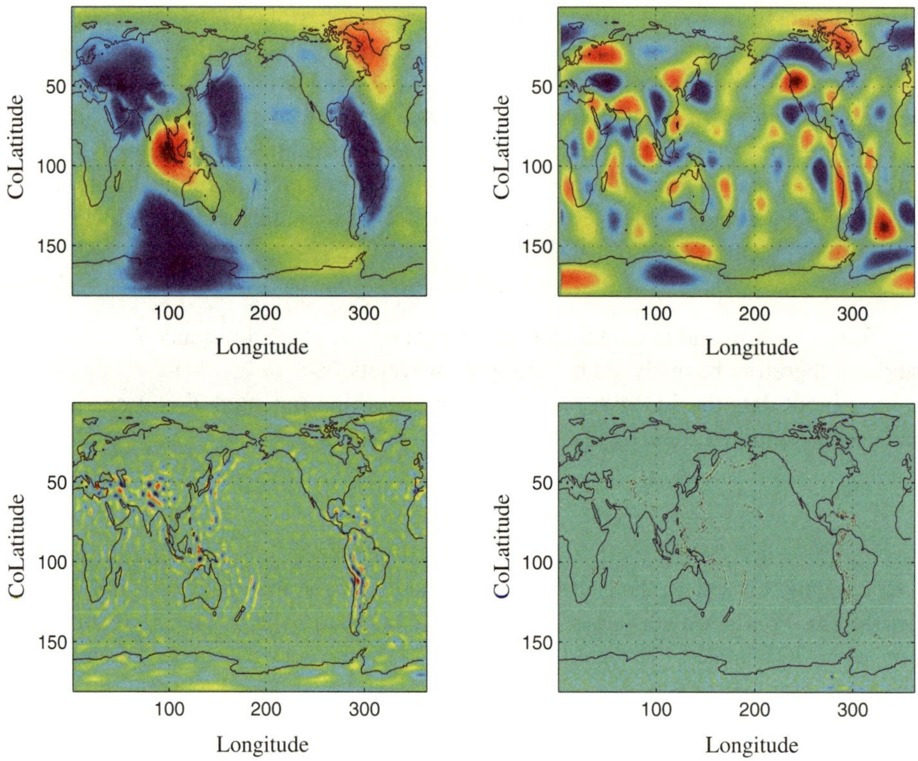

Figure 2.51. MRA of the Earth's gravitational field. From top left to bottom right: complete field, scales 3, 5, 7

Basically, in the simple form presented here, the spherical wavelet analysis is a method for a sub-band coding of a signal given on the sphere. Though this is very useful for many practical problems, the general acceptance in the geo-scientific community is still pending. The main obstacle for the application of spherical wavelet techniques is the lack of numerical algorithms, which have an efficiency comparable to Mallat's algorithm in the Euclidean case.

The spherical wavelet theory was mainly developed by the *Geomathematics Group* of the University in Kaiserslautern. It reaches far beyond the ideas and techniques explained here. Profound references to the theory of spherical wavelets are [27], or [22]. The applicability of the spherical wavelet technique to geodetic problems has been discussed in a larger number of papers. The publications [16] and [17] deal with the approximation of data given on the sphere. The recovery of the Earth's gravity field from terrestrial and satellite observations by means of spherical wavelets is treated in the papers [35], [37] and [36].

2.11.2 Triangulation based wavelets

Classically, wavelet constructions have been employed in Euclidean spaces (the real line $\mathbb{R}$ or the plane $\mathbb{R}^2$). Since most data are given on a finite interval or a finite rectangle the wavelet theory has been extended to wavelets on an interval.

However, the underlying geometry is still Euclidean. The wavelet theory on curved surfaces is still at its beginning. Besides the spherical wavelets, which make a special use of the spherical harmonics expansion on the sphere, here a different technique, based on a uniform approximation of the sphere by regular polyhedra will be presented.

In the classical setting wavelets are built with the help of a scaling functions. Both wavelets and scaling functions fulfill so-called scaling equations, i.e. wavelet and scaling functions on a certain scale j can be expressed as linear combinations of the same scaling function on the next finer level $j-1$. As in the Euclidean case also for the sphere the wavelet will be derived from a multi-resolution analysis (MRA).

Definition 2.11.7. Let $L_2(\sigma)$ be the Hilbert space of all square integrable functions on the unit sphere σ. A nested sequence of closed subspaces V_j is called a *multi-resolution analysis* (MRA) of $L_2(\sigma)$, if

$$\{0\} \subset \cdots \subset V_{j+1} \subset V_j \subset V_{j-1} \subset \cdots \subset L_2(\sigma), \tag{2.285}$$

$$\overline{\bigcup_{j\in\mathbb{Z}} V_j} = L_2(\sigma) \tag{2.286}$$

holds and for each j scaling functions $\varphi_{j,k},\ k \in \mathcal{K}(j)$ exist, such that

$$\operatorname{span}\{\varphi_{j,k} \mid k \in \mathcal{K}(j)\} = V_j \tag{2.287}$$

is valid and for the index sets $\mathcal{K}(j)$

$$\mathcal{K}(j) \subset \mathcal{K}(j-1) \tag{2.288}$$

holds.

In contrast to the classical case the scaling functions $\varphi_{j,k}$ do not have to be translates of one particular mother function φ_j. Relation (2.285) implies the existence of scaling coefficients $h_{j,k,l}$ such that

$$\varphi_{j,k} = \sum_{l\in\mathcal{K}(j-1)} h_{j,k,l}\varphi_{j-1,l} \tag{2.289}$$

is valid. As in the Euclidean case the wavelets code the difference between the signal representation on two consecutive resolution levels. In more detail: Wavelets are base function of the wavelet spaces W_j, which themselves are defined as the orthogonal complement of V_j in V_{j-1}:

$$V_j \oplus W_j = V_{j-1}. \tag{2.290}$$

Definition 2.11.8. Consider the set of functions

$$W := \{\psi_{j,m} \mid j \in \mathbb{Z}, m \in \mathcal{M}(j)\}, \tag{2.291}$$

where $\mathcal{M}(j) \subset \mathcal{K}(j-1)$ is an index set again. If

1. W is a base of $L_2(\sigma)$ and
2. the set $\{\psi_{j,m} \mid m \in \mathcal{M}(j)\}$ is a base of W_j,

then the functions $\psi_{j,m}$ are called a *spherical wavelet base*.

Since $W_j \subset V_{j-1}$ there have to exist coefficients $g_{j,m,l}$ with

$$\psi_{j,m} = \sum_{l \in \mathcal{K}(j-1)} g_{j,m,l}\varphi_{j-1,l}\,, \quad m \in \mathcal{M}(j). \tag{2.292}$$

So far the orthogonality of the wavelet has not been mentioned. In fact it is difficult or even impossible to construct orthogonal wavelets on the sphere with a support smaller than the whole surface of the sphere. Therefore our considerations will be restricted to biorthogonal wavelets on the sphere.

The dual scaling functions $\tilde{\varphi}_{l,m}$ are defined by the condition

$$\langle \varphi_{j,k}, \tilde{\varphi}_{j,l} \rangle = \delta_{k,l}. \tag{2.293}$$

For a dual scaling function the scaling equation

$$\tilde{\varphi}_{j,k} = \sum_{l \in \mathcal{K}(j-1)} \tilde{h}_{j,k,l}\tilde{\varphi}_{j-1,l} \tag{2.294}$$

holds. Additionally, for a given set of spherical wavelets $\psi_{j,k}$ there exist dual wavelets $\tilde{\psi}_{n,m}$, which are biorthogonal to each other,

$$\langle \psi_{j,k}, \tilde{\psi}_{n,m} \rangle = \delta_{j,n}\delta_{k,m}. \tag{2.295}$$

This implies $\langle \tilde{\psi}_{j,k}, \varphi_{n,m} \rangle = \langle \psi_{j,k}, \tilde{\varphi}_{n,m} \rangle = 0$. The essence of biorthogonality is the use of dual bases for the decomposition and for the reconstruction of a signal f:

$$f = \sum_{j,m} \langle \tilde{\psi}_{j,m}, f \rangle \psi_{j,m} = \sum_{j,m} \langle \psi_{j,m}, f \rangle \tilde{\psi}_{j,m}. \tag{2.296}$$

Once the scaling coefficients $h_{j,k,l}$ and $g_{j,m,l}$ are given, an analog of Mallat's algorithm can be constructed. Let

$$c_k^{(n)} := \langle f, \tilde{\varphi}_{n,k} \rangle \tag{2.297}$$

be the Fourier coefficients of a signal f with respect to the base of the scaling space V_n. The Fourier coefficients $c_k^{(n+1)}$ of the projection $P_{n+1}f$ into the coarser space V_{n+1}

and the Fourier coefficients $d_k^{(n+1)}$ of the projection $Q_{n+1} f$ of f into the difference space W_{n+1} can be computed recursively according to

$$c_k^{(n+1)} = \sum_{l \in \mathcal{K}(n)} \tilde{h}_{n,k,l} c_l^{(n)}, \tag{2.298}$$

$$d^{(n+1)} = \sum_{l \in \mathcal{K}(n)} \tilde{g}_{n,k,l} c_l^{(n)}. \tag{2.299}$$

The reconstruction step then is as follows:

$$c_k^{(n)} = \sum_{l \in \mathcal{K}(n+1)} h_{n+1,k,l} c_l^{(n+1)} + \sum_{l \in \mathcal{K}(n+1)} g_{n+1,k,l} d_l^{(n+1)}. \tag{2.300}$$

In order to get a clearer picture of the situation, a simple example of biorthogonal wavelets will be given now.

Example 2.11.4. In preparation of the construction of biorthogonal wavelets on the sphere biorthogonal wavelets on the real line, which are related to the Haar wavelet will be discussed here. Suppose

$$\varphi_{j,k}(x) := \chi_{[2^j k, 2^j k + 2^j]}(x), \tag{2.301}$$

where χ_M is the characteristic function of the set M, are the primal scaling functions. Obviously, the dual scaling functions are

$$\tilde{\varphi}_{j,l} := \frac{1}{2^j} \chi_{[2^j k, 2^j k + 2^j]}(x), \tag{2.302}$$

which leads to the orthogonality condition

$$\langle \varphi_{j,l}, \ddot{\varphi}_{j,m} \rangle = \delta_{l,m}.$$

It is not difficult to see that the primal wavelets are

$$\psi_{j,k}(x) = \varphi_{j-1,2k}(x) - \varphi_{j-1,2k+1}, \tag{2.303}$$

and the dual wavelets are

$$\tilde{\psi}_{j,k}(x) = \tilde{\varphi}_{j-1,2k}(x) - \tilde{\varphi}_{j-1,2k+1}. \tag{2.304}$$

Both, the primal and the dual scaling function and the primal and dual wavelet are shown in Figure 2.52.

Evidently, the scaling coefficients are

$$h_{j,k.0} = h_{j,k,1} = 1, \quad g_{j,k.0} = 1, \quad g_{j,k,1} = -1, \tag{2.305}$$

$$\tilde{h}_{j,k.0} = \tilde{h}_{j,k,1} = 1, \quad \tilde{g}_{j,k.0} = 1, \quad \tilde{g}_{j,k,1} = -1. \tag{2.306}$$

All other coefficients are equal to zero.

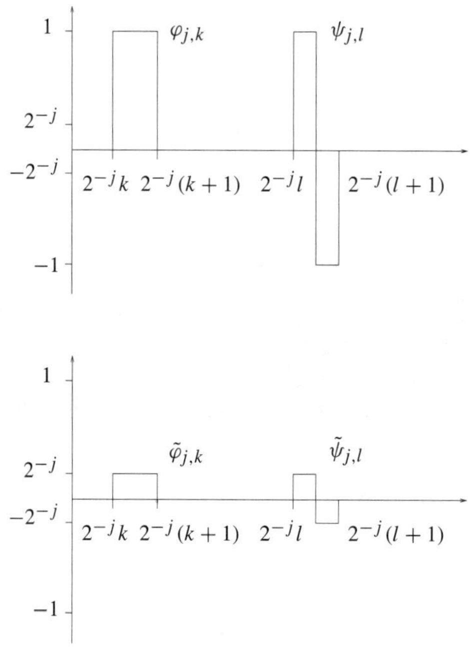

Figure 2.52. Primal scaling function and wavelet (top); dual scaling function and wavelet (bottom)

Haar wavelets on the sphere. This simple example is the starting point for the construction of a Haar wavelet base on the sphere. In order to achieve this, a nested sequence of triangulation of the sphere is necessary. The construction of such a nested sequence starts from a subdivision of the sphere into four spherical triangles, as it is indicated in Figure 2.53, where one of the four spherical triangles is marked by solid lines.

To generate the next finer grid, the midpoints of each side of each triangle are connected by lines. In this way each triangle is divided in four smaller triangles, as it is shown in Figure 2.54.

So in each refinement level j the surface of the sphere σ is the union of the triangles $T_{j,k}$:

$$\sigma = \bigcup_{k \in \mathcal{K}(j)} T_{j,k}. \tag{2.307}$$

For each j and k the triangle $T_{j,k}$ is the union of four *child triangles* $T_{j-1,l}$:

$$T_{j,k} = \bigcup_{l=0}^{3} T_{j-1,l}. \tag{2.308}$$

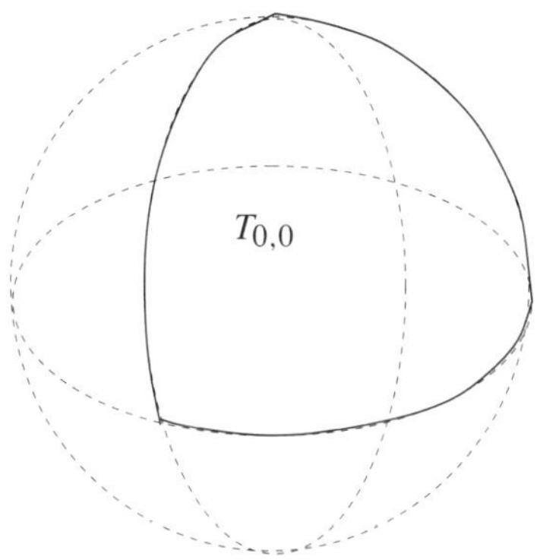

Figure 2.53. Subdivision of the sphere into four spherical triangles

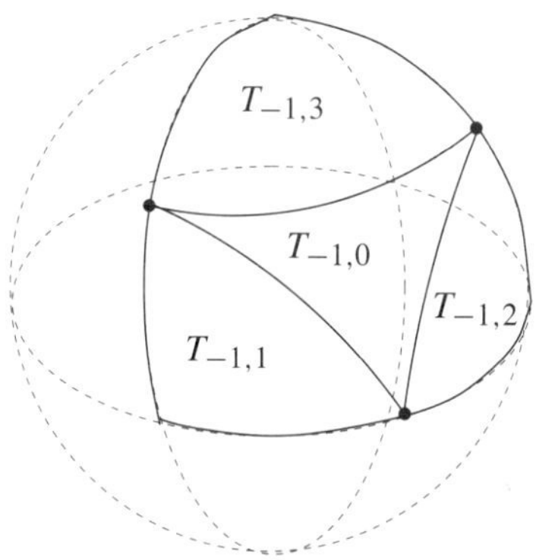

Figure 2.54. Refinement of the subdivision of the sphere

On a certain level j the spherical Haar scaling functions $\varphi_{j,k}$ are defined as

$$\varphi_{j,k} := \chi_{T_{j,k}}, \tag{2.309}$$

with χ being the characteristic function of the triangle $T_{j,k}$. Then the dual scaling function $\tilde{\varphi}_{j,k}$ is

$$\tilde{\varphi}_{j,k} = \frac{1}{\mu(T_{j,k})} \chi_{T_{j,k}}, \tag{2.310}$$

where $\mu(T_{j,k})$ is the Lebesgue measure of the triangle $T_{j,k}$. In Figure 2.55 a spherical Haar scaling function and its dual are displayed.

For each spherical Haar scaling function $\varphi_{j,k}$ there are three independent spherical Haar wavelets defined by

$$\psi_{j,k}^{(m)} := 2\Big(\varphi_{j-1,m} - \frac{I_{j-1,m}}{I_{j-1,0}} \varphi_{j-1,0}\Big), \quad m = 1, 2, 3, \tag{2.311}$$

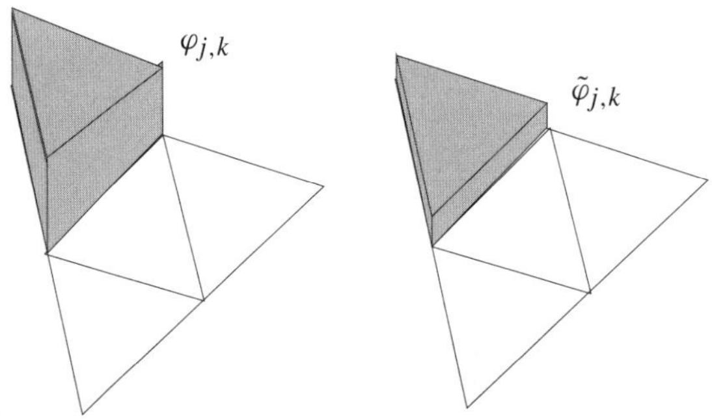

Figure 2.55. Spherical Haar scaling function and its dual.

with $I_{j,k}$ being the integral of the scaling function

$$I_{j,k} := \int_\sigma \varphi_{j,k}\, d\sigma = \mu(T_{j,k}). \tag{2.312}$$

In Figure 2.56 the spherical Haar scaling function $\varphi_{j,k}$ and one of the corresponding spherical Haar wavelets $\psi_{j,k}^{(m)}$ are displayed.

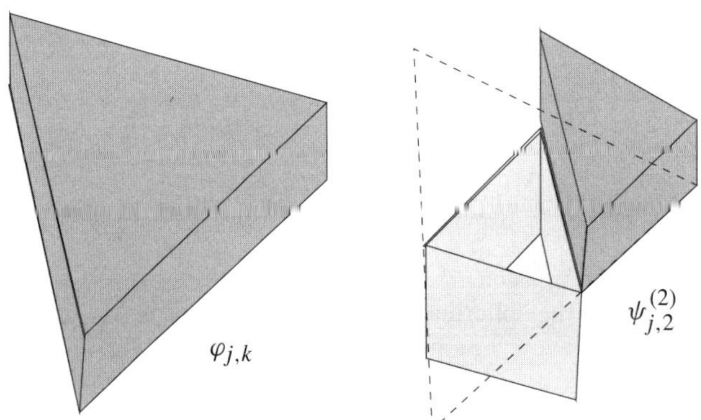

Figure 2.56. Spherical Haar scaling function and one of its spherical Haar wavelets

Quite similar the dual spherical Haar wavelet is defined by

$$\tilde{\psi}_{j,k}^{(m)} := \frac{1}{2}(\tilde{\varphi}_{j-1,m} - \tilde{\varphi}_{j,k}), \quad m = 1, 2, 3. \tag{2.313}$$

In Figure 2.57 the dual spherical Haar scaling function $\tilde{\varphi}_{j,k}$ and one of the corresponding dual spherical Haar wavelets $\tilde{\psi}_{j,k}^{(m)}$ are shown.

Obviously, the following relations are true

$$\langle \varphi_{j,k}, \tilde{\varphi}_{j,l} \rangle = \delta_{k,l},$$

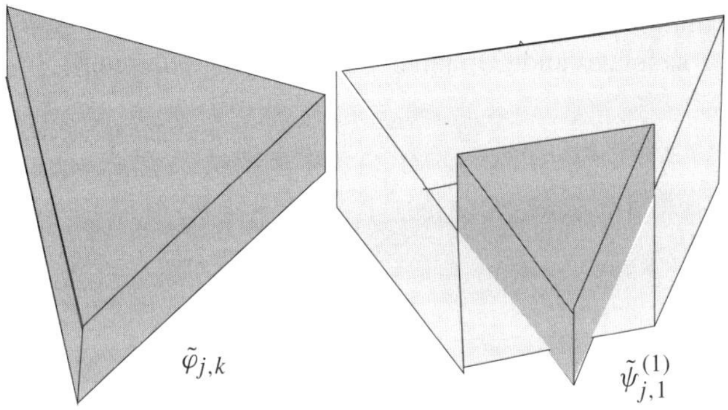

Figure 2.57. Dual spherical Haar scaling function and one of its dual spherical Haar wavelets

$$
\begin{aligned}
\langle \psi_{j,k}^{(l)}, \tilde{\psi}_{n,m}^{(p)} \rangle &= \delta_{j,n}\delta_{k,m} \cdot \langle \psi_{j,k}^{(l)}, \tilde{\psi}_{j,k}^{(p)} \rangle \\
&= \delta_{j,n}\delta_{k,m} \left\langle \left(\varphi_{j-1,l} - \frac{I_{j-1,l}}{I_{j-1,0}} \varphi_{j-1,0} \right), (\tilde{\varphi}_{j-1,p} - \tilde{\varphi}_{j,k}) \right\rangle \\
&= \delta_{j,n}\delta_{k,m} \left(\langle \varphi_{j-1,l}, \tilde{\varphi}_{j-1,p} \rangle - \langle \varphi_{j-1,l}, \tilde{\varphi}_{j,k} \rangle + \frac{I_{j-1,l}}{I_{j-1,0}} \langle \varphi_{j-1,0}, \tilde{\varphi}_{j,k} \rangle \right) \\
&= \delta_{j,n}\delta_{k,m} \left(\delta_{l,p} - \frac{I_{j-1,j}}{\mu(T_{j,k})} + \frac{I_{j-1,l}}{I_{j-1,0}} \frac{I_{j-1,0}}{\mu(T_{j,k})} \right) \\
&= \delta_{j,n}\delta_{k,m}\delta_{l,p},
\end{aligned}
\tag{2.314}
$$

$$
\begin{aligned}
\langle \varphi_{j,k}, \psi_{j,k}^{(l)} \rangle &= \left\langle \varphi_{j,k}, \varphi_{j-1,l} - \frac{I_{j-1,l}}{I_{j-1,0}} \varphi_{j-1,0} \right\rangle \\
&= I_{j-1,l} - \frac{I_{j-1,l}}{I_{j-1,0}} I_{j-1,0} \\
&= 0,
\end{aligned}
\tag{2.315}
$$

and

$$
\begin{aligned}
\langle \tilde{\varphi}_{j,k}, \tilde{\psi}_{j,k}^{(l)} \rangle &= \langle \tilde{\varphi}_{j,k}, \tilde{\varphi}_{j-1,l} - \tilde{\varphi}_{j,k} \rangle \\
&= \frac{1}{\mu(T_{j,k})\mu(T_{j-1,l})} \langle \varphi_{j,k}, \varphi_{j-1,l} \rangle - \frac{1}{\mu(T_{j,k})^2} \langle \varphi_{j,k}, \varphi_{j,k} \rangle \\
&= \frac{\mu(T_{j-1,l})}{\mu(T_{j,k})\mu(T_{j-1,l})} - \frac{\mu(T_{j,k})}{\mu(T_{j,k})^2} \\
&= 0.
\end{aligned}
\tag{2.316}
$$

Hence the functions $\varphi_{j,k}$, $\psi_{j,k}^{(l)}$. $\tilde{\varphi}_{j,k}$, $\tilde{\psi}_{j,k}^{(l)}$ indeed form a biorthogonal wavelet base on the sphere.

More information about triangulation based wavelets on the sphere and their applications in Geosciences and Computer vision can be found in [92] and [100].

3 Applications

3.1 Pattern recognition

3.1.1 Polar motion

The rotation axis of the Earth is neither fixed in inertial space nor is it with respect to the Earth's body. The gravitative forces of the Sun and Moon acting on the equatorial bulge of the Earth are changing the orientation of the rotation axis in inertial space. These changes are called *precession* and *nutation* and can be predicted with a very high accuracy.

Additionally, there is a small movement of the Earth's rotation axis with respect to its crust, which is called *polar motion*. Both nutation and polar motion are the superposition of Earth's response to external forces and free nutation of the Earth. Nutation is primarily the forced response of the Earth and can be predicted by geophysical and orbital models. The polar motion represents the forced and the free response of the Earth to external forces in almost equal parts. Again, the forced part can be predicted but the free part can only be determined by Space Geodesy methods. Moon and Sun and the planets exert gravitational forces on the equatorial bulge. Since the rotating Earth behaves like a gyro, it reacts to these forces by a clockwise movement of its rotation axis. This movement consist of two constituents:

- precession,
- nutation.

Definition 3.1.1. *Luni-solar precession* is the circular motion of the celestial pole with a period of 25 800 years and an amplitude equal to the obliquity of the ecliptic of $23^\circ.5$. The precession causes a westerly movement of the equinox of about $50''.3$ per year.

Planetary precession consist of a $0^\circ.5$ per year rotation of the ecliptic resulting in an easterly motion of the equinox by about $12''.5$ per century and an decrease of the obliquity of the ecliptic by about $47''$ per century.

Definition 3.1.2. The combined effect of luni-solar and planetary precession is called *general precession* or simply *precession*.

Definition 3.1.3. The short periodic motion of the pole superimposed on the precession with oscillations of 1 day to 18.6 years (the main period) and a maximum amplitude of $9''.2$ is called *nutation*.

Besides the movement of the Earth's rotation axis in space there is an additional variation of the rotation axis relative to the Earth's crust. This motion is primarily due to the elastic properties of the Earth and the exchange of angular momentums between the solid Earth, the oceans and the atmosphere.

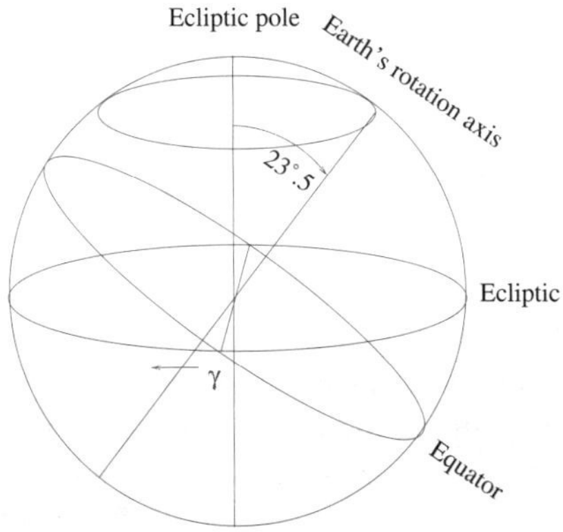

Figure 3.1. Precession

Definition 3.1.4. Polar motion is the rotation of the true celestial pole as defined by the precession and nutation models with respect to the z-axis of a conventionally chosen terrestrial reference system.

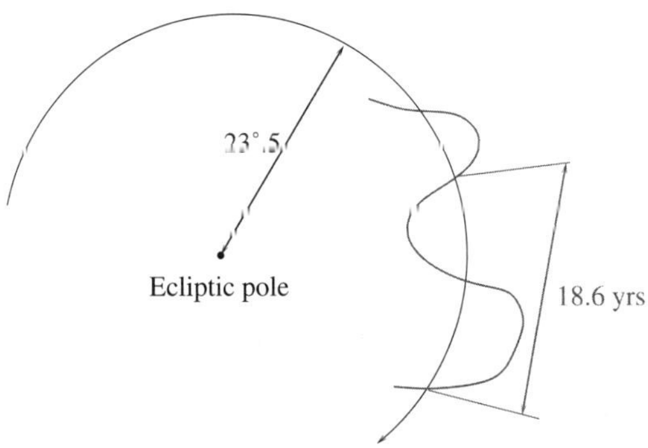

Figure 3.2. Nutation

Polar motion consists of a free and a forced oscillation. The free oscillation is counterclockwise with a period of 430 days (*Chandler period*) and an amplitude of 3–6 m.

The forced component again consists of two parts. The first part is excited by the tidal forces and therefore has a diurnal period, with an amplitude of one order of magnitude smaller than the free oscillation. The second part has an annual period since it is excited by the annual changes in the atmosphere. Its amplitude is about as large as the free oscillation.

Polar motion cannot be predicted by models, it has to be observed by space techniques. The accuracy of those observation has achieved a very high level, accounting for 0.2–0.5 milliarcseconds which is equivalent to 6–15 mm at the Earth's surface. Polar motion values can be downloaded from the *International Earth Rotation Service* (IERS) as tables of daily values of pole coordinates.

In Figure 3.3 a part of the polar motion time series between 1925.0 and 1976.15 with a sampling rate of 0.05 yr is displayed.

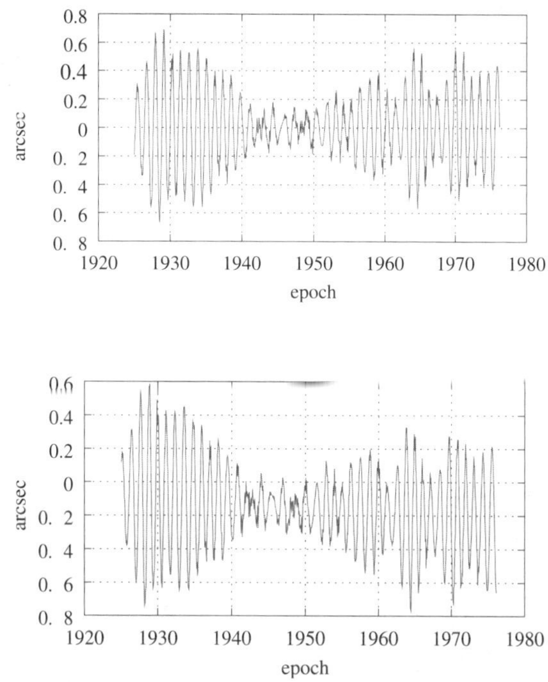

Figure 3.3. x- and y-component of polar motion

The results of the analysis of these time series with the Morlet wavelet is shown in Figure 3.4. For this analysis the value of the Morlet parameter ω_0 was chosen as $\omega_0 = 5.0$.

Three features of the polar motion are clearly visible from the wavelet spectra:

1. the forced annual oscillation,
2. the Chandler wobble with a period of 435 days,
3. the attenuation of the energy of the polar motion between 1940.0 and 1950.0.

The attenuation is due to the three-decade oscillation, the so-called Markowitz wobble. Since the time span of the polar motion is only 50 years, the Markowitz wobble does not show up as a separate frequency band in the wavelet spectrum.

These three effects cannot be separated in a Fourier spectrum. In Figure 3.5 the results of the short-time Fourier transformation of the x-component of the polar motion

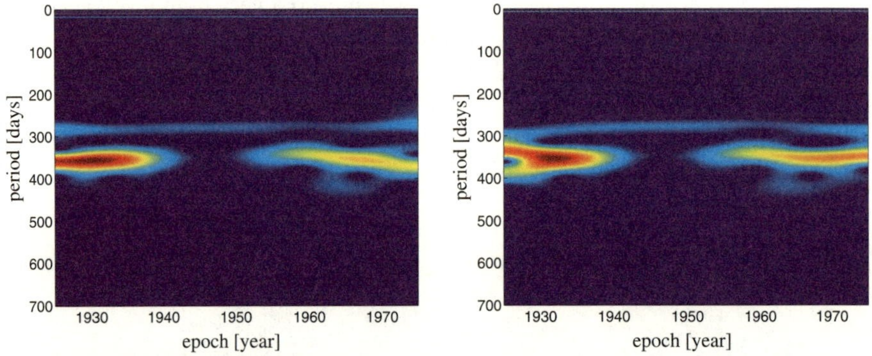

Figure 3.4. Wavelet spectrum of the x- and y-component of polar motion

both for a narrow and for a wide window are displayed. It is clearly visible that the short-time Fourier transformation does not separate annual oscillation and Chandler wobble as clearly as the wavelet transformation.

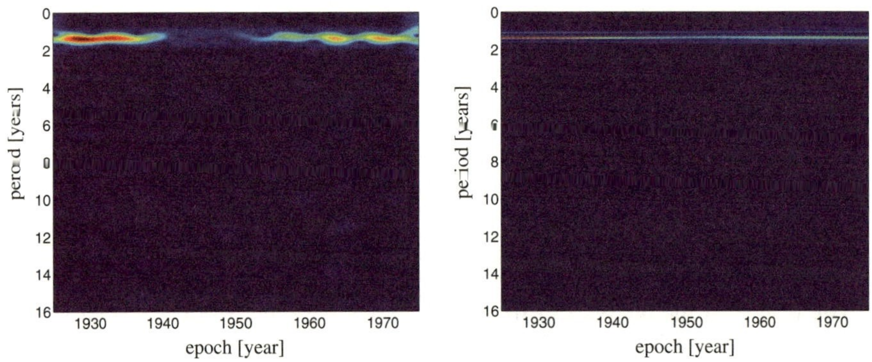

Figure 3.5. Short-time Fourier spectrum of the x-component of polar motion; narrow window (left) and wide window (right)

The time series of the coordinates of the instantaneous position of the Earth's rotation axis is frequently analyzed by different kinds of wavelets. Such studies aim at the detection of time varying features of the polar motion. Contributions, which belong to this group of investigations are [85], [86], [89] and [84]. In the papers [90] and [91] a special emphasis is given to the analysis of short-periodic variations in the Earth's polar motion. In a more general setting the time series of the polar motion can be considered as realizations of stochastic processes. A systematic extension of the theory of stochastic processes to wavelet bases can be found in [83] and [87] and [88]. The localization properties of wavelets enable the extension of the Wiener–Kolmogorov prediction to cases with a stationary signal but an non-stationary noise. These developments are reported in [56] and [55].

3.1.2 Atmospheric turbulence

Fully developed turbulence occurs in incompressible flows at high Reynolds numbers. It shows features of self-similarity and is therefore closely related to fractals. Consequently, the discussion of the application of wavelets for the study of turbulence will be started with the introduction of some concepts of fractal theory. The turbulence analysis aims at the identification of places where the velocity field suddenly changes its regularity. The regularity of a function f can be measured by its Hölder exponent h:

$$|f(x_0 + l) - f(x_0)| \sim C l^{h(x_0)}. \tag{3.1}$$

The larger the Hölder exponent $h(x_0)$ the more regular is the function f at the place x_0. The Hausdorff dimension is a concept for the characterization of fractal sets:

Definition 3.1.5. Let $h \subset \mathbb{R}^m$ be an arbitrary subset of $\mathbb{R}^m$. Let $N(\varepsilon)$ be the number of balls of radius ε which are necessary to cover h completely. The number

$$D(h) := \lim_{\varepsilon \to 0} \frac{\ln N(\varepsilon)}{\ln \varepsilon} \tag{3.2}$$

is called the *Hausdorff dimension* of h.

It is easy to see that for simple sets h the Hausdorff dimension coincides with the usual definition of its dimension. For more complicated sets, the Hausdorff dimension can have a non-integer value. For this reason sets with a fractal Hausdorff dimension are called *fractals*.

Intuitively,

$$N_h(l) \sim l^{-D(h)} \tag{3.3}$$

gives the number of balls of diameter l, which are necessary to cover the set of all points, where the velocity field has the Lipschitz regularity h.

The connection between the Lipschitz regularity and wavelet analysis is given by the following theorem:

Theorem 3.1.1. *Suppose that ψ is a wavelet with n vanishing moments, having a compact support. Let θ be a function with*

$$\int_{\mathbb{R}} \theta(t)\, dt \neq 0, \quad \psi = (-1)^n \theta^{(n)}(t). \tag{3.4}$$

Let $f \in L_1([a, b])$. If there exists $s_0 > 0$ such that $|Wf(u, s)|$ has no local maximum for $u \in [a, b]$ and $s < s_0$, then f is uniformly Lipschitz n on $[a + \varepsilon, b - \varepsilon]$, for any $\varepsilon > 0$.

Proof. [71]. □

In plain words the theorem states that a function f is Lipschitz $\alpha_0 < n$ at x_0 if there is a sequence $\{(u_p, s_p)\}_p$ of wavelet maxima points with

$$\lim_{p\to\infty} u_p = x_0, \qquad \lim_{p\to\infty} s_p = 0. \tag{3.5}$$

Consequently, if the local maxima of the wavelet spectrum are connected by lines, the prolongation of these lines towards the scale $s = 0$ end at a point x_0, where f is Lipschitz $\alpha_0 < n$. For a fixed scale s the set of local maxima of the wavelet spectrum $|Wf(u, s)|$ can be considered as a covering of the set of singularity points of f by wavelets of the scale s. At these maxima locations

$$|Wf(u, s)| \sim s^{\alpha_0 + 1/2} \tag{3.6}$$

holds.

Definition 3.1.6. Let $\{u_p(s)\}$ be the positions of all local maxima of $|Wf(u, s)|$ at the fixed scale s. The function

$$\mathcal{Z}(q, s) = \sum_p |Wf(u_p, s)|^q \tag{3.7}$$

is called the *partition function*.

Of course, $\mathcal{Z}$ decays for the scale s tending to zero. The scaling exponent $\tau(q)$

$$\tau(q) := \liminf_{s} \frac{\ln Z(q, s)}{\ln s} \tag{3.8}$$

measures the decay of $\mathcal{Z}$. The scaling exponent τ and the Hausdorff dimension D are closely related to each other.

Theorem 3.1.2. *Let $\Lambda = [\alpha_{min}, \alpha_{max}]$ be the support of $D(\alpha)$. Let ψ be a wavelet with $n > \alpha_{max}$ vanishing moments. Then it holds*

$$\tau(q) = \min_{q\in\Lambda} \left(q \left(\alpha + \frac{1}{2} \right) - D(\alpha) \right). \tag{3.9}$$

Proof. [3]. □

In numerical calculations $\tau(q)$ is computed by evaluating the sum $\mathcal{Z}(s, q)$. Solving (3.9) for $D(\alpha)$ gives the unknown Hausdorff dimension.

In Figure 3.6 the typical behavior of the Hausdorff dimension as a function of α is displayed.

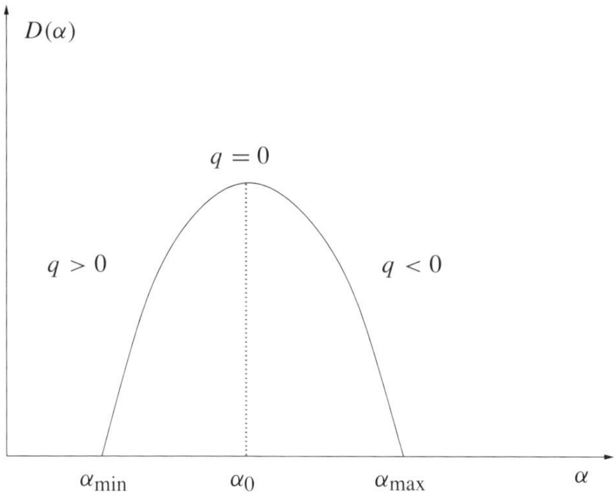

Figure 3.6. Hausdorff dimension as function of Hölder exponent α

Hence the computation of the Hausdorff dimension consists of four steps:

1. Compute $Wf(u, s)$ and the modulus maxima at each scale s. Concatenate the maxima across scales.

2. Compute the partition function
$$\mathcal{Z}(q, s) = \sum_p |Wf(u_p, s)|^q .$$

3. Compute $\tau(q)$ as a linear regression of $\ln \mathcal{Z}(q, s)$ with respect to $\ln s$.

4. Compute Hausdorff dimension as
$$D(\alpha) = \min_{q \in \mathbb{R}} \left(q(\alpha + 1/2) - \tau(q) \right) . \tag{3.10}$$

The statistical theory of turbulence was introduced by the contributions of Kolmogorov [59] and Obuchov [76]. This theory applies the tools of the theory of stationary stochastic processes to understand the partition of energy, at different scales, in the solution of the Navier–Stokes equation. Only the inertial zone, which lies between the smallest scales (where dynamic energy is dissipated in heat) and the largest scales (where exterior forces supply energy), is treated in the statistical theory of turbulence. In this inertial zone it is supposed that energy is neither produced nor dissipated but only transferred from one scale to another according to a constant rate ε. Additionally, it is assumed that the statistical properties of the velocity components of the transport are invariant and isotropic in the probabilistic sense. The main result of Kolmogorov's statistical theory is

$$E\{|u_i(\boldsymbol{x} + \Delta\boldsymbol{x}, t) - u_i(\boldsymbol{x}, t)|^n\} = K_n \cdot (E\{\varepsilon\})^{\frac{n}{3}} \cdot |\Delta\boldsymbol{x}|^{\frac{n}{3}}, \tag{3.11}$$

where $u_i, i = 1, 2, 3$ are the velocity components, ε is energy dissipation rate

$$\varepsilon = \frac{\nu}{2}\left(\frac{\partial u_i}{\partial x_j} + \frac{\partial u_j}{\partial x_i}\right)^2, \tag{3.12}$$

ν is the kinematic viscosity and K_n is an universal constant. The pendant to the scaling law in the space domain is the 5/3 -law

$$S(\omega) \sim \varepsilon^{2/3}|\omega|^{-5/3} \tag{3.13}$$

for the power density.

For $n = 2$ the scaling law (3.11) has been confirmed by many experiments. However, it is not accurate for $n > 2$ as evidenced by many other experiments. The deviation from the scaling law (3.11) can be attributed to the fact the dissipation of turbulent energy occurs only in a small fraction of the fluid volume. This intermittency results in a $\varepsilon(\boldsymbol{x})$ with a turn-on turn-off behavior. This means that turbulence has features which change not only in scale but also in space. Hence the Fourier power spectrum does not elucidate the multi-fractal structure of fully developed turbulence. Therefore it is a natural idea to use wavelets in order to unfold turbulence signals in scale and space.

In general, turbulence studies means the identification of sets with small Hölder exponents. This leads to the following interpretation of the scaling law (3.11): The velocity field is statistically homogeneous with a Lipschitz regularity 1/3. For purpose of isolating singular structures a link between the so-called structure function

$$D_f^q(x) := E\{|f(x_0 + x) - f(x_0)|^q\} \tag{3.14}$$

and the Hölder exponent is used. Due to Parisi and Fritsch 1984 [77] this relationship is

$$D_f^q(r) \sim r^{\inf_{\alpha>0}[p\alpha+1-D(\alpha)]}, \tag{3.15}$$

where $D(\alpha)$ is the Hausdorff dimension of the set

$$\{x \mid h(x) = \alpha\}. \tag{3.16}$$

Practical computations of the Hausdorff dimension $D(\alpha)$ show that it has its maximum at $\alpha = 1/3$ as it is predicted by Kolmogorov's theory. Nevertheless, the support of $D(\alpha)$ is not reduced to $\{1/3\}$, which means that a turbulent flow is not a statistically homogeneous field.

Practical applications of wavelets for the study of atmospheric turbulence can be found in [8], [49] and [46].

3.1.3 Fault scarps from seafloor bathymetry

A wavelet decomposition of bathymetric data reveals structures which are easily overlooked in the raw data. Additionally, isolated features such as fault scarps can be extracted by an wavelet analysis.

Usually most of the bathymetry data are collected from ship traverses. These tracks are densely spaced in along-track direction but the tracks are often several kilometers apart. Hence these data are basically one-dimensional spatial series. Those series can be successfully analyzed by one-dimensional wavelet analysis. In [64] a 1 600 km long profile near to Hawaii is analyzed. For this analysis a Morlet wavelet is used and the returned wavelet spectrum shows clearly two distinct zones:

- a small zone with large wavelet coefficients on long scales,
- the rest of the profile showing dominating wavelet coefficients on the short scales.

The anomalous small zone is interpreted as the site of a short-lived, abandoned spreading center.

Up to now only few areas of the seafloor have been mapped by a swath bathymetric survey, meaning that only in these few areas two-dimensional bathymetry data are available. Nevertheless, a two-dimensional analysis can give a much deeper insight than only a track-wise analysis. A typical two-dimensional bathymetry image is shown in Figure 3.7. It results from a bathymetric survey of the Mid-Atlantic Ridge.

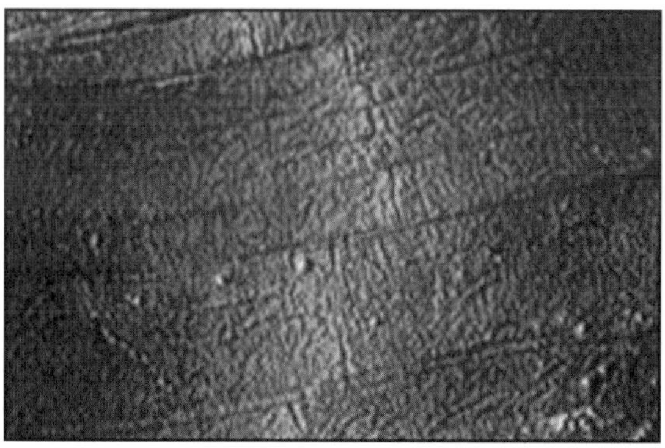

Figure 3.7. Bathymetry of a section of the Mid-Atlantic Ridge

The deep central median valley and the surrounding faulted blocks are clearly visible in the image. These faulted blocks tend to be long linear ridges running parallel to the axis of spreading. To find the location of the faults is one goal of the bathymetric survey. Therefore one tries to extract relatively long linear features with a preferred orientation from the image. For this purpose a wavelet has to be constructed, which is sensitive to exactly those features. This goal can be achieved by the construction of a two-dimensional wavelet $\psi(x_1, x_2)$ with a quick decay in x_1 and a slow decay in x_2 direction. Then the bathymetry data $d(\boldsymbol{x})$ can be subjected a continuous two-dimensional wavelet transform

$$\mathcal{W}^{\mathrm{eu}}\{\boldsymbol{d}\}(a, \vartheta, \boldsymbol{b}) = \frac{1}{\sqrt{c_\psi}} \int_{\mathbb{R}^2} \frac{1}{a} \psi\left(\frac{1}{a}\boldsymbol{R}^\top(\vartheta)(\boldsymbol{x}-\boldsymbol{b})\right) \cdot \boldsymbol{d}(\boldsymbol{x})\, d\boldsymbol{x} \tag{3.17}$$

with a fixed orientation angle ϑ. The orientation angle has to be chosen so that it coincides with the orientation of the mid-valley of the Mid-Atlantic Ridge.

As an example for a wavelet with different scales in different coordinate directions the tensor product of the Mexican hat wavelet

$$\psi(x_1) = (1 - x_1^2)e^{-\frac{x_1^2}{2}} \tag{3.18}$$

and the scaling function of the linear spline wavelet

$$\varphi(x_2) = \begin{cases} \frac{1}{4}(x_2+2), & -2 \le x_2 < 0 \\ \frac{1}{4}(2-x_2), & 0 \le x_2 < 2 \\ 0, & \text{otherwise} \end{cases} \tag{3.19}$$

can be used. The resulting wavelet is displayed in Figure 3.8.

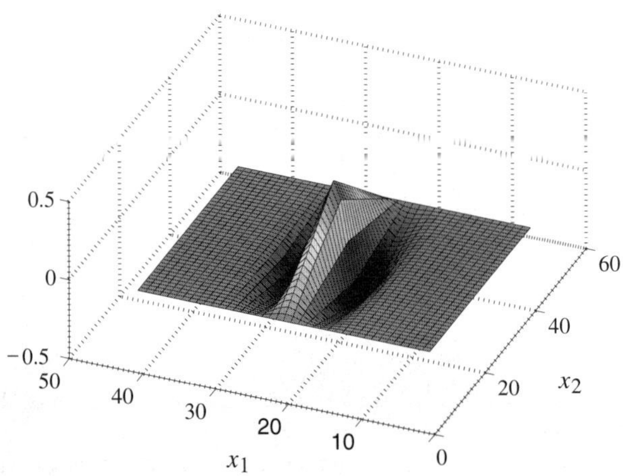

Figure 3.8. Two-dimensional wavelet for the analysis of Sea-floor faults

With the help of this wavelet the bathymetry data can be analyzed twice. Once with the rotation angle $\vartheta = 85°$ and another time with the rotation angle $\vartheta = -5°$. In the first case the faults running across the Mid-Atlantic Ridge in the second case the faults running parallel to the Mid-Atlantic Ridge are emphasized. A threshold operation applied to the wavelet spectra extracts the across and the parallel faults.

In Figure 3.9 the wavelet spectra and the results of thresholding these spectra are shown.

From Figure 3.9 it is clearly visible that in the case of $\vartheta = 85^\circ$ the faults running across the Mid Atlantic Ridge are extracted, the choice of $\vartheta = -5^\circ$ reveals the faults running parallel the Mid-Atlantic Ridge.

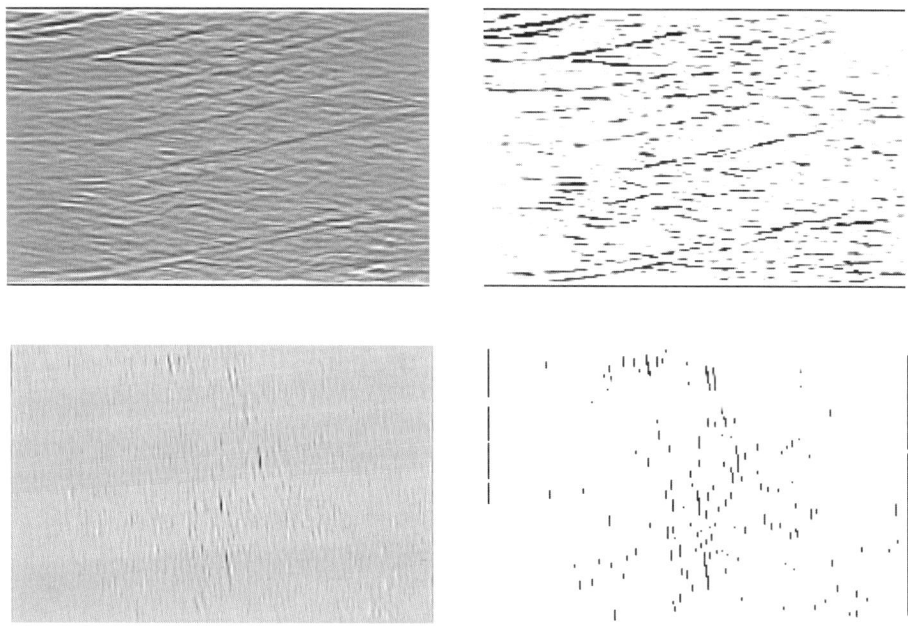

Figure 3.9. Wavelet spectra for rotation angles $\vartheta = 85^\circ$ (top left) and $\vartheta = -5^\circ$ (bottom left) and corresponding thresholded spectra (right)

Similar investigations are reported in [67]. In this paper bathymetric data of a 100 km $\times$ 70 km section of the Mid-Atlantic Ridge are analyzed. With a linear B-spline wavelet the fault scarps parallel to the Mid-Atlantic Ridge were extracted from the bathymetry data. The obtained information about preferred fault spacing, fault orientation and densities can help to understand the complex process the geophysical process of ocean floor spreading.

3.1.4 Seismic reflection horizons

The source of seismic waves can be natural or artificial. The natural sources of seismic waves are earthquakes. Since earthquakes are not readily available a continuous recording of seismographs and waiting for earthquakes is highly inefficient. For this reason the use of explosives can generate widely control-able artificial sources of seismic waves. The artificially generated seismic waves are recorded at an array of geophones located at a regular grid around the place, where the waves are generated. Basically, there are two different sets of seismic data acquired using artificial sources, one is obtained by refraction the other by reflection.

- In the refraction method the arrival time of wave as a function of the distance from the sources is measured.
- In the reflection method the travel time of the reflected signal from the source to the reflection horizon and back to the receiver is measured.

The relationship between the travel time of the reflected signal and the elastic properties of the soil is given by the one-dimensional wave equation

$$\rho(x)\frac{\partial^2 u}{\partial t^2} = \frac{\partial}{\partial x}\left[E(x)\frac{\partial u}{\partial x}\right], \tag{3.20}$$

where ρ is the density E an elastic constant and $u(x,t)$ the longitudinal displacement due to compression waves. If we assume that the medium consists of a stack of homogeneous layers as shown in Figure 3.10, the equation simplifies to

$$\rho_i \frac{\partial^2 u_i}{\partial t^2} = E_i \frac{\partial^2 u_i}{\partial x_2}, \tag{3.21}$$

where E_i and ρ_i are constants. Using arbitrary functions $f_i(t)$, the solution of (3.21) can be written as

$$u_i = f_i(t \pm x/c_i), \quad c_i = \sqrt{E_i/\rho_i}. \tag{3.22}$$

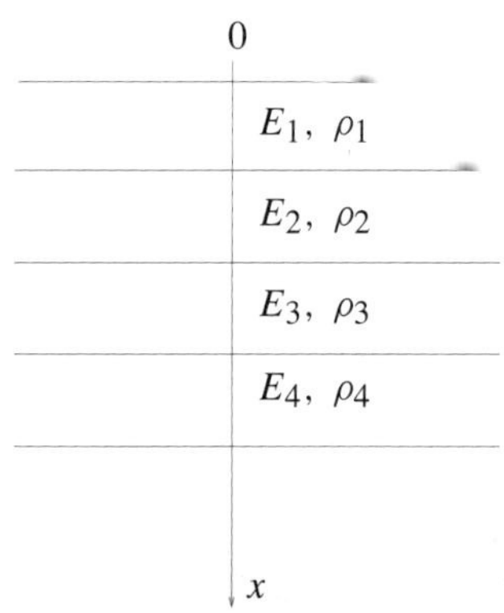

Figure 3.10. Medium of homogeneous layers

Without any loss of generality the thicknesses Δx_i of the layers can be chosen in such a way that the travel-time

$$\Delta t_i = \frac{\Delta x_i}{c_i} = \delta t = \text{const} \tag{3.23}$$

through every layer is the same. Let us now consider the reflection and transmission at the interface from the i-th to the $(i+1)$-th layer. The situation is sketched in Figure 3.11.

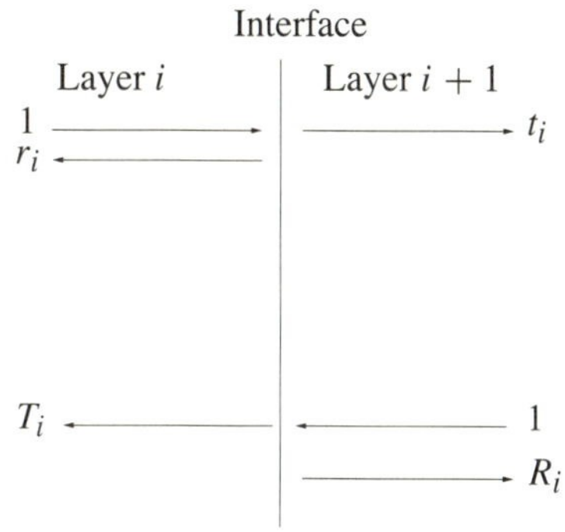

Figure 3.11. Interface between layers i and $(i+1)$

Applying the condition of continuity of displacement across the interface

$$u_i = u_{i+1}, \tag{3.24}$$

and the condition of continuity of traction

$$E_i \frac{\partial u_i}{\partial x} = E_{i+1} \frac{\partial u_{i+1}}{\partial x}, \tag{3.25}$$

the following relations between the amplitudes of the transmitted and reflected pulses are obtained:

$$1 + r_i = t_i, \tag{3.26}$$

$$1 + R_i = T_i, \tag{3.27}$$

$$r_i = -R_i = \frac{c_i \rho_i - c_{i+1} \rho_{i+1}}{c_i \rho_i + c_{i+1} \rho_{i+1}}. \tag{3.28}$$

If now an impulse at $x = 0$ is sent to the layered medium the geophones receive the reflections from many interfaces as it is shown in Figure 3.12.

This means, the impulses, which are reflected at the different interfaces arrive at the receiver with a time spacing of $2\Delta t$. A receiver placed directly at the point where the impulse was injected to the material would register a time series of the form shown in Figure 3.13.

In practical applications not only one geophones but a whole array of geophones placed at different distances from the seismic source are used. These geophones receive besides the direct wave also the reflected wave as it is indicated at the left side of Figure 3.14.

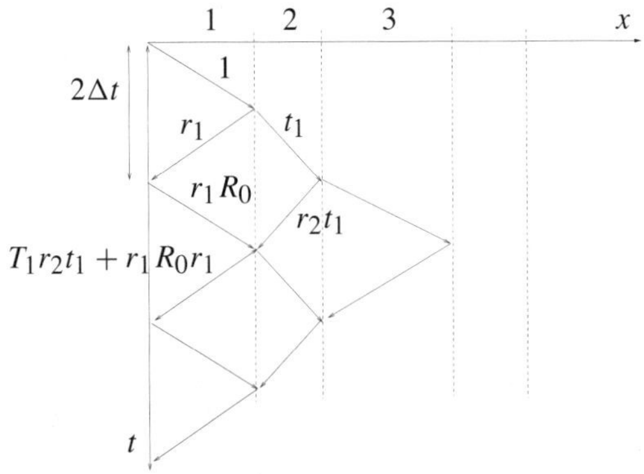

Figure 3.12. Transmission and reflection at interfaces

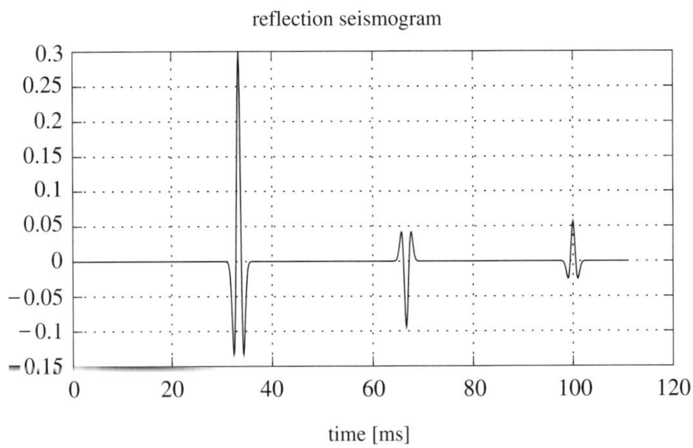

Figure 3.13. Direct registration of the reflected impulses

If the seismograms of the registered signals at the individual geophones are stacked according to the distance to the seismic source the typical form of a reflection seismogram as it is shown at the right side of Figure 3.14 is obtained.

Obviously, in the seismogram the signature of the reflection horizons is given by a succession of features on a precisely defined scale. This leads in a natural way to the application of wavelets for the analysis of seismic data. Information like lengths of an event, its tip, frequency and amplitude are essential for geophysical interpretation of seismic data. In Figure 3.15 a section of a seismogram is displayed as a gray-coded image.

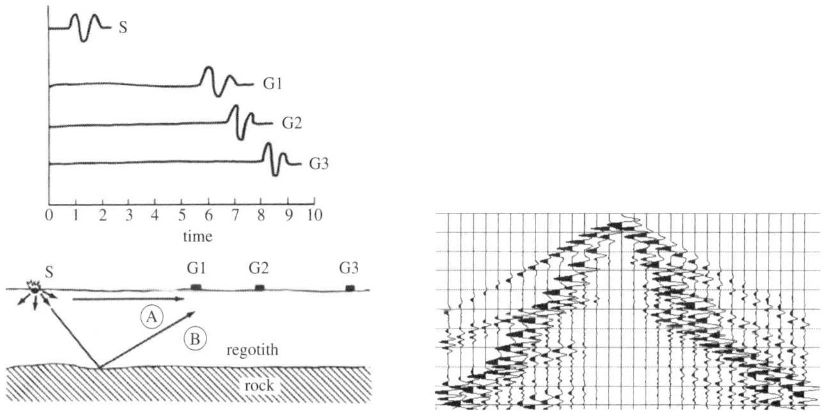

Figure 3.14. Array of geophones (left) and the registered seismogram (right)

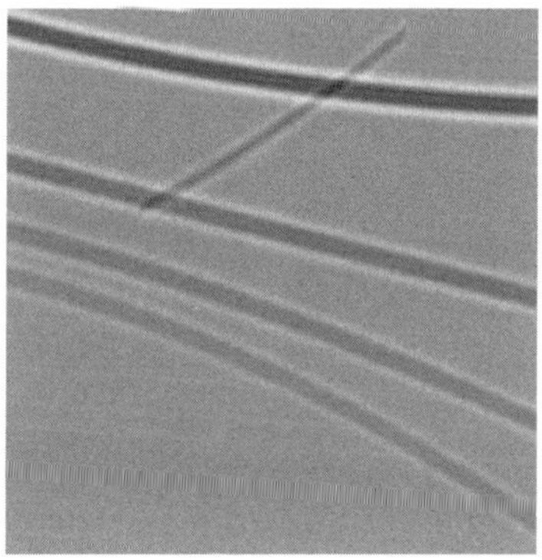

Figure 3.15. Gray-coded section of a seismogram

The inversion of seismograms aims at an identification of the individual seismic reflection horizons, their lengths and their dip-angles. Since the scale of the reflected signal is approximatively known a wavelet analysis of each individual record at exactly this scale reveals the location of the reflection horizon as the occurrence of large values in the wavelet spectrum. An subsequent thresholding operation can suppress those parts in the spectrum which are not related to reflection horizons and will leave only the reflection incidences.

This process is indicated in Figure 3.16. For the wavelet analysis the Mexican-hat wavelet with the scale parameter $a = 15.0$ was used. A hard-thresholding was applied by setting to zero all spectral lines with a modulus smaller than 40 % of the maximum modulus. Obviously by this strategy only the reflection horizons with a strong reflection are detected. If also the weak reflections should be detected a milder thresholding would be necessary. If this procedure is repeated for every recording and if the results are stacked again a picture as is displayed in Figure 3.17 is obtained.

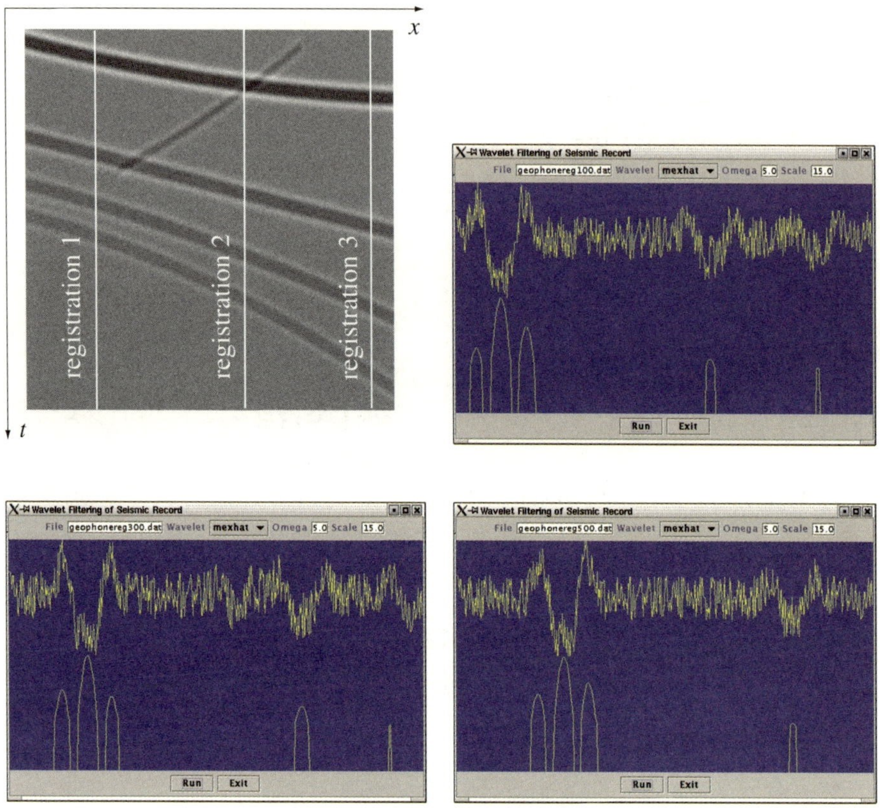

Figure 3.16. Seismogram with three selected recordings (top left) and the recordings with their thresholded wavelet spectra (top right to bottom right)

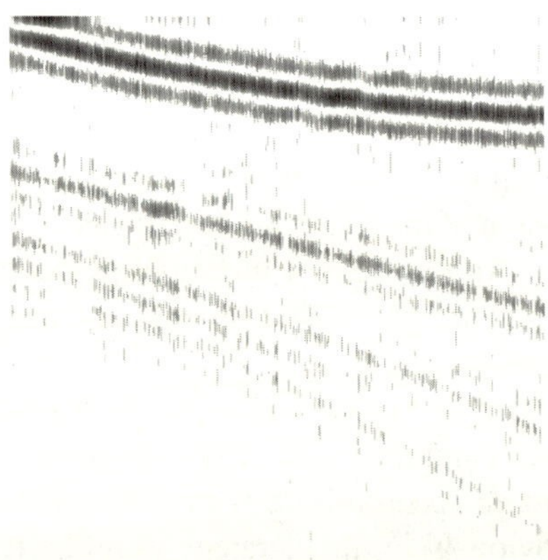

Figure 3.17. Reflection horizons detected by wavelet analysis

One of the first application of wavelets for seismic studies is reported in [61]. Examples for the wavelet analysis of normal incidence seismic data, i.e. of seismic data where source and receiver are at the same location are given in [78]. Similar ideas, without relating the directly to wavelet theory, are used in the paper [66] and [102].

3.1.5 GPS cycle-slip detection

The NAVSTAR GPS (*NAV*igation *S*ystem with *T*ime and *R*anging *G*lobal *P*ositioning *S*ystem) is a satellite based navigation system providing precise three dimensional positions and time information. Among the observables of the GPS system carrier phase observations have the highest accuracy potential. Carrier phase observations are generated by comparison of the carrier received from the satellite with a copy of this carrier generated in the receiver. Because of the travel time of the signal the received and the generated carrier differ by a time lag, which corresponds to the travel time of the signal between the satellite and the receiver. Due to the strict periodicity of both the received and the generated carrier the time lag can only be determined up to an unknown integer number of carrier cycles.

Hence a GPS carrier phase observation can be described by the following equation:

$$\Phi_i(t) = -\left(\frac{f_i}{c}\right)(\rho(t) - I_i(t) + T(t) + f_i(\delta^S - \delta^R)) + N_i, \tag{3.29}$$

where

- f_i is the nominal L_i frequency,
- ρ is the geometrical distance between satellite and receiver,
- I_i is the frequency dependent path lengthening due to ionospheric refraction,
- T is the frequency independent path lengthening due to tropospheric refraction,
- δ^S, δ^R are the satellite clock and the receiver clock errors respectively,
- N_i is the integer phase ambiguity,
- the index i refers to the two frequencies L_1, L_2 implemented in the GPS system.

The values of N_i do not change as long as the connection between satellite and receiver is not interrupted. If as a result of a bad signal to noise ratio this connection is disturbed, the receiver has to lock the satellite again. A new acquisition also changes the value of the integer ambiguities N_i. This effect is called *cycle -slip* and is sketched in Figure 3.18.

Therefore one of the basic steps in GPS data processing is the detection and correction of cycle slips. A large number of different techniques has been developed for this purpose. One of the frequently used method is the analysis of the so-called phase double differences. For the double differences observations of two receivers i, j to two satellites p, q are used. The situation is displayed in Figure 3.19.

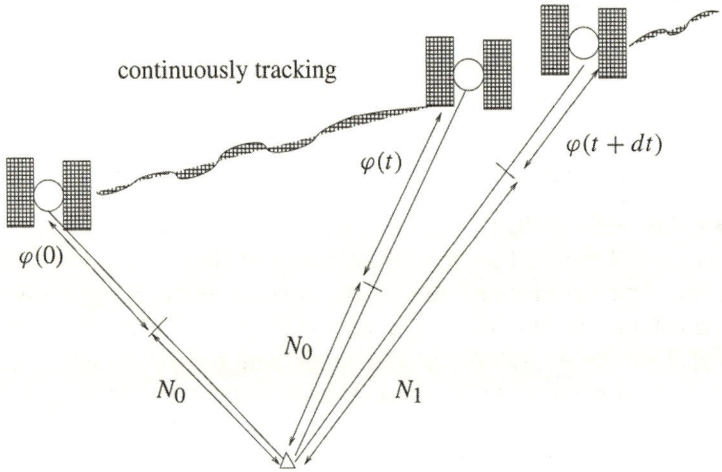

Figure 3.18. Cycle slip due to loss-of-lock

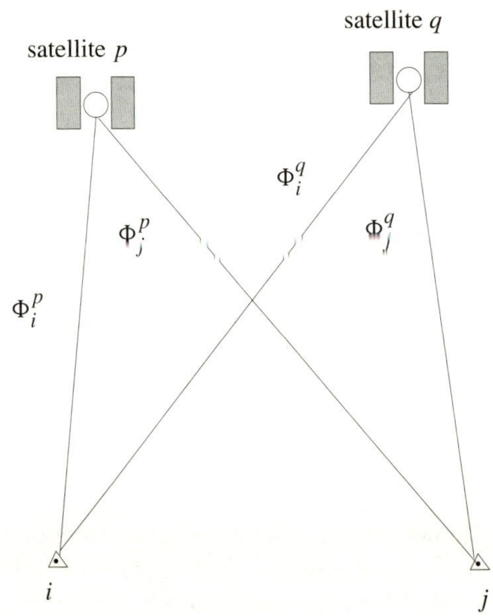

Figure 3.19. Configuration of double differences observations

The phase double difference is defined by

$$\nabla\Delta\Phi_{ij}^{pq} := (\Phi_i^p - \Phi_j^p) - (\Phi_i^q - \Phi_j^q). \tag{3.30}$$

Inserting (3.29) in this definition one obtains

$$\begin{aligned}\nabla\Delta\Phi_{ij}^{pq} = -\left(\frac{f}{c}\right)\big[&(\rho_i^p - \rho_j^p) - (\rho_i^q - \rho_j^q)\\ &- (I_i^p - I_j^p) - (I_i^q - I_j^q) + (T_i^p - T_j^p) - (T_i^q - T_j^q)\\ &+ (N_i^p - N_j^p) - (N_i^q - N_j^q)\big].\end{aligned} \tag{3.31}$$

Obviously, the satellite and the receiver clock errors cancel out in the double differences. If the distance between the receivers is smaller than about 30 km, the signals registered at both receivers have approximatively the same path lengthening due to ionospheric and tropospheric influences. This means, the tropospheric and ionospheric terms can be neglected, leading to

$$\nabla\Delta\Phi_{ij}^{pq} = -\left(\frac{f}{c}\right)\big[(\rho_i^p - \rho_j^p) - (\rho_i^q - \rho_j^q) + (N_i^p - N_j^p) - (N_i^q - N_j^q)\big]. \tag{3.32}$$

Consequently, the change in $\nabla\Delta\Phi_{ij}^{pq}$ is only due to the change in the satellite–receiver geometry. Since the satellites move slowly – they need about 12 hours for one revolution around the Earth – the change in $\nabla\Delta\Phi_{ij}^{pq}$ is very smooth. Sudden jumps in $\nabla\Delta\Phi_{ij}^{pq}$ indicate the occurrence of a cycle slip. In Figure 3.20 two hours of L1 phase double differences, recorded with a sampling rate of 15 seconds are displayed.

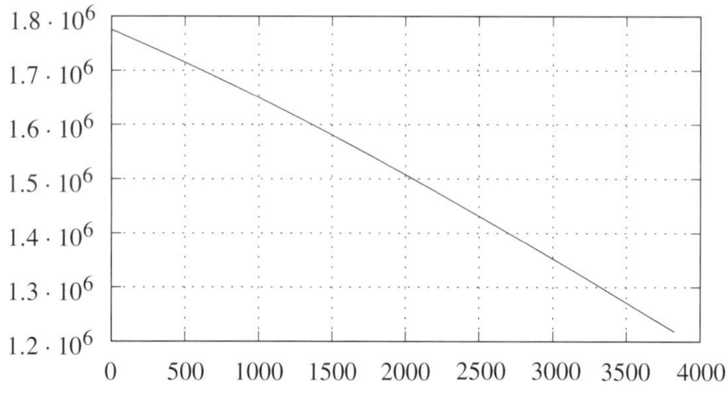

Figure 3.20. Double differences with cycle slip at epoch 100

Because the modulus of $\nabla\Delta\Phi_{ij}^{pq}$ is quite large the jump at epoch 100 is not visible and it is also difficult to detect by statistical tests. For this reason the smooth trend is eliminated by fitting a low order polynomial to $\nabla\Delta\Phi_{ij}^{pq}$. The residuals after the fit of a cubic polynomial are displayed in Figure 3.21. In the residuals the cycle slip is clearly visible as a sudden change in the sign of the residuals. This change of sign can easily be detected by statistical tests.

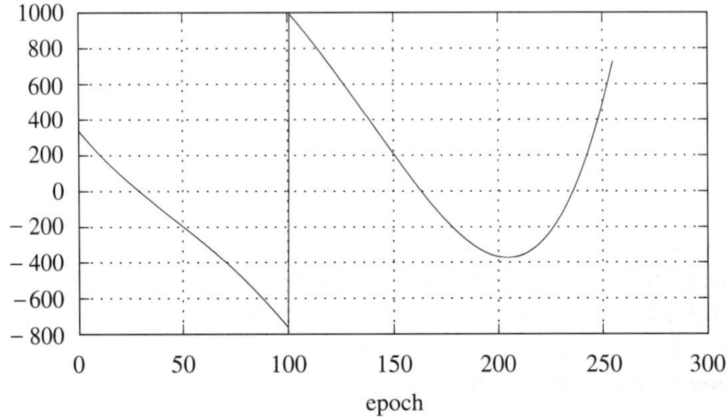

Figure 3.21. Residuals of a cubic polynomial fitted to phase double differences

The essence of cycle slip detection by analysis of the double differences is a two-step algorithm,

1. a low-pass filter by extracting a polynomial trend and

2. a high-pass filter by extraction short wavelength features in the residuals.

Therefore the cycle slip detection algorithm can be considered as a band-pass filter. Since wavelet transformation implements also a band pass filter a direct link between cycle slip detection and wavelet transformation is given. The basic idea of this link is to identify cycle slips as large entries in the wavelet spectrum of $\nabla\Delta\Phi_{ij}^{pq}$. For this purpose a multi-resolution analysis based on the Daubechies wavelet of order 4 was created and the projection of $\nabla\Delta\Phi_{ij}^{pq}$ into the Wavelet space W_1 was computed. The corresponding wavelet coefficients $d_k^{(1)}$ are displayed in Figure 3.22 in a logarithmic scale.

In this spectrum the cycle slip clearly shows up as an isolated spike which is very easy to detect. Used in this way, wavelet transformation is a very efficient and fast tool for cycle slip detection.

In [14] the a different method for the detection of cycle slips with wavelets is discussed. The idea behind the detection of cycle slips by wavelet transformation is the observation that the wavelet spectrum has large values at places where the signal locally has a similar shape as the analyzing wavelet. Therefore a wavelet has to be constructed which reflects the typical signal structure of a cycle slip. Entries in the spectrum with large values are interpreted as places, where cycle slips occurred.

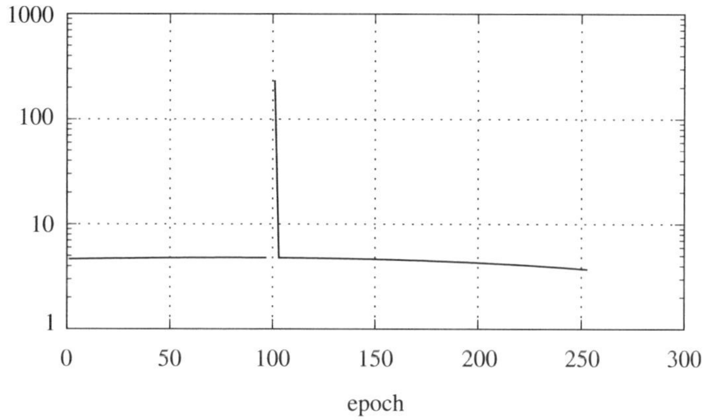

Figure 3.22. Daubechies 4 wavelet spectrum of double differences

For the cycle slip detection the phase observations on both frequencies are combined

$$
\begin{aligned}
PP &= \Phi_1 - \left(\frac{f_1}{f_2}\right)\Phi_2 \\
&= \left(\frac{f_1}{c}\right)(I_1(t) - I_2(t)) + \left[N_1 - \left(\frac{f_1}{f_2}\right)N_2\right].
\end{aligned} \tag{3.33}
$$

Since the ionospheric refraction I_i chances slowly with time, a cycle slip will give a jump in the *PP* combination. A typical *PP* combination containing a cycle slip is displayed in Figure 3.23.

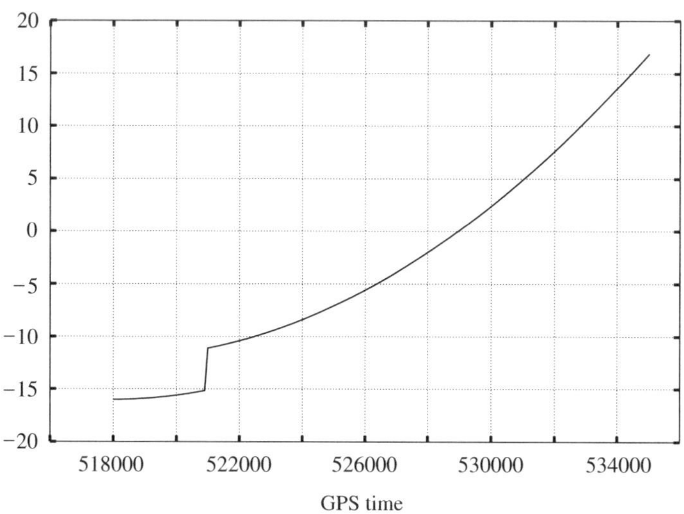

Figure 3.23. Cycle slip in *PP* combination

For the analysis of the *PP* combination the authors developed a special wavelet $\psi(t)$ given by its discrete values

$$\psi(t) = \begin{cases} 1.0, & t = -2 \\ -1.0, & t = -1 \\ -2.0, & t = 0 \\ 2.0, & t = 1 \\ 1.0, & t = 2 \\ 0, & \text{for all other } t \in \mathbb{Z}. \end{cases} \tag{3.34}$$

Between these discrete points the wavelet is interpolated linearly. The wavelet is displayed in Figure 3.24.

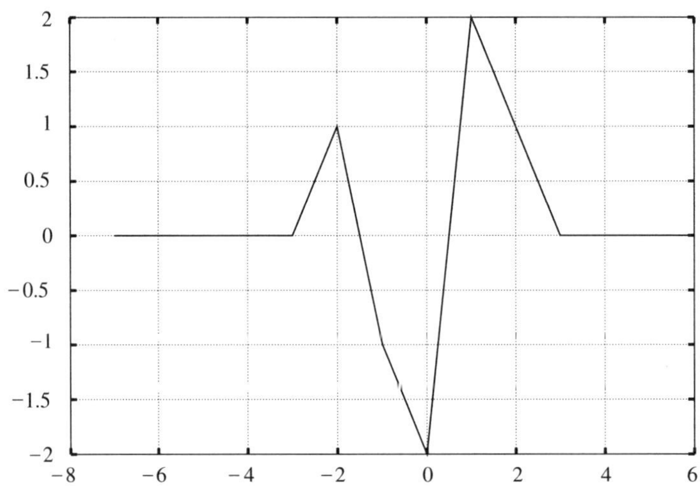

Figure 3.24. Empirical wavelet for the detection of cycle slips

The wavelet spectrum of the *PP* combination with respect to the wavelet (3.34) was computed with respect to the single scale $\lambda = 1$ and the result is plotted in Figure 3.25.

The cycle slip is clearly visible from its typical signature in the wavelet spectrum. The slow variation of the *PP* combination, due to the variation of the ionospheric influence is completely suppressed.

Despite the fact that the analysis was really carried out using a wavelet there is one essential difference to a real wavelet analysis. Typical for a wavelet analysis is the fact that the signal is analyzed on different scales. In the cycle slip example the analysis was carried out only on the finest scale.

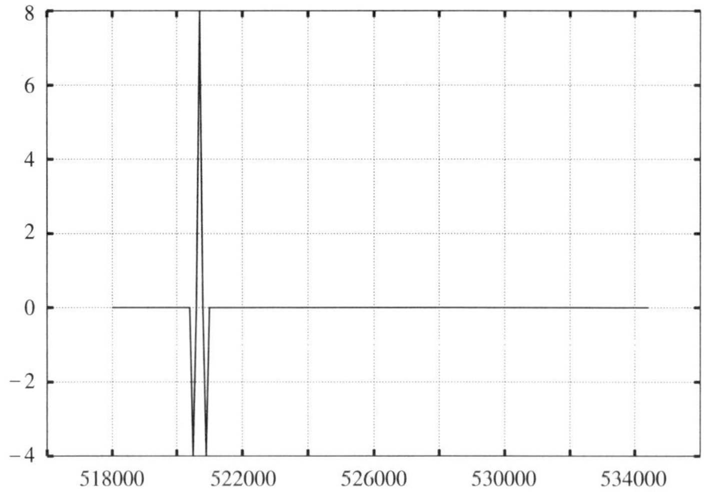

Figure 3.25. Wavelet spectrum of the *PP* combination

3.1.6 Edge detection in images

Frequently, a pattern recognition in images aims at a detection of sharp edges. On first sight edges could be understood as places in the image, where the image intensity has a sharp transition. But this definition is not satisfactory since also textures have sharp transitions in image density without being connected to an edge. A clear distinction between edges and textures requires the inclusion of the scale of the phenomena into the consideration. This leads in a natural way to a wavelet analysis of the image. The edges can be identified as the modulus maxima of the wavelet spectrum of the image. Edge detection algorithms, which are based on this idea are described in [80] and [11].

Maxima of wavelet spectra of images. The standard algorithm for edge detection in images is the search for maxima of the gray value gradient

$$\nabla f(\boldsymbol{x}) = \left(\frac{\partial f}{\partial x_1}(\boldsymbol{x}), \frac{\partial f}{\partial x_2}(\boldsymbol{x}) \right). \tag{3.35}$$

The vector $\nabla f(\boldsymbol{x})$ points into the direction of the maximum change of f. A point $f(\boldsymbol{x})$ is considered an edge, if it is a local maximum of $\|\nabla f\|$ in the direction $\nabla f(\boldsymbol{x})$, i.e. if

$$\|\nabla f(\boldsymbol{x})\| \geq \|\nabla f(\boldsymbol{y})\| \tag{3.36}$$

holds for

$$\boldsymbol{y} = \boldsymbol{x} + \lambda \nabla f(\boldsymbol{x}) \tag{3.37}$$

and $\lambda \in \mathbb{R}$ small enough. In this way all sharp transitions in image density will be detected independent of their scale. In order to implement a multi-scale version of this edge-detection approach, the data has to be convoluted in advance with a smoothing kernel $\Theta(i, j)$, which is dilated to the respective scale. Usually, this smoothing kernel is symmetric $\Theta(\boldsymbol{x}) = \Theta(-\boldsymbol{x})$ and its partial derivatives

$$\psi^1(\boldsymbol{x}) := -\frac{\partial\Theta(x_1, x_2)}{\partial x_1}, \quad \psi^2(\boldsymbol{x}) := -\frac{\partial\Theta(x_1, x_2)}{\partial x_2} \tag{3.38}$$

are wavelets. For this reason the concatenation of smoothing and gradient computation can be interpreted as wavelet transform:

$$\begin{aligned}
\nabla(\Theta * f)(\boldsymbol{u}) &= \left(\frac{\partial}{\partial x_1}(\Theta * f)(\boldsymbol{u}), \frac{\partial}{\partial x_2}(\Theta * f)(\boldsymbol{u})\right) \\
&= \left(\left(\frac{\partial\Theta}{\partial x_1} * f\right)(\boldsymbol{u}), \left(\frac{\partial\Theta}{\partial x_2} * f\right)(\boldsymbol{u})\right) \\
&= \left(\left(\frac{\partial\Theta}{\partial x_1}(-\bullet) * f\right)(\boldsymbol{u}), \left(\frac{\partial\Theta}{\partial x_2}(-\bullet) * f\right)(\boldsymbol{u})\right) & (3.39) \\
&= \left((-\psi^1(-\bullet) * f)(\boldsymbol{u}), (-\psi^2(-\bullet) * f)(\boldsymbol{u})\right) & (3.40) \\
&= \left(-\langle\psi^1(\bullet - \boldsymbol{u}), f\rangle, -\langle\psi^2(\bullet - \boldsymbol{u}), f\rangle\right) \\
&= \left(-\mathcal{W}^1\{f\}(1, \boldsymbol{u}), -\mathcal{W}^2\{f\}(1, \boldsymbol{u})\right).
\end{aligned}$$

The modulus of the wavelet spectrum is defined by

$$\mathcal{M}f(1, \boldsymbol{u}) := \sqrt{|\mathcal{W}^1\{f\}(1, \boldsymbol{u})|^2 + |\mathcal{W}^2\{f\}(1, \boldsymbol{u})|^2}. \tag{3.41}$$

This derivation coincides with the intuitive understanding of a wavelet transform as a band-pass filter. The concatenation of a smoothing as a low-pass filter, and the differentiation, as a high-pass filter, yields a band-pass filter. Band-pass filtering is typical for a wavelet transformation. Hence the maximum modulus of the wavelet spectrum reveals places where on the preselected scale a sharp transition in the image intensity occurs.

So far the edge detection has only be considered on the scale $a = 1$. For an edge detection on other scales the smoothing kernel has to be scaled:

$$\Theta_j(\boldsymbol{x}) := 2^{-j}\Theta(2^{-j}\boldsymbol{x}). \tag{3.42}$$

On this scale we have

$$\begin{aligned}\nabla(\Theta_j * f)(\boldsymbol{u}) &= \left(\frac{\partial}{\partial x_1}(\Theta_j * f)(\boldsymbol{u}), \frac{\partial}{\partial x_2}(\Theta_j * f)(\boldsymbol{u})\right)\\ &= \left(\left(\frac{\partial \Theta_j}{\partial x_1} * f\right)(\boldsymbol{u}), \left(\frac{\partial \Theta_j}{\partial x_2} * f\right)(\boldsymbol{u})\right)\\ &= 2^{-j}\left(2^{-j}(-\psi^1 * f)(\boldsymbol{u}), 2^{-j}(-\psi^2 * f)(\boldsymbol{u})\right) \qquad (3.43)\\ &= 2^{-j}\left(-\langle \psi^1(2^{-j}(\bullet - \boldsymbol{u})), f\rangle, -\langle \psi^2(2^{-j}(\bullet - \boldsymbol{u})), f\rangle\right)\\ &= 2^{-j}\left(-\mathcal{W}^1\{f\}(2^{-j}, \boldsymbol{u}), -\mathcal{W}^2\{f\}(2^{-j}, \boldsymbol{u})\right)\end{aligned}$$

and

$$\mathcal{M} f(2^{-j}, \boldsymbol{u}) := 2^{-j}\sqrt{|\mathcal{W}^1\{f\}(2^{-j}, \boldsymbol{u})|^2 + |\mathcal{W}^2\{f\}(2^{-j}, \boldsymbol{u})|^2}. \qquad (3.44)$$

In this way between edges and elements of a texture in an image can be distinguished. On the smallest scale $j = 0$ the modulus maxima of the wavelet spectrum

$$\mathcal{M} f(1, \boldsymbol{u}) := \sqrt{|\mathcal{W}^1\{f\}(1, \boldsymbol{u})|^2 + |\mathcal{W}^2\{f\}(1, \boldsymbol{u})|^2} \qquad (3.45)$$

reveals both the edges and the texture. On suitable larger scale $j > 0$ only the edges show up in the modulus maxima of the wavelet spectrum

$$\mathcal{M} f(2^{-j}, \boldsymbol{u}) := \sqrt{|\mathcal{W}^1\{f\}(2^{-j}, \boldsymbol{u})|^2 + |\mathcal{W}^2\{f\}(2^{-j}, \boldsymbol{u})|^2}. \qquad (3.46)$$

Example 3.1.1. In order to demonstrate the effect of a wavelet transformation on the edge detection a synthetical image was constructed, which shows a figure as well as a texture. This image is shown in Figure 3.26. The big square in the left upper corner represents the figure and the 9 small squares in the right lower corner represent the pattern.

Figure 3.26 was subjected to a wavelet transformation on the scales $j = 0$ and $j = 3$ and the wavelet spectrum maxima are displayed im Figure 3.27.

It is clearly visible that on the smallest scale both the texture and the edges are extracted. On a larger scale the texture is suppressed and only the edges show up.

The application of wavelet techniques for edge detection in photogrammetry and remote sensing is still in its initial phase. Examples for the application of wavelets for fast stereo vision can be found in [101] and [65].

Figure 3.26. Test image containing a figure and a texture

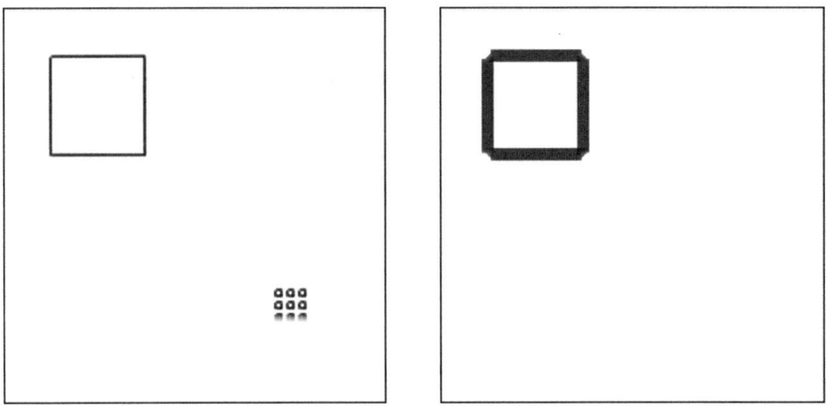

Figure 3.27. Wavelet modulus maxima on scale $j = 0$ (left) and on scale $j = 3$ (right)

3.2 Data compression and denoising

3.2.1 Wavelet filters and estimation

Frequently, a signal is hidden in some background noise. An estimation of signal is computed by attenuating the noise while amplifying the signal. For decades this technique is predominated by linear estimators because of their simplicity. In this section some examples of wavelet based non-linear estimators with an improved performance will be presented.

Let

$$f : [\Omega, \mathcal{A}, P] \to \mathbb{R}^N \tag{3.47}$$

be a random vector which is considered the signal. The random vector

$$n : [\Omega, \mathcal{A}, P] \to \mathbb{R}^N \tag{3.48}$$

is considered the noise. The observations which are available are the sum of signal and noise:

$$x := f + n. \tag{3.49}$$

Usually, f, n are assumed to be uncorrelated. The goal is to construct an estimation $\hat{f}$ of f which is linear in the observations

$$\hat{f} := Lx, \quad L \in \mathcal{L}(\mathbb{R}^N, \mathbb{R}^N) \tag{3.50}$$

in such a way that the average loss of $\hat{f}$ compared to f is minimal. In most applications the loss is measured as the square norm of the difference between $\hat{f}$ and f:

$$r(L, f) := E\{\|Lx - f\|^2\}. \tag{3.51}$$

Definition 3.2.1. A linear estimator L^* is called *Wiener estimator*, if it minimizes the loss among all linear estimators:

$$r(L^*, f) = \inf_{L \in \mathcal{L}(\mathbb{R}^N, \mathbb{R}^N)} r(L, f). \tag{3.52}$$

Theorem 3.2.1. *A linear estimator L^* is a Wiener estimator if and only if*

$$E\{(f - Lx) \cdot x^\top\} = 0 \tag{3.53}$$

holds.

Proof. The loss function

$$r(L, f) = E\{\langle Lx - f, Lx - f\rangle\}$$

is a quadratic form in L. Therefore it reaches its minimum for that L^* that generates a vanishing first derivative:

$$0 = \frac{\partial E\{\langle Lx - f, Lx - f\rangle\}}{\partial L} = -2E\{(Lx - f)x^\top\}.$$ □

The theorem provides a characterization of the Wiener estimator independent of the statistical properties of signal and noise. From this characterization an explicit expression for the Wiener estimator L^* can be derived if additional information about the statistical properties are supplied.

Theorem 3.2.2. *Without any loss of generality let us assume that both the signal* $\boldsymbol{f}$ *and the noise* $\boldsymbol{n}$ *have a zero mean*

$$0 = E\{\boldsymbol{f}\} = E\{\boldsymbol{n}\}. \tag{3.54}$$

Let

$$\boldsymbol{C}_{xx} = E\{\boldsymbol{x}\boldsymbol{x}^\top\}, \quad \boldsymbol{C}_{fx} = E\{\boldsymbol{f}\boldsymbol{x}^\top\} \tag{3.55}$$

be the auto- and the cross-covariance matrix, respectively. Then the Wiener estimator is

$$L^* = \boldsymbol{C}_{fx} \cdot \boldsymbol{C}_{xx}^{-1}. \tag{3.56}$$

Proof.

$$\begin{aligned} E\{(L^*\boldsymbol{x} - \boldsymbol{f})\boldsymbol{x}^\top\} &= E\{(\boldsymbol{C}_{xx}^{-1}\boldsymbol{C}_{fx}\boldsymbol{x} - \boldsymbol{f})\boldsymbol{x}^\top\} \\ &= \boldsymbol{C}_{fx}\boldsymbol{C}_{xx}^{-1}\boldsymbol{C}_{xx} - \boldsymbol{C}_{fx} \\ &= 0. \end{aligned}$$

□

If the observations $\boldsymbol{x}$, as well as the signal $\boldsymbol{f}$ and the noise $\boldsymbol{n}$ are represented with respect to the canonical base of $\mathbb{R}^N$, the linear equations (3.56) have to be solved for the construction of the Wiener estimator. This can be avoided if a special base, adapted to signal and noise, the so-called Karhunen–Loève base is used. Let

$$\boldsymbol{y} : [\Omega, \mathcal{A}, P] \to \mathbb{R}^N \tag{3.57}$$

be a random vector with vanishing mean and the covariance matrix

$$\boldsymbol{C}_{yy} := E\{\boldsymbol{y}\boldsymbol{y}^\top\}. \tag{3.58}$$

Definition 3.2.2. An orthogonal base $\{\boldsymbol{g}_k\}_{0 \le k < N}$ is called a *Karhunen–Loève* base of $\boldsymbol{y}$ if

$$\boldsymbol{C}_{yy}\boldsymbol{g}_k = \sigma_k^2 \boldsymbol{g}_k, \quad k = 0, \ldots, N-1 \tag{3.59}$$

holds.

Theorem 3.2.3. *If there exist a Karhunen–Loève basis* $\{\boldsymbol{g}_k\}$ *which diagonalizes both the covariance matrices* $\boldsymbol{C}_{ff}$ *and* $\boldsymbol{C}_{nn}$, *i.e.*

$$\boldsymbol{C}_{ff}\boldsymbol{g}_k = \beta_k^2 \boldsymbol{g}_k, \quad \boldsymbol{C}_{nn}\boldsymbol{g}_k = \sigma_k^2 \boldsymbol{g}_k, \quad k = 0, \ldots, N-1, \tag{3.60}$$

then

$$L^* = \boldsymbol{G} \operatorname{diag}\left(\frac{\beta_0^2}{\beta_0^2 + \sigma_0^2}, \ldots, \frac{\beta_{N-1}^2}{\beta_{N-1}^2 + \sigma_{N-1}^2} \right) \boldsymbol{G}^\top \tag{3.61}$$

holds, where $\boldsymbol{G}$ *is the matrix with the base vectors* $\boldsymbol{g}_k$ *in its rows.*

Proof.

$$
\begin{aligned}
L^* &= \boldsymbol{C}_{xf} \cdot \boldsymbol{C}_{xx}^{-1} \\
&= \boldsymbol{C}_{ff} \cdot \boldsymbol{C}_{xx}^{-1} \\
&= \boldsymbol{G}\,\mathrm{diag}(\beta_0^2, \ldots, \beta_{N-1}^2)\boldsymbol{G}^\top \Big(\boldsymbol{G}\,\mathrm{diag}(\beta_0^2, \ldots, \beta_{N-1}^2)\boldsymbol{G}^\top \\
&\qquad + \boldsymbol{G}\,\mathrm{diag}(\sigma_0^2, \ldots, \sigma_{N-1}^2)\boldsymbol{G}^\top\Big)^{-1} \\
&= \boldsymbol{G}\,\mathrm{diag}(\beta_0^2, \ldots, \beta_{N-1}^2)\,\mathrm{diag}\left(\frac{1}{\beta_0^2 + \sigma_0^2}, \ldots, \frac{1}{\beta_{N-1}^2 + \sigma_{N-1}^2}\right)\boldsymbol{G}^\top \\
&= \boldsymbol{G}\,\mathrm{diag}\left(\frac{\beta_0^2}{\beta_0^2 + \sigma_0^2}, \ldots, \frac{\beta_{N-1}^2}{\beta_{N-1}^2 + \sigma_{N-1}^2}\right)\boldsymbol{G}^\top.
\end{aligned}
$$

□

Remark. As a consequence of the theorem an inversion-free representation of the estimation $\hat{\boldsymbol{f}}$ can be given:

$$
\begin{aligned}
\hat{\boldsymbol{f}} &= L^*\boldsymbol{x} \\
&= \boldsymbol{G}\,\mathrm{diag}\left(\frac{\beta_0^2}{\beta_0^2 + \sigma_0^2}, \ldots, \frac{\beta_{N-1}^2}{\beta_{N-1}^2 + \sigma_{N-1}^2}\right)\boldsymbol{G}^\top\boldsymbol{x} \qquad (3.62) \\
&= \sum_{k=0}^{N-1} \frac{\beta_k^2}{\beta_k^2 + \sigma_k^2} \langle \boldsymbol{g}_k, \boldsymbol{x}\rangle \boldsymbol{g}_k.
\end{aligned}
$$

The interpretation of this representation is that the Wiener estimator works as an attenuation of each signal component $\langle \boldsymbol{x}, \boldsymbol{g}_k\rangle \mathbf{g}_k$ by a factor, which depends on the signal-to-noise ratio β_k^2/σ_k^2. In this way signal components with a bad signal-to-noise ratio are suppressed.

The disadvantage of this diagonal Wiener estimator is that even signal components with a bad signal-to-noise ratio will be still present in the estimation. In many cases, a thresholding estimation

$$
\hat{\boldsymbol{f}} = \sum_{k=0}^{N-1} \rho_T(\langle \boldsymbol{g}_k, \boldsymbol{x}\rangle)\boldsymbol{g}_k \qquad (3.63)
$$

with an appropriate threshold function ρ_T will be superior to (3.62). Standard choices for the threshold function are the so-called hard thresholding

$$
\rho_T(x) := \begin{cases} x, & |x| > T \\ 0, & |x| \le T \end{cases} \qquad (3.64)
$$

and the so-called soft thresholding

$$
\rho_T(x) := \begin{cases} x - T, & x > T \\ x + T, & x < -T \\ 0, & |x| \le T. \end{cases} \qquad (3.65)
$$

The threshold T is generally chosen in such a way that it is with a high probability larger than the maximum noise coefficient $\langle \boldsymbol{n}, \boldsymbol{g}_k \rangle$.

Thresholding has particular interesting properties if it is performed with respect to a wavelet base. An extensive study of this technique can be found in [2]. The observations are represented with respect to an orthogonal wavelet base

$$\boldsymbol{x} = \sum_{j=1}^{J} \sum_{m=0}^{2^{J-j}-1} \langle \boldsymbol{x}, \psi_{j,m} \rangle \psi_{j,m} + \langle \boldsymbol{x}, \varphi_{J,0} \rangle \varphi_{J,0}. \tag{3.66}$$

with $N = 2^J$. A thresholding estimator in this wavelet base can be written as

$$\hat{\boldsymbol{f}} = \sum_{j=1}^{J} \sum_{m=0}^{2^{J-j}-1} \rho_T(\langle \boldsymbol{x}, \psi_{j,m} \rangle) \psi_{j,m} + \rho_T(\langle \boldsymbol{x}, \varphi_{J,0} \rangle) \varphi_{J,0}, \tag{3.67}$$

where ρ_T can be the hard thresholding (3.64) or the soft thresholding (3.65). Generally speaking, in a wavelet base large amplitude coefficients correspond to transient signal variations. This means a wavelet thresholding keeps only transients coming from the signal without adding others due to noise.

In Figure 3.28 the effect of the hard wavelet thresholding is demonstrated for a saw-tooth like signal with superimposed noise.

Obviously, in this estimation the noise is widely suppressed while the sharp edges of the transient signal parts are not smoothed.

If the same signal is represented with respect to a Fourier base $\boldsymbol{g}_k(t) = e^{-\imath kt}$ and if the same threshold estimation technique is applied, the removal of the noise also smoothes the sharp edges of the signal. This effect is illustrated in Figure 3.29.

The hard thresholding is automatically a kind of an adaptive smoothing that depends on the regularity of the signal: The wavelet coefficients of the observations are the sum of two components

$$\langle \boldsymbol{x}, \psi_{j,m} \rangle = \langle \boldsymbol{f}, \psi_{j,m} \rangle + \langle \boldsymbol{n}, \psi_{j,m} \rangle, \tag{3.68}$$

the wavelet coefficient of the signal and the wavelet coefficient of the noise. In general, small wavelet coefficients $\langle \boldsymbol{x}, \psi_{j,m} \rangle$ occur in regions, where the signal $\boldsymbol{f}$ is smooth. Setting these coefficients to zero means locally smoothing the observations. Since the signal itself is already smooth in these regions thresholding means eliminating the noise locally.

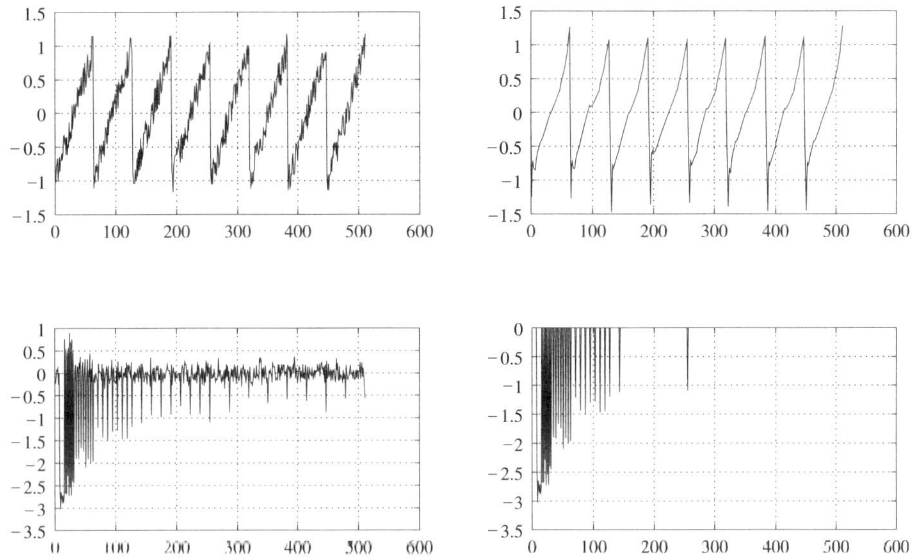

Figure 3.28. Transient signal with noise (top left), its Daubechies 4 spectrum (bottom left), the hard thresholded spectrum (bottom right) and the estimated signal (top right)

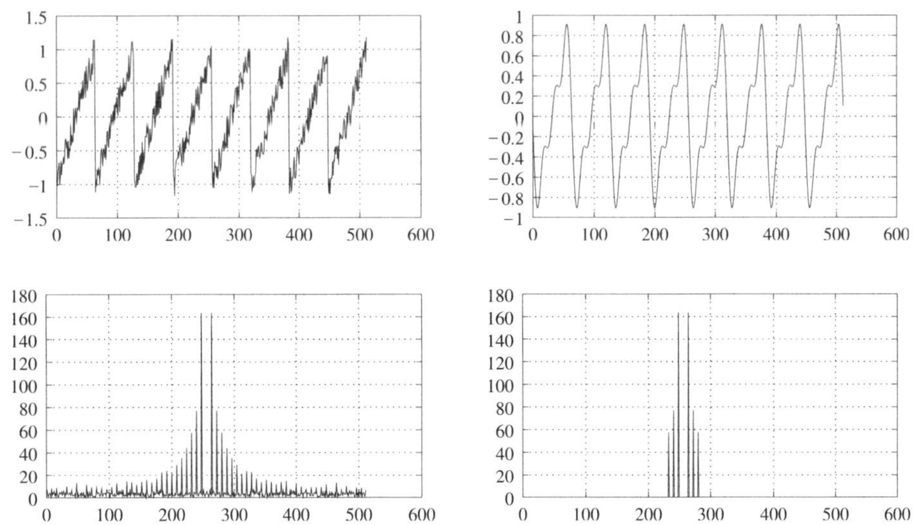

Figure 3.29. Transient signal with noise (top left), its Fourier spectrum (bottom left), the hard thresholded Fourier spectrum (bottom right) and the estimated signal (top right)

Large wavelet coefficients $\langle \boldsymbol{x}, \psi_{j,m} \rangle$ occur in the neighborhood of sharp signal transitions. These coefficients are kept under the thresholding policy. For this reason the smoothing of these sharp variations is avoided.

The assumption that for a piecewise smooth signal the noise manifests itself mainly on the finest wavelet scale leads to a simple method for the estimation of the noise variance. Under this assumption for a signal of length N on the finest scale

$$\langle \boldsymbol{x}, \psi_{1,m} \rangle \approx \langle \boldsymbol{n}, \psi_{1,m} \rangle \tag{3.69}$$

holds. Let us denote the median of the absolute values these coefficients by M.

$$M := \mathrm{Med}\{|\langle \boldsymbol{x}, \psi_{1,m} \rangle|\}_{m=0,\dots,N/2}. \tag{3.70}$$

If the length $N = 2^J$ is large enough, the values $\langle \boldsymbol{x}, \psi_{0,m} \rangle$ can be assumed to be normal distributed and it holds

$$E\{M\} \approx 0.6745\sigma \tag{3.71}$$

(cf. [57]). Therefore the noise variance can be estimated by

$$\hat{\sigma}^2 = \left(\frac{M}{0.6745}\right)^2. \tag{3.72}$$

This procedure is robust against outliers, which are produced by the few large wavelet coefficients in the vicinity of sharp signal transitions. A detailed study of noise variance estimation based on this idea can be found in [20].

Donoho denoising. A more efficient noise suppression can be obtained by a more sophisticated thresholding policy, such as the so-called Donoho-threshold. In [19] the following threshold policy is proposed:

$$\hat{d}_{m,k} = \mathrm{sgn}(d_{m,k})(|d_{m,k}| - \tau)_+, \quad \tau := \sqrt{2\log(2^J)}\sigma \tag{3.73}$$

with $(x)_+$ defined by

$$(x)_+ := \begin{cases} x, & x > 0 \\ 0, & x \leq 0. \end{cases} \tag{3.74}$$

This thresholding policy sets all wavelet coefficients to zero which are smaller than the bound τ. Finally, an estimation $\hat{f}$ of f is constructed from the thresholded wavelet coefficients

$$\hat{f}(t) = \hat{c}_{J,0}\varphi_J(t) + \sum_{m=1}^{J} \sum_{k=0}^{2^{J-m}-1} \hat{d}_{m,k}\psi_{m,k}(t), \tag{3.75}$$

which is hopefully free of noise but still contains the sharp features of the signal.

As an example we consider the signal displayed in Figure 3.30 which imitates one signal discussed in [19]. The signal consists of a large-scale sinusoidal component, a singular spike at 3.8 and a superimposed noise. The logarithm of the modulus of its wavelet spectrum is shown in Figure 3.31.

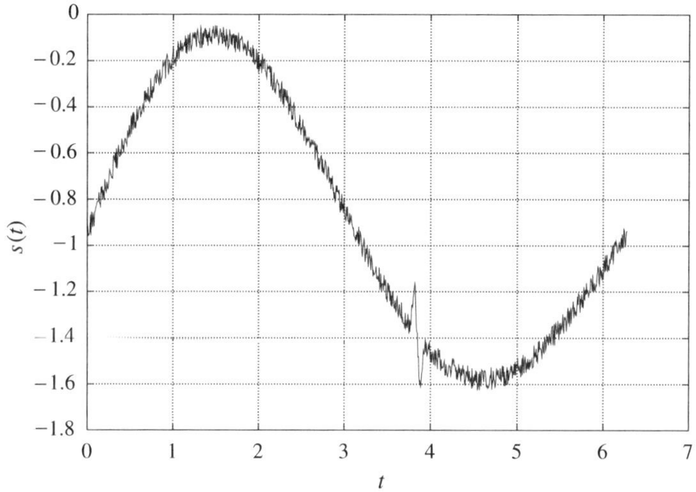

Figure 3.30. Noisy signal with isolated spikes

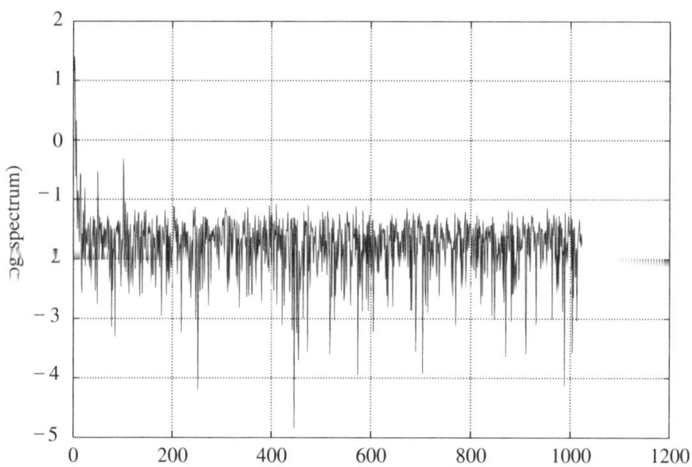

Figure 3.31. Logarithm of wavelet spectrum

Obviously, the majority of the wavelet spectrum is smaller than 10^{-1}. This part of the spectrum represents the noise contained in the data. If a thresholding is applied, only 13 out of 1024 wavelet coefficients remain different from zero. The signal reconstructed from this 13 nonzero coefficient is displayed in Figure 3.32.

The reconstructed signal is practically free of noise but the sharp features are still preserved. They are not blurred during the denoising process. In comparison to the wavelet denoising the result of the usual low-pass filtering is displayed in Figure 3.33.

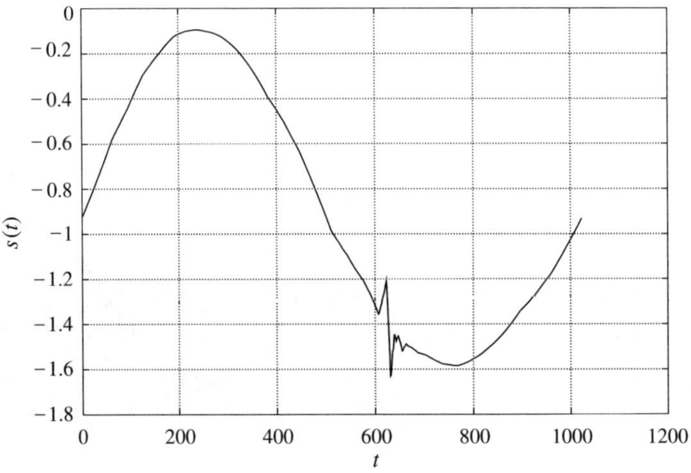

Figure 3.32. Reconstructed signal

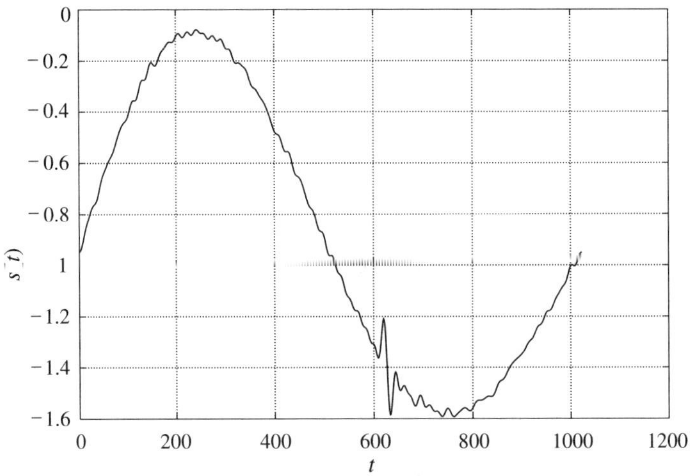

Figure 3.33. Low-pass filtered signal

Here the noise is strongly reduced but not completely eliminated. The spike in the data is still visible. If an even stronger low-pass filter is applied the result can be seen in Figure 3.34.

3.2.2 Deconvolution thresholding

The inversion of noisy data is often instable and amplifies the data noise considerably. In this section data will be investigated, which are degraded by white Gaussian noise.

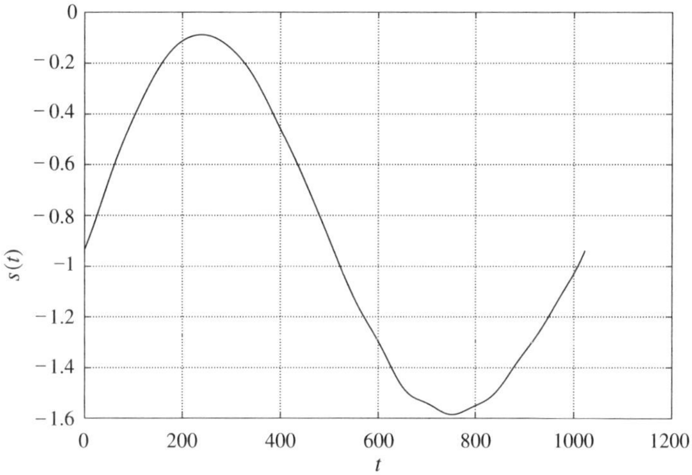

Figure 3.34. Stronger low-pass filter

It will turn out that threshold estimators with respect to a base which diagonalizes the covariance of the noise are an appropriate deconvolution technique.

It is assumed that the data y

$$y = k * f + n \tag{3.76}$$

are observed. The observation is the convolution of the signal f with the kernel k plus a superimposed noise n. The noise n is assumed to be white and Gaussian with the variance σ^2. The goal is an estimation of f from the noisy and convoluted data y.

It is not guaranteed that the convolution operator has a unique inverse. For this reason, the pseudo-inverse k^- will be used for the deconvolution.

Lemma 3.2.1. *Let y be an observation of a signal defined on* $[0, N]$. *Let*

$$\mathcal{B} := \{g_m(t) := N^{-1/2} e^{-\imath 2\pi mt/N} \mid m \in \{0\} \cup \mathbb{N}\}. \tag{3.77}$$

Then the pseudo-inverse k^- is given by

$$k^-(t) = \sum_{m=0}^{\infty} \lambda_m^- g_k(t), \tag{3.78}$$

where

$$\lambda_m^- := \begin{cases} \langle k * g_m, g_m \rangle^{-1}, & \langle k * g_m, g_m \rangle \neq 0 \\ 0, & \langle k * g_m, g_m \rangle = 0. \end{cases} \tag{3.79}$$

Proof. Let $V = \{k * f \mid f \in L_2[0, N]\}$ be the image of the convolution operator k and $V^\perp$ the orthogonal complement of V in $L_2[0, N]$. The set $\mathcal{B}$ is a complete orthonormal base of $L_2[0, N]$. For a $y \in L_2[0, N]$ the pseudo-inverse k^- is defined by

$$\|k^- * y\|^2 = \min\{\|f\|^2 \mid k * f = y\}.$$

For any solution f of $k * f = y$

$$\langle y, g_m\rangle = \langle k * f, g_m\rangle = \langle k * g_m, f\rangle = \langle k * g_m, g_m\rangle \cdot \langle g_m, f\rangle,$$

$m \in \{0\} \cup \mathbb{N}$ holds. Hence any solution of the equation $k * f = y$ is generated by

$$f = \sum_{m=0}^{\infty} \lambda_m \langle y, g_m\rangle g_m,$$

with

$$\lambda_m := \begin{cases} \langle k * g_m, g_m\rangle^{-1}, & \langle k * g_m, g_m\rangle \neq 0 \\ \text{arbitrary value}, & \langle k * g_m, g_m\rangle = 0. \end{cases}$$

Due to the orthogonality of the base functions g_m the relation

$$\|f\|^2 = \sum_{m=0}^{\infty} (\lambda_m)^2 \langle y, g_m\rangle^2$$

holds. This leads to

$$\min \|f\|^2 = \sum_{m=0}^{\infty} (\lambda_m^-)^2 \langle y, g_m\rangle^2.$$

Hence

$$\begin{aligned} f &= \sum_{m=0}^{\infty} \lambda_m^- \langle y, g_m\rangle g_m \\ &= \sum_{m=0}^{\infty} \sum_{l=0}^{\infty} \lambda_m^- \langle y, g_m\rangle \delta_{ml} g_l \\ &= \sum_{m=0}^{\infty} \sum_{l=0}^{\infty} \lambda_m^- \langle y, g_m\rangle (g_m * g_l) \\ &= \Big(\sum_{m=0}^{\infty} \langle y, g_m\rangle g_m\Big) * \Big(\sum_{l=0}^{\infty} \lambda_l^- g_l\Big) \\ &= k^- * y \end{aligned}$$

holds. □

The estimation $\hat{f}$ is obtained by the application of the pseudo-inverse k^- to the data y

$$\hat{f} = k^- * y = (k^- * k) * f + k^- * n. \tag{3.80}$$

The pseudo-inverse does not provide a complete estimation of f. It only estimates the projection $F = P_V f$ of the signal f onto the image V of k. Since the pseudo-inverse k^- is linear, the transformed noise $Z := k^- * n$ is Gaussian again with the covariance operator

$$K = \sigma^2 k^- * k^-. \tag{3.81}$$

Because the exponential functions g_k are eigenfunctions of every convolution operator, the base (3.77) diagonalizes both the covariance operator of $F := (k^- * k) * f$ and Z. Hence it forms a Karhunen–Loève base for the estimation problem (3.80).

Lemma 3.2.2. *Let C_{ff} be the covariance operator of f and C_{nn} be the covariance operator of n. The quantities β_m, σ_m are defined by*

$$\beta_m^2 = E\{\langle f, g_m\rangle^2\} = \langle C_{ff} g_m, g_m\rangle \tag{3.82}$$

$$\sigma_m^2 = E\{\langle n, g_m\rangle^2\} = \langle C_{nn} g_m, g_m\rangle = \sigma^2 \tag{3.83}$$

Then the Wiener estimator is given by

$$\hat{F} = \sum_{m\in I} \frac{\beta_m^2}{\beta_m^2 + (\lambda_m^-)^2\sigma^2} \langle \hat{f}, g_m\rangle g_m, \tag{3.84}$$

where $I := \{m \in \{0\} \cup \mathbb{N} \mid \lambda_m^- \neq 0\}$.

Proof. With the notations

$$F := (k^- * k) * f, \quad Z := k^- * n,$$

the deconvolution problem (3.80) can be reformulated as denoising problem

$$\hat{f} = F + Z.$$

The covariance operators of X and Z are given by

$$C_{FF} = k^- * k^- * k * k * C_{ff}, \quad C_{ZZ} = k^- * k^- * C_{nn} = \sigma^2 k^- * k^-.$$

Hence

$$E\{|\langle F, g_m\rangle|^2\} = (\lambda_m^-)^2 \langle k * g_m, g_m\rangle^2 \beta_m^2$$

and

$$E\{|\langle Z, g_m\rangle|^2\} = (\lambda_m^-)^2\sigma^2$$

holds. Since $\{g_k\}$ is a Karhunen–Loève base the Wiener estimator is given by

$$\begin{aligned}
\hat{F} &= \sum_{m=0}^{\infty} \frac{E\{\langle F, g_m\rangle^2\}}{E\{\langle F, g_m\rangle^2\} + E\{\langle Z, g_m\rangle^2\}} \langle \hat{f}, g_m\rangle g_m \\
&= \sum_{m=0}^{\infty} \frac{(\lambda_m^-)^2 \langle k * g_m, g_m\rangle^2 \beta_m^2}{(\lambda_m^-)^2 \langle k * g_m, g_m\rangle^2 \beta_m^2 + (\lambda_m^-)^2\sigma^2} \langle \hat{f}, g_m\rangle g_m. \\
&= \sum_{m\in I} \frac{\beta_m^2}{\beta_m^2 + (\lambda_m^-)^2\sigma^2} \langle \hat{f}, g_m\rangle g_m.
\end{aligned}$$

□

Many applications in geo-sciences involve convolution kernels which are compact. This means that the image space coincides with the whole space

$$V = L_2[0, N] \tag{3.85}$$

but the eigenvalues tend to zero

$$\lim_{m\to\infty} \lambda_m = 0. \tag{3.86}$$

Consequently k is a smoothing kernel. A simple inversion would lead to

$$\hat{f} = \sum_{m=0}^{\infty} \frac{1}{\lambda_m} \langle y, g_m \rangle g_m, \tag{3.87}$$

which clearly is an unbounded and therefore instable deconvolution operator. If for compact kernel the estimation formula (3.84) is used, the simple relation

$$\hat{F} = \sum_{m=0}^{\infty} \frac{\beta_m^2}{\beta_m^2 + \lambda_m^{-2}\sigma^2} \langle \hat{f}, g_m \rangle g_m. \tag{3.88}$$

is obtained. Here the amplification $1/\lambda_m$ is counterbalanced by a regularization parameter σ^2 resulting in a bounded and therefore stable deconvolution operator.

As in the denoising case the Wiener deconvolution has the disadvantage that signal components with a bad signal to-noise ratio will remain in the solution. This can be avoided by the application of a thresholding algorithm.

$$\hat{F} = \sum_{k=0}^{\infty} \frac{\beta_k^2}{\beta_k^2 + \lambda_k^{-2}\sigma^2} \rho_T(\langle \hat{f}, g_k \rangle) g_k. \tag{3.89}$$

Formula (3.89) describes the threshold deconvolution in a exponential base. This exponential base has the advantage that it diagonalizes the convolution operator. On the other hand this base is optimally suited for stationary signals. Transient signals are better approximated with respect to a wavelet base. The price which has to be paid for working in a wavelet base it that the convolution operator is not longer diagonal with respect to this base. Fortunately, only a few of the off-diagonal elements are really large and a suitable threshold will set them to zero. In this way the advantages of representing also transient signal components and leading to a diagonal estimator can be combined.

In a wavelet base $\mathcal{B} := \{\psi_{jm} \mid j = 1, \ldots, J, m = 0, \ldots, N/2^j\}$ the observation equation (3.76) can be written as

$$\langle y, \psi_{pq} \rangle = \sum_{j=1}^{J} \sum_{m=0}^{N/2^{-j}} \langle k * \psi_{jm}, \psi_{pq} \rangle f_{jm} + \langle n, \psi_{pq} \rangle. \tag{3.90}$$

In many cases the matrix $(\langle k * \psi_{jm}, \psi_{pq}\rangle)$ is sparse and strongly diagonal dominant. Therefore it can be replaced by a thresholded matrix

$$\langle y, \psi_{pq}\rangle = \sum_{j=1}^{J} \sum_{m=0}^{N/2^{-j}} \rho_T(\langle k * \psi_{jm}, \psi_{pq}\rangle) f_{jm} + \langle n, \psi_{pq}\rangle. \tag{3.91}$$

For a suitable choice of the matrix and the threshold T the thresholded matrix is diagonal

$$\rho_T(\langle k * \psi_{jm}, \psi_{pq}\rangle) = \operatorname{diag}(\langle k * \psi_{jm}, \psi_{jm}\rangle). \tag{3.92}$$

This leads to the following threshold deconvolution:

$$\hat{f} = \sum_{j=1}^{J} \sum_{m=0}^{N/2^{-j}} \frac{\langle y, \psi_{jm}\rangle}{\langle k * \psi_{jm}, \psi_{jm}\rangle} \psi_{jm}. \tag{3.93}$$

Example 3.2.1. Let us consider the convolution kernel

$$k(t) = \frac{1}{2n+1} \sum_{l=-n}^{n} \delta(t-l), \tag{3.94}$$

which is the formula expression for a smoothing average operator of length $2n+1$. For every convolution kernel the functions g_k are eigenfunctions. The corresponding eigenvalues λ_k can be derived from following equation:

$$\begin{aligned} k * g_k &= \frac{1}{2n+1} \sum_{l=-n}^{n} e^{-\imath 2\pi kl/N} g_k \\ &= \frac{1}{2n+1} \Big(1 + 2 \sum_{l=1}^{n} \cos(2\pi kl/N)\Big) g_k. \end{aligned}$$

Obviously the eigenvalues are given by

$$\lambda_k = \frac{1}{2n+1} \Big(1 + 2 \sum_{l=1}^{n} \cos(2\pi kl/N)\Big). \tag{3.95}$$

For $n = 4$ the eigenvalue spectrum $\{\lambda_l\}$ of k is plotted in Figure 3.35.

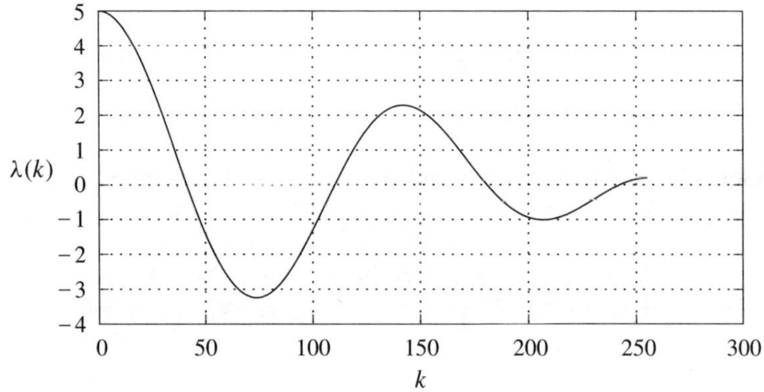

Figure 3.35. Spectrum of the convolution operator

As input data a sawtooth signal as it is shown in Figure 3.36 is used.

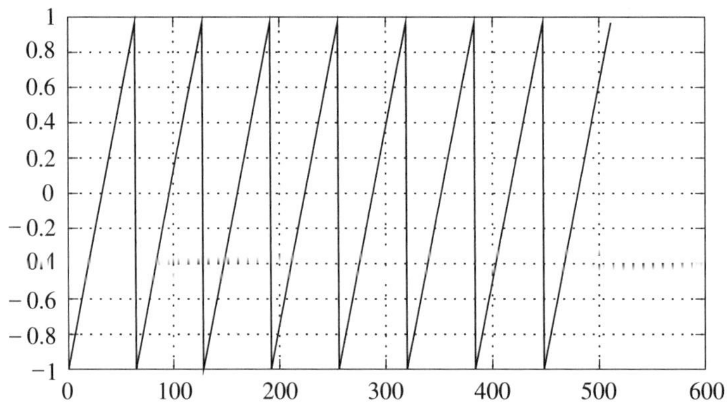

Figure 3.36. Signal f to be recovered

After convolving with the kernel (3.94) and adding noise, the resulting observations are displayed in Figure 3.37.

The wavelet spectrum of the observations $\langle y, \psi_{jm}\rangle$ can be computed by the application of Mallat's algorithm. This spectrum is shown in Figure 3.38

The matrix $K := (\langle k * \psi_{jm}, \psi_{jm}\rangle)$ is sparse but not exactly diagonal, as can be seen in Figure 3.39.

Instead of approximating the matrix K by its diagonal $K \approx \operatorname{diag}(\langle k * \psi_{jm}, \psi_{jm}\rangle)$ a thresholding policy is applied both to the matrix and to the observation spectrum

$$\tilde{K} = \rho_T(\langle k * \psi_{jm}, \psi_{jm}\rangle), \quad \tilde{b}_{jm} = \rho_T(\langle y, \psi_{jm}\rangle). \tag{3.96}$$

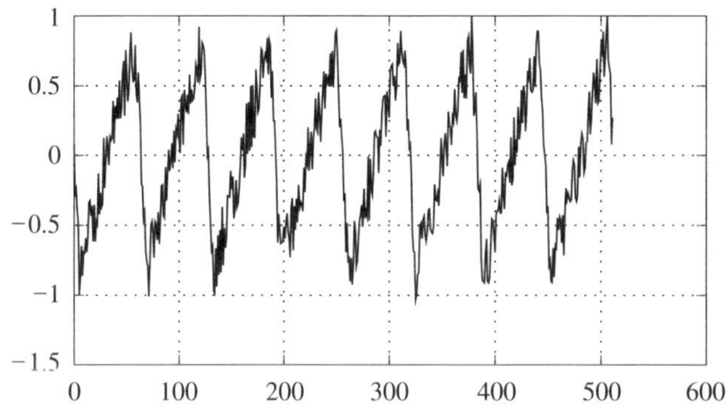

Figure 3.37. Observations

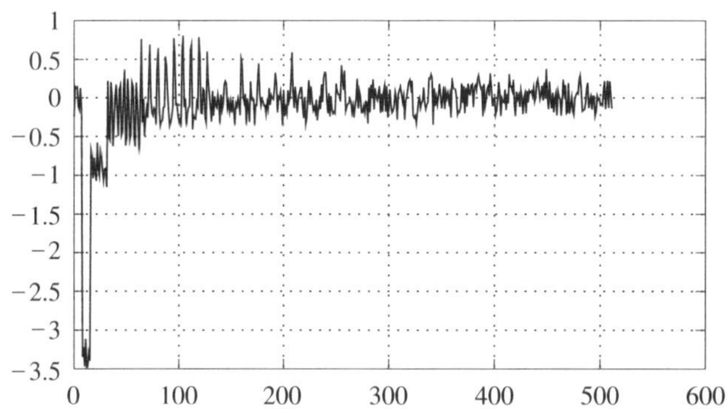

Figure 3.38. Wavelet spectrum of observations

The solution of this thresholded system is considered the wavelet spectrum of the signal to be recovered

$$(\langle f, \psi_{jm} \rangle) := \tilde{K}^{-1}(\tilde{b}_{jm}). \tag{3.97}$$

The wavelet reconstruction of the estimated signal spectrum $(\langle f, \psi_{jm} \rangle)$ yields an estimation for the unknown signal f. This estimation is displayed in Figure 3.40.

Clearly, the noise level has strongly reduced without mollifying the sharp peaks. On the other hand the thresholding of matrix elements leads to some undesired artefacts in the estimation of the signal. Here, with a proper choice of the thresholds a balance between denoising, deconvolution and the generation of artefacts has to be found.

More examples for the application of wavelet deconvolution techniques in Physical Geodesy are listed in [39].

As it was already demonstrated in the previous example, the thresholding in the denoising approach can lead to artefacts in the solution. In order to avoid this one

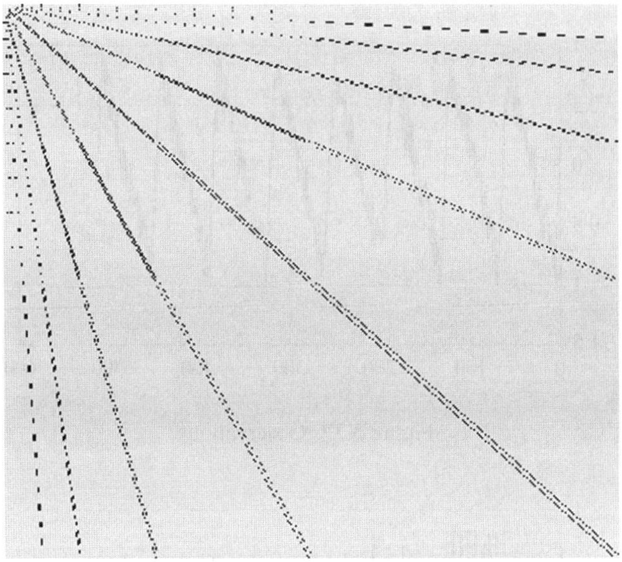

Figure 3.39. System matrix of wavelet deconvolution

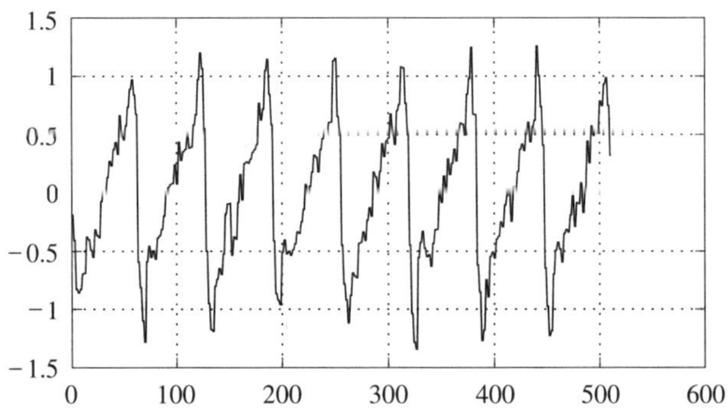

Figure 3.40. Result of thresholded wavelet deconvolution

could represent the observation and the signal to be estimated with respect to different tight frames. These frames can be chosen in such a way, that the deconvolution matrix gets diagonal. This second tight frame can be derived from the first one as the solution of the following equation:

$$k * w_{nm} = \kappa_{nm} \psi_{nm}, \quad \|w_{nm}\| = 1. \tag{3.98}$$

If the first frame ψ_{nm} is a wavelet frame, the resulting frame w_{nm} is called a vaguelette frame. With respect to this wavelet-vaguelette pair the deconvolution equations get diagonal,

$$\langle f, \psi_{nm} \rangle = \kappa_{nm} \langle y, w_{nm} \rangle. \tag{3.99}$$

As for the wavelet also for the vaguelette Mallat's algorithm can be used for decomposition and reconstruction. Hence, if once the wavelet-vaguelette pair of a convolution kernel k has been developed, the wavelet-vaguelette decomposition is a highly efficient algorithm for deconvolution problems.

Results about the application of this wavelet-vaguelette decomposition for the solution of geodetic integral equations are given in [41], [51] and [52].

3.2.3 Image compression

One of the most successful applications of discrete wavelet transformation is image compression. The so-called pyramidal algorithm [10], which is widely used in image processing can be considered a predecessor of the multi-resolution analysis in wavelet theory. The multi-resolution analysis and Mallat's algorithm for fast discrete wavelet transform are the cornerstones for the application of wavelets in image processing.

A digital grey-coded image can be represented by a $n \times n$ matrix $F = (f_{ij})$ with integer entries f_{ij}. The value f_{ij} is the grey value of the pixel in the i-th row and the j-th column. In general

$$f_{ij} \in \{0, 1, 2, \ldots, 255\} \tag{3.100}$$

holds. This means that the storage requirement of such an image is $S = n^2$ byte. The compression aims at an alternative representation of the image requiring only $S_{\text{comp}} < S$ bytes. The ratio

$$k := \frac{S}{S_{\text{comp}}} \tag{3.101}$$

is called compression rate. Using the scaling function φ of the Haar wavelet a grey-coded image has the representation

$$f(x, y) = \sum_{i,j=0}^{n-1} f_{ij} \varphi(x - i) \cdot \varphi(y - j). \tag{3.102}$$

This means the image is represented with respect to the tensor Haar scaling function base. The idea of image compression is the representation of the image with respect to an alternative base and to choose this base in such a way that in the new representation as many coefficients as possible are so small that they can be neglected without influencing the visual quality of the image.

Hence a typical algorithm for image compression by wavelets consists of three steps:

- Computation of wavelet spectrum.

- Thresholding.

- Compact coding of the remaining non-zero wavelet coefficients.

The coding itself consist again of two steps

- the quantization and
- the coding of the quantized coefficient values.

In the original image representation all coefficients can have only values out of a finite set (in most cases only the integers between 0 and 255). Due to the wavelet transform the image representation with respect to a wavelet base has real coefficients. Quantization means to map these real coefficients on a finite set. The simplest form of quantization is the uniform quantizer

$$Q(x) = \begin{cases} 0, & |x| < \Delta/2 \\ \operatorname{sgn}(x)k\Delta, & (k-1/2)\Delta \leq |x| \leq (k+1/2)\Delta, \end{cases} \tag{3.103}$$

i.e., all values in an interval of the length Δ are quantized by the mean value of the interval. In general the quantized value is an integer.

The last step, the coding, is an intelligent method to store these integers in such a way that frequently occurring numbers are coded by a low number of bits. An efficient coding algorithm is the so-called Huffmann coding:

Example 3.2.2. Suppose that the following values are obtained from quantization:

$$9, 123, 17, 63, 129, 17, 123, 123, 52.$$

In this sequence the number 123 occurs three times, 17 two times and the integers 9, 129, 52 one time each. Therefore, 123 gets the shortest binary code, 17 the next-shortest and so on. This leads to the following coding table

123	0
17	10
63	1100
9	1101
129	1110
52	1111

Consequently, the coded sequence is

$$11010101100111010001111.$$

Obviously, the original sequence needs 1 byte per entry accounting for 72 bits for the whole sequence. The Huffmann coded sequence needs only 23 bits.

The choice of a proper base, which leads to a large number of small coefficients and an efficient coding leads to a high compression rate. Currently, the JPEG compression standard [104] is widely used.

Fourier compression (JPEG). The JPEG image compressing uses block-wise cosine function as base functions. The image is decomposed in blocks of $L \times L$ pixels. In each block the image is represented with respect to the orthogonal base

$$\varphi_{k,j}(n,m) = \frac{1}{\gamma_k \gamma_j} \frac{2}{L} \cos\left(\frac{k\pi}{L}(n+\frac{1}{2})\right) \cos\left(\frac{j\pi}{L}(m+\frac{1}{2})\right) \tag{3.104}$$

with

$$\gamma_i = \begin{cases} 1/\sqrt{2}, & i = 0 \\ 1, & \text{otherwise.} \end{cases} \tag{3.105}$$

In the JPEG standard the block-size L is set to $L = 8$. Pictures of N^2 pixels are divided in $N^2/64$ blocks of 8×8 pixels each. The grey-values $f_{8I+i,8J+j}$ in the block (I, J) have the following representation with respect to the base (3.104):

$$f_{8I+i,8J+j} = \frac{1}{16} \sum_{n=0}^{8} \sum_{m=0}^{8} c_{n,m}^{(I,J)} \varphi_{n,m}(i,j) \tag{3.106}$$

with

$$c_{n,m}^{(I,J)} = \gamma_n \gamma_m \sum_{i=0}^{8} \sum_{j=0}^{8} f_{8I+i,8J+j} \varphi_{n,m}(i,j). \tag{3.107}$$

Usually, the block-cosine spectrum $c_{n,m}^{(I,J)}$ contains a comparatively small number of coefficients with a large modulus and a large number of very small coefficients. The biggest part of the information about the image is carried by the small number of large coefficients. The majority of the coefficients is small and contributes only marginally to the information about the image. Therefore, applying a threshold strategy, the number of coefficients which are necessary to reproduce the main information about the image can be dramatically reduced.

Example 3.2.3. In order to demonstrate the JPEG compression principle, the JPEG algorithm will be applied to the famous *Lena* image shown in Figure 3.41. The histogram of the block-cosine coefficients is shown in Figure 3.42. It displays for every class the ratio of the number of coefficients belonging to this class and the total number of coefficients. Though the modulus of the coefficients can reach the value of 7000 about 90 % of the coefficients are smaller than 200. Therefore a threshold strategy can be applied: All coefficients which are smaller than 1 % of the value of the largest coefficients are neglected. This lead to a compression ratio of

$$k = 64.1. \tag{3.108}$$

It has to be mentioned, that no quantization and no coding was applied. An additional quantization and coding would still increase the compression ratio considerably.

The image, which was reconstructed from the remaining coefficients is shown in Figure 3.43. It is clearly visible that the JPEG compression is not loss-free. The recon-

Figure 3.41. Test image *Lena*

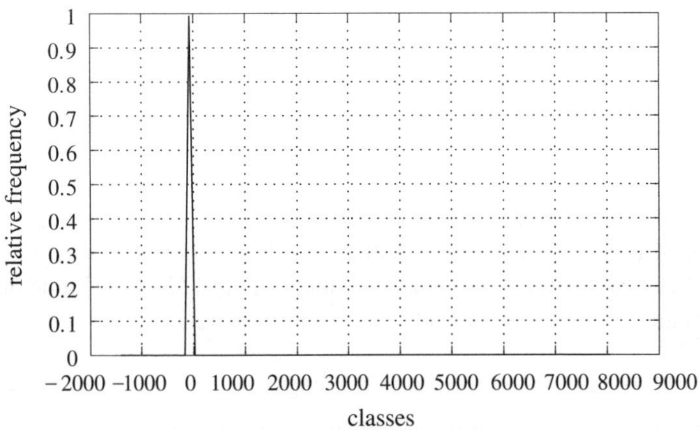

Figure 3.42. Histogram of block-cosine coefficients of Lena image

Figure 3.43. Reconstructed image after thresholding the block-cosine spectrum

structed image has small artefacts. Especially at Lena's shoulder the block structure of the JPEG compression algorithm is visible as a mosaic of small squares.

Wavelet compression. A larger compression rate as for the JPEG algorithm can be obtained if the image base is chosen in such a way that the histogram of the coefficients with respect to this base is even stronger concentrated around zero.

For this purpose orthogonal wavelets with a compact support can be chosen as a base. The localization property of wavelets promises a stronger concentration and the use of Mallat's algorithm provides a very efficient tool for coding and decoding the image.

Example 3.2.4. As a test image again the Lena image was used. The Daubechies wavelet of order 6 was chosen as wavelet base. The histogram of the wavelet coefficients of the test image is displayed in Figure 3.44.

If the wavelet histogram in Figure 3.44 is compared to the block-cosine histogram in Figure 3.42 two effects can be noticed:

1. The magnitude of the largest coefficients has increased from 7 000 to 20 000.

2. The concentration of the small coefficients has increased from 90% to 98%.

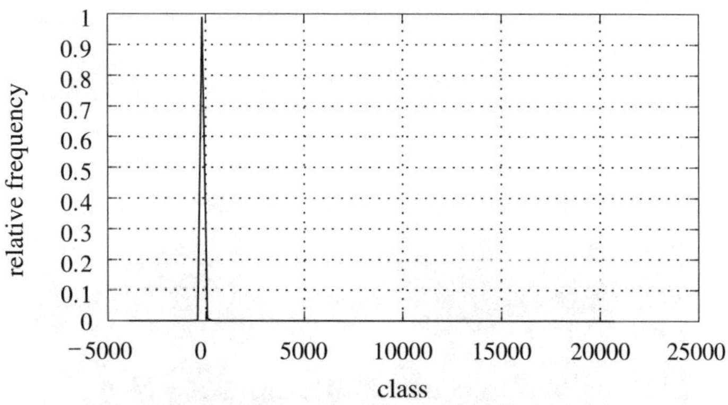

Figure 3.44. Histogram of Daubechies 6 wavelet spectrum of Lena image

Hence a higher compression rate by thresholding should be possible. But also the quality of the reconstructed image can be improved by keeping the same compression rate as for the JPEG compression.

Applying the threshold policy to neglect all coefficients which are smaller than 1% of the maximal coefficients leads to a compression rate of

$$k = 102.2, \tag{3.109}$$

which is larger than the compression ratio of the JPEG algorithm. The reconstructed image is displayed in Figure 3.45.

A visual comparison of both images, given in Figure 3.46 shows that the compression loss of the wavelet compression is smaller, even if the compression rate is higher. No artefacts are visible. Especially, there is no block pattern in the reconstructed image.

The reason for the better performance of the wavelet compression is that sharp gray-value contrast in the image can be better detected by wavelets than by a cosine base.

The difference between JPEG and wavelet compression can be shown more clearly if test images with simple geometrical figures are used. Two images showing a square and a circle in the center are used as test images. The images are shown in Figure 3.47.

Both images are compressed using JPEG and wavelet compression. The reconstructions of the compressed square images are displayed in Figure 3.48.

Figure 3.45. Reconstructed image after thresholding the wavelet spectrum

Figure 3.46. Comparison of JPEG and wavelet compression

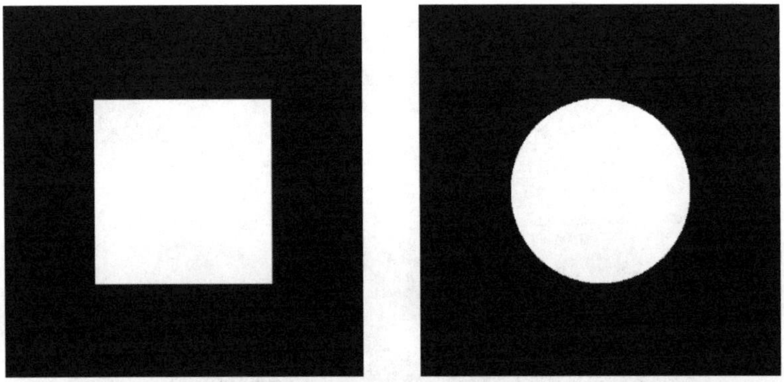

Figure 3.47. Test images

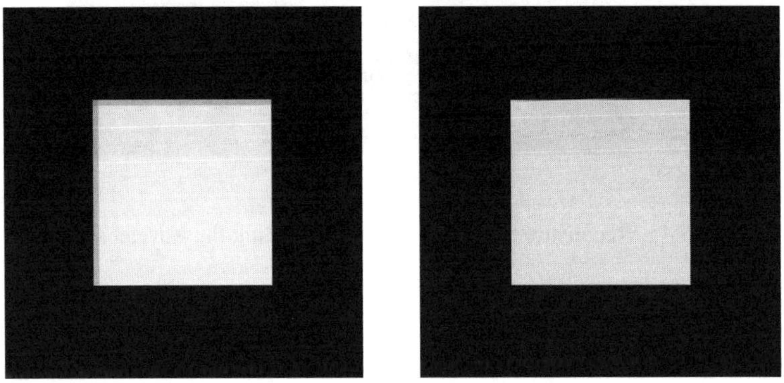

Figure 3.48. JPEG (left) and wavelet (right) compressed square

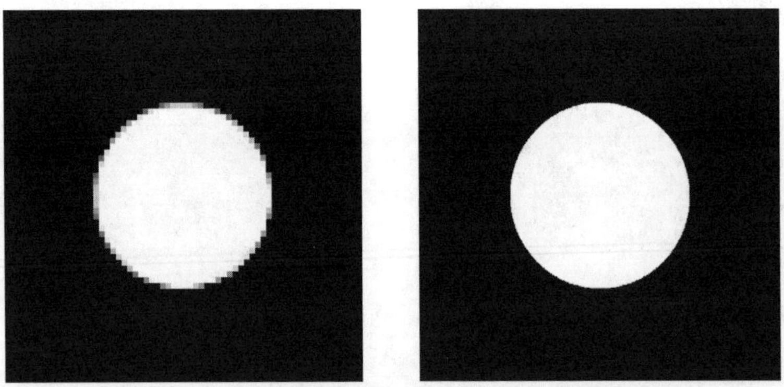

Figure 3.49. JPEG (left) and wavelet (right) compressed circle

The compression rates of both images are quite different: $k = 233.1$ for JPEG and $k = 4166.6$ for the wavelet compression. While, the wavelet compressed square is almost perfectly reconstructed, the JPEG compressed square shows clearly visible artefacts on the upper and on the left side. This tendency becomes even more apparent, if the circle image is considered. The comparison of both compression results is given in Figure 3.49.

3.3 Sub-band coding, filtering and prediction

3.3.1 QMF filter design and wavelets

As it has already been mentioned in Section 2.2, a filter bank splits a signal x into two parts: a low frequency part Hx and a high frequency part Gx. A similar situation occurs in the case of a multi-resolution analysis. There a function $f \in V_0$ is split in a smoothed part $Pf \in V_1$ and the difference part $Qf \in W_1$. Now the question arises, under which conditions a filter design is equivalent to a wavelet analysis.

In order to point out the similarities between wavelet multi-resolution analysis and digital filtering, some concepts of digital signal processing theory should be recalled here: Four digital filters H, G, H^*, G^* defined according to (1.48) with their z-transforms according to (1.49) are called a *PR filter bank*, if

$$H(-z)H^*(z) + G(-z)G^*(z) = 0 \tag{3.110}$$

$$H(z)H^*(z) + G(z)G^*(z) = 2 \tag{3.111}$$

holds. The filter bank is called *orthogonal* if additionally

$$H(z) = H^*(z^{-1}) \tag{3.112}$$

is valid. The filter bank is called a *quadrature mirror filter* (QMF), if

$$|H(z)|^2 + |G(z)|^2 = 1 \tag{3.113}$$

holds. Additionally, a filter H is called a *conjugate mirror filter*, if

$$|H(z)|^2 + |H(-z)|^2 = 2 \tag{3.114}$$

is true.

The decomposition operators H, G and the reconstruction operators H^*, G^* of Mallat's algorithm can be related to the combination of linear convolution filters $\bar{H}$, $\bar{G}$, $\bar{G}^*$, $\bar{H}^*$ combined with dyadic down- and up-sampling. Define the convolution filters $\bar{H}, \bar{G}, \bar{G}^*, \bar{H}^*$ by

$$(\bar{H}x)_n := \sum_{k\in\mathbb{Z}} \bar{h}_k x_{n-k}, \quad \bar{h}_k = h_{-k}, \tag{3.115}$$

$$(\bar{G}x)_n := \sum_{k\in\mathbb{Z}} \bar{g}_k x_{n-k}, \quad \bar{g}_k = g_{-k}, \tag{3.116}$$

$$(\bar{H}^*x)_n := \sum_{k\in\mathbb{Z}} h_k x_{n-k}, \tag{3.117}$$

$$(\bar{G}^*x)_n := \sum_{k\in\mathbb{Z}} g_k x_{n-k}. \tag{3.118}$$

Then the decomposition steps of Mallat's algorithm can be identified with the filters $\bar{H}, \bar{G}$ followed by a down-sampling by the factor 2:

$$(Hc)_n = \sum_{k\in\mathbb{Z}} h_{k-2n} c_k = \sum_{k\in\mathbb{Z}} \bar{h}_{2n-k} c_k = (\bar{H}c)_{2n}, \tag{3.119}$$

$$(Gc)_n = \sum_{k\in\mathbb{Z}} g_{k-2n} c_k = \sum_{k\in\mathbb{Z}} \bar{g}_{2n-k} c_k = (\bar{G}c)_{2n}, \tag{3.120}$$

$$(H^*c)_n = \sum_{k\in\mathbb{Z}} h_{n-2k} c_k = \sum_{k\in\mathbb{Z}} h_{n-k} \check{c}_k = (\bar{H}^*\check{c})_n, \tag{3.121}$$

$$(G^*c)_n = \sum_{k\in\mathbb{Z}} g_{n-2k} c_k = \sum_{k\in\mathbb{Z}} g_{n-k} \check{c}_k = (\bar{G}^*\check{c})_n, \tag{3.122}$$

where

$$\check{c}_n = \begin{cases} c_{n/2}, & n \text{ even} \\ 0, & n \text{ odd} \end{cases} \tag{3.123}$$

is the sequence which is obtained by up-sampling the sequence c_k by the factor 2.

The consequence of this relationship is that all what can be said about the convolution filters applies also to Mallat's decomposition and reconstruction filters.

A multi-resolution analysis of $L_2(\mathbb{R})$ can be generated by the choice of a scaling function φ. The coefficients h_k of its scaling equation (2.67) define a digital convolution filter by

$$(Hx)_k := \sum_n h_n x_{k-n}. \tag{3.124}$$

Lemma 3.3.1. *The digital convolution filter* (3.124), *defined by the scaling coefficients of the scaling function* φ *of an MRA of* $L_2(\mathbb{R})$, *is a conjugate mirror filter, if the shifted versions of the scaling functions are orthogonal to each other, i.e. if*

$$\langle \varphi(\bullet), \varphi(\bullet - k)\rangle = \delta_{0k} \tag{3.125}$$

holds.

Proof. For the transfer function (2.71) of the scaling equation holds

$$|h(\omega)|^2 + |h(\omega+\pi)|^2 = 1.$$

The z-transform H of the digital filter defined by the scaling coefficients h_k is related to the transfer function h by

$$h(\omega) = \frac{\sqrt{2}}{2}\sum_{k\in\mathbb{Z}} h_k e^{-\imath k\omega} = \frac{\sqrt{2}}{2}\sum_{k\in\mathbb{Z}} h_k (e^{\imath\omega})^{-k} = \frac{\sqrt{2}}{2} H(e^{\imath\omega}).$$

Hence

$$\begin{aligned} 2 &= 2(|h(\omega)|^2 + |h(\omega+\pi)|^2) = |H(e^{\imath\omega})|^2 + ||H(-e^{\imath\omega})|^2 \\ &= |H(z)|^2 + |H(-z)|^2, \qquad z = e^{\imath\omega}. \end{aligned}$$

This proves that the convolution filter is a conjugate mirror filter. □

Remark. If for a signal $f \in V_0$ with the representation

$$f(x) = \sum_{k\in\mathbb{Z}} c_k^{(0)} \varphi(x-k) \tag{3.126}$$

the validity of $f(k) = c_k^{(0)}$ can be imputed, then the smoothing step

$$c_k^{(1)} = \sum_{n\in\mathbb{Z}} h_n c_{n+k}^{(0)} \tag{3.127}$$

of Mallat's algorithm is considered a low-pass filter. In contrast to the standard case of digital signal processing it is not a convolution but a correlation filter. Defining related filter coefficients by $\bar{h}_k := h_{-k}$ then Mallat's smoothing step can also be expressed as a convolution filter

$$c_k^{(1)} = \sum_{n\in\mathbb{Z}} \bar{h}_{-n} c_{n+k}^{(0)} = \sum_{n\in\mathbb{Z}} \bar{h}_n c_{k-n}^{(0)}. \tag{3.128}$$

For its z-transform follows

$$\bar{H}(z) = \sum_{k\in\mathbb{Z}} \bar{h}_k z^{-k} = \sum_{k\in\mathbb{Z}} h_k \left(\frac{1}{z}\right)^{-k} = H\left(\frac{1}{z}\right).$$

Hence, with $z = 1/\zeta$,

$$2 = |H(\zeta)|^2 + |H(-\zeta)|^2 = |\bar{H}\left(\frac{1}{\zeta}\right)|^2 + |\bar{H}\left(-\frac{1}{\zeta}\right)|^2 = |\bar{H}(z)|^2 + |\bar{H}(-z)|^2 \tag{3.129}$$

holds, and Mallat's smoothing step is also a conjugate mirror filter.

Another important class of digital filters are the so-called *quadrature mirror filters* (QMF). They are characterized by the property that their filter characteristics ad up to 1. Unfortunately the smoothing and the differencing filter of Mallat's algorithm do not form a QMF filter in general. Only in the simplest possible case of the Haar wavelet this property holds.

Example 3.3.1. The coefficients of the scaling equations for the Haar wavelet are

$$h_0 = h_1 = \frac{1}{\sqrt{2}}, \qquad g_0 = -g_{-1} = -\frac{1}{\sqrt{2}}.$$

With the introduction of new filter coefficients

$$\bar{h}_k = h_{-k}, \quad k = -1, 0, \qquad \bar{g}_k = g_{-k}, \quad k = 0, 1$$

the smoothing and the differencing step of Mallat's algorithm can be expressed as convolution filters followed by a down-sampling by the factor 2:

$$(Hc)_n = \sum_k h_k c_{2n+k} = \sum_k \bar{h}_k c_{2n-k} = (\bar{h} * c)(2n) = D(\bar{h} * c).$$

The same holds for G where D means the down-sampling operator

$$(Dx)_n := x_{2n}.$$

The z-transforms of the convolution filters $\bar{h}, \bar{g}$ are

$$\begin{aligned} \bar{h}(z) &= \sum_k \bar{h}_k z^{-k} = h_0 + h_1 z \\ \bar{g}(z) &= \sum_k \bar{g}_k z^{-k} = g_0 + g_{-1} z^{-1}. \end{aligned}$$

If a sequence $\{x_n\}$ has the z-transform $X(z)$, then the down-sampled sequence $D\{x_n\}$ has the z-transform

$$\begin{aligned} DX(z) &= \sum_k (Dx)_k z^{-k} \\ &= \sum_k x_{2k} z^{-k} \\ &= \frac{1}{2}\Big(\sum_k x_k z^{-k/2} + \sum_k x_k (-1)^k z^{-k/2}\Big) \\ &= \frac{1}{2}\left(X(\sqrt{z}) + X(-\sqrt{z})\right). \end{aligned}$$

Hence Mallat's smoothing and differencing filters have the z-transforms

$$H(z) = \frac{1}{2}\left(h_0 + h_1\sqrt{z} + h_0 - h_1\sqrt{z}\right) = h_0$$

$$G(z) = \frac{1}{2}\left(g_{-1}\frac{1}{\sqrt{z}} + g_0 - g_{-1}\frac{1}{\sqrt{z}}\right) = g_0.$$

Applying the properties

$$g_0 = h_1, \quad h_0^2 + h_1^2 = 1,$$

the relation

$$|H(z)|^2 + |G(z)|^2 = h_0^2 + g_0^2 = h_0^2 + h_1^2 = 1$$

follows and consequently the Haar wavelet establishes a QMF filter.

While the smoothing filter H and the differencing filter G of Mallat's algorithm constitute a QMF filter only in the trivial case of the *Haar* wavelet, any orthogonal wavelet generates a PR filter bank.

Theorem 3.3.1. *The filters $\bar{H}$, $\bar{G}$, $\bar{H}^*$, $\bar{G}^*$* (3.115)–(3.118) *generated by an orthogonal wavelet form a PR filter bank.*

Proof. The z-transform of $\bar{H}$ and $\bar{G}$ are

$$\bar{H}(z) = \sum_{k\in\mathbb{Z}} \bar{h}_k z^{-k} = \sum_{k\in\mathbb{Z}} h_k z^k$$

$$\bar{G}(z) = \sum_{k\in\mathbb{Z}} \bar{g}_k z^{-k} = \sum_{k\in\mathbb{Z}} g_k z^k = \sum_{k\in\mathbb{Z}} (-1)^k h_{1-k} z^k = -z \sum_{k\in\mathbb{Z}} (-1)^k h_k z^{-k}$$
$$= -z\bar{H}\left(-\frac{1}{z}\right).$$

For the $\bar{H}^*$ and the $\bar{G}^*$ filters the z-transforms are

$$\bar{H}^*(z) = \sum_{k\in\mathbb{Z}} h_k z^{-k} = \bar{H}\left(\frac{1}{z}\right)$$

$$\bar{G}^*(z) = \sum_{k\in\mathbb{Z}} g_k z^{-k} = \sum_{k\in\mathbb{Z}} (-1)^k h_{1-k} z^{-k} = -\frac{1}{z} \sum_{k\in\mathbb{Z}} (-1)^k h_k z^k$$
$$= -\frac{1}{z}\bar{H}(-z).$$

As the next preliminary step the orthogonality of the scaling function

$$\sum_{k\in\mathbb{Z}} h_{2m+k} h_k = \delta_{0m}$$

is used to compute

$$\begin{aligned}\bar{H}(z)\bar{H}\left(\frac{1}{z}\right) &= \Big(\sum_{k\in\mathbb{Z}} h_k z^k\Big)\Big(\sum_{l\in\mathbb{Z}} h_l z^{-l}\Big)\\ &= \sum_{k\in\mathbb{Z}}\sum_{l\in\mathbb{Z}} h_k h_l z^{k-l}\\ &= \sum_{m\in\mathbb{Z}}\sum_{l\in\mathbb{Z}} h_{m+l} h_l z^m.\end{aligned}$$

This leads to

$$\begin{aligned}\bar{H}(-z)\bar{H}^*(z)+\bar{G}(-z)\bar{G}^*(z) &= \bar{H}(-z)\bar{H}\left(\frac{1}{z}\right)+(z\bar{H}\left(\frac{1}{z}\right)\left(-\frac{1}{z}\bar{H}(-z)\right)\\ &= 0\\ \bar{H}(z)\bar{H}^*(z)+\bar{G}(z)\bar{G}^*(z) &= \bar{H}(z)\bar{H}\left(\frac{1}{z}\right)+\left(-z\bar{H}\left(-\frac{1}{z}\right)\left(-\frac{1}{z}\bar{H}(-z)\right)\right)\\ &= \bar{H}(z)\bar{H}\left(\frac{1}{z}\right)+\bar{H}(-z)\bar{H}\left(-\frac{1}{z}\right)\\ &= \sum_{m\in\mathbb{Z}}\sum_{l\in\mathbb{Z}} h_{m+l}h_l z^m+\sum_{m\in\mathbb{Z}}\sum_{l\in\mathbb{Z}}(-1)^m h_{m+l}h_l z^m\\ &= \sum_{m\in\mathbb{Z}} 2\sum_{l\in\mathbb{Z}} h_{2m+l}h_l z^{2m}\\ &= 2\sum_{m\in\mathbb{Z}} \delta_{0m} z^{2m}\\ &= 2.\end{aligned}$$

Consequently, an orthogonal wavelet produces a perfect reconstruction filter bank. □

All in all, it can be summarized that Mallat's algorithm for an orthogonal wavelet is equivalent to a perfect reconstruction filter bank. In general, the reversal is not true. The coefficients of the low- and the high-pass filter of a perfect reconstruction filter bank in general cannot be identified with the scaling coefficients of an orthogonal scaling function and its corresponding wavelet. For a perfect reconstruction filter bank there are four filters H, G, H^*, G^*, which are connected to each other by two conditions (3.110), (3.111). Hence there are two degrees of freedom. In wavelet theory the four filters H, G, H^*, G^* are descendants of the single filter H with the filter coefficients being identical to the scaling coefficients of the underlying orthogonal scaling function φ. This leads to the property that the results of the of the smoothing and the differentiating step are orthogonal to each other. A similar result is not true for a general perfect reconstruction filter.

Concluding it can be said, that orthogonal wavelets induce perfect reconstruction filter banks with additional desirable properties.

3.3.2 Prediction of stationary signals with superimposed non-stationary noise

Collocation is widely used for the treatment of geodetic measurements. The collocation theory is based on the stationarity assumption. In practice this assumption is only piecewise fulfilled, since the variance of the data errors differs in different areas.

A first solution of the non-stationary collocation problem is given in [82]. In this paper the following extension for the well-known Wiener–Kolmogorov equation to the non-stationary case is found:

$$(C_{\xi\xi} + C_{nn})a = C_{\xi\eta} f, \quad \widehat{\langle f, \xi\rangle} = \langle a, \eta\rangle. \tag{3.130}$$

Here η is a Hilbert space valued random variable, representing the observed data with the covariance operator $C_{\eta\eta}$ and ξ is a Hilbert space valued random variable, representing the unknown solution with the cross-covariance operator $C_{\xi\eta}$. The data η contains white noise with a piecewise varying variance $w(t)$, i.e.

$$C_{nn} = w(t)\mathbf{I}, \tag{3.131}$$

with $\mathbf{I}$ being the identity operator. Since, in contrast to the classical Wiener–Kolmogorov equations w is not constant, the equations cannot be transformed into the frequency-domain and solved by FFT. Therefore a wavelet solution will be given here which exhibits the following features:

- The piecewise constancy of the noise-variance will be optimally reflected by the underlying Haar wavelets.
- In contrast to the FFT-case the matrix is no longer diagonal but at least sparse.

Wavelet solution of Wiener–Kolmogorov equations. The equation (3.130) is an operator equation, which in general can only be solved approximately. One standard technique for the numerical solution of operator equations is Galerkins's method. The basic idea of this method is to look for a solution of (3.130) not in the whole Hilbert space H but in a finite-dimensional subspace $H_n \subset H$. With $\{e_i\}$ being a base of H_n this leads to the so-called *Galerkin equations*

$$\sum_{j=1}^{n} \langle e_i, (C_{\xi\xi} + C_{nn})e_j\rangle \alpha_j = \langle C_{\xi\eta} f, e_i\rangle, \quad i = 1, \dots n, \tag{3.132}$$

or, in more detail,

$$\sum_{j=1}^{n} [\langle e_i, C_{\xi\xi} e_j\rangle + \langle e_i, w(t) e_j\rangle]\alpha_j = \langle C_{\xi\eta} f, e_i\rangle, \quad i = 1, \dots n. \tag{3.133}$$

The question remains how the base functions e_i can be properly chosen. One criterion can be the nature of w. The quantity w is the variance of the data-noise. The data-noise is supposed to have a piecewise constant variance. Hence w can be optimally represented by Haar-wavelets

$$w(t) = w_N \varphi_{N,0}(t) + \sum_{n=0}^{N} \sum_{m=0}^{2^{N-n}-1} w_{n,m} \psi_{n,m}(t), \tag{3.134}$$

with

$$\varphi_{N,0}(t) = 2^{-\frac{N}{2}} \varphi(2^{-N} t), \tag{3.135}$$

$$\psi_{n,m}(t) = 2^{-\frac{n}{2}} \psi(2^{-n} t - m) \tag{3.136}$$

and

$$\varphi(t) = \begin{cases} 1, & 0 \le t < 1 \\ 0, & \text{else} \end{cases}, \tag{3.137}$$

$$\psi(t) = \begin{cases} 1, & 0 \le t < \frac{1}{2} \\ -1, & \frac{1}{2} \le t < 1 \\ 0, & \text{else} \end{cases}. \tag{3.138}$$

This leads to the following form of the Galerkin equations

$$\sum_{j=1}^{n} \Big[\langle e_i, C_{\xi\xi} e_j \rangle + w_N \langle e_i, \varphi_{N,0} e_j \rangle + \sum_{n=0}^{N} \sum_{m=0}^{2^{N-n}-1} w_{n,m} \langle e_i, \psi_{n,m} e_j \rangle \Big] \alpha_j = \langle e_i, C_{\xi\eta} f \rangle. \tag{3.139}$$

The Galerkin equations constitute a system of linear equations for the unknown coefficients α_i of the approximate solution a_n. For an efficient solution of these equations it is necessary to find closed expressions for the matrix coefficients

$$a_{ij} = \langle e_i, C_{\xi\xi} e_j \rangle + w_N \langle e_i, \varphi_{N,0} e_j \rangle + \sum_{n=0}^{N} \sum_{m=0}^{2^{N-n}-1} w_{n,m} \langle e_i, \psi_{n,m} e_j \rangle \tag{3.140}$$

of those linear equations. For this purpose the base functions e_i should be as simple as possible. The simplest is the usage of Haar wavelets. This choice generates the following form of the Galerkin equations:

$$\begin{aligned} &\Big[\langle \varphi_{N,0}, C_{\xi\xi} \varphi_{N,0} \rangle + w_N \langle \varphi_{N,0}, \varphi_{N,0}^2 \rangle + \sum_{n=0}^{N} \sum_{m=0}^{2^{N-n}-1} w_{n,m} \langle \varphi_{N,0}, \psi_{n,m} \varphi_{N,0} \rangle \Big] \alpha_{N,0} \\ &+ \Big[\sum_{p=0}^{N} \sum_{q=0}^{2^{N-p}-1} \Big[\langle \varphi_{N,0}, C_{\xi\xi} \psi_{p,q} \rangle + w_N \langle \varphi_{N,0}, \varphi_{N,0} \psi_{p,q} \rangle \\ &+ \sum_{n=0}^{N} \sum_{m=0}^{2^{N-n}-1} w_{n,m} \langle \varphi_{N,0}, \psi_{n,m} \psi_{p,q} \rangle \Big] \Big] \alpha_{p,q} = \langle \varphi_{N,0}, C_{\xi\eta} f \rangle, \end{aligned} \tag{3.141}$$

$$\Big[\langle\psi_{r,s}, C_{\xi\xi}\varphi_{N,0}\rangle + w_N\langle\psi_{r,s}, \varphi_{N,0}^2\rangle + \sum_{n=0}^{N}\sum_{m=0}^{2^{N-n}-1} w_{n,m}\langle\psi_{r,s}, \psi_{n,m}\varphi_{N,0}\rangle\Big]\alpha_{N,0}$$
$$+\Big[\sum_{p=0}^{N}\sum_{q=0}^{2^{N-p}-1}\Big[\langle\psi_{r,s}, C_{\xi\xi}\psi_{p,q}\rangle + w_N\langle\psi_{r,s}, \varphi_{N,0}\psi_{p,q}\rangle \tag{3.142}$$
$$+\sum_{n=0}^{N}\sum_{m=0}^{2^{N-n}-1} w_{n,m}\langle\psi_{r,s}, \psi_{n,m}\psi_{p,q}\rangle\Big]\Big]\alpha_{p,q} = \langle\psi_{r,s}, C_{\xi\eta}f\rangle,$$
$$r = 0,\dots N-1, \quad s = 0\dots 2^{N-r}-1.$$

In matrix notation these equations have the following form

$$\begin{pmatrix} a_{N,0|N,0} & \cdots & a_{N,0|1,2^{N-1}-1} \\ a_{N-1,0|N,0} & \cdots & a_{N-1,0|1,2^{N-1}-1} \\ \vdots & \ddots & \vdots \\ a_{1,2^{N-1}-1|N,0} & \cdots & a_{1,2^{N-1}-1|1,2^{N-1}-1} \end{pmatrix} \times \begin{pmatrix} \alpha_{N,0} \\ \alpha_{N-1,0} \\ \vdots \\ \alpha_{1,2^{N-1}-1} \end{pmatrix}$$
$$= \begin{pmatrix} (\varphi_{N,0}, C_{\xi\eta}f) \\ (\psi_{N-1,0}, C_{\xi\eta}f) \\ \vdots \\ (\psi_{1,2^{N-1}-1}, C_{\xi\eta}f) \end{pmatrix}, \tag{3.143}$$

with

$$a_{N,0|N,0} = \langle\varphi_{N,0}, C_{\xi\xi}\varphi_{N,0}\rangle + w_N\langle\varphi_{N,0}, \varphi_{N,0}^2\rangle$$
$$+\sum_{n=0}^{N}\sum_{m=0}^{2^{N-n}-1} w_{n,m}\langle\varphi_{N,0}, \psi_{n,m}\varphi_{N,0}\rangle, \tag{3.144}$$

$$a_{N,0|p,q} = \langle\varphi_{N,0}, C_{\xi\xi}\psi_{p,q}\rangle + w_N\langle\varphi_{N,0}, \varphi_{N,0}\psi_{p,q}\rangle$$
$$+\sum_{n=0}^{N}\sum_{m=0}^{2^{N-n}-1} w_{n,m}\langle\varphi_{N,0}, \psi_{n,m}\psi_{p,q}\rangle, \tag{3.145}$$

$$a_{r,s|N,0} = \langle\psi_{r,s}, C_{\xi\xi}\varphi_{N,0}\rangle + w_N(\psi_{r,s}, \varphi_{N,0}^2\rangle$$
$$+\sum_{n=0}^{N}\sum_{m=0}^{2^{N-n}-1} w_{n,m}\langle\psi_{r,s}, \psi_{n,m}\varphi_{N,0}\rangle, \tag{3.146}$$

$$a_{r,s|p,q} = \langle\psi_{r,s}, C_{\xi\xi}\psi_{p,q}\rangle + w_N\langle\psi_{r,s}, \psi_{p,q}\varphi_{N,0}\rangle$$
$$+\sum_{n=0}^{N}\sum_{m=0}^{2^{N-n}-1} w_{n,m}\langle\psi_{r,s}, \psi_{n,m}\psi_{p,q}\rangle. \tag{3.147}$$

These expressions can be simplified using the relations

$$\langle \varphi_{N,0}, \varphi_{N,0}^2 \rangle = 2^{-\frac{3N}{2}} \int_0^{2^N} dt = 2^{-\frac{N}{2}}, \tag{3.148}$$

$$\begin{aligned} \langle \varphi_{N,0}, \psi_{n,m} \varphi_{N,0} \rangle &= 2^{-N} \int_0^{2^N} \psi_{n,m}(t)\, dt \\ &= 2^{-N} 2^{-\frac{n}{2}} \int_0^{2^N} \psi(2^{-n}t - m)\, dt \\ &= 2^{-N} 2^{\frac{n}{2}} \int_{-m}^{2^{N-n}-m} \psi(z)\, dz \\ &= 0, \end{aligned} \tag{3.149}$$

$$\begin{aligned} \langle \varphi_{N,0}, \psi_{n,m} \psi_{p,q} \rangle &= 2^{-\frac{N}{2}} \int_0^{2^N} \psi_{n,m}(t) \psi_{p,q}(t)\, dt \\ &= 2^{-\frac{N}{2}} \delta_{np} \delta_{mq}, \end{aligned} \tag{3.150}$$

leading to

$$a_{N,0|N,0} = [\langle \varphi_{N,0}, C_{\xi\xi} \varphi_{N,0} \rangle + w_N 2^{-\frac{N}{2}}], \tag{3.151}$$

$$a_{N,0|p,q} = [\langle \varphi_{N,0}, C_{\xi\xi} \psi_{p,q} \rangle + 2^{-\frac{N}{2}} w_{p,q}], \tag{3.152}$$

$$a_{r,s|N,0} = [\langle \psi_{r,s}, C_{\xi\xi} \psi_{N,0} \rangle + 2^{-\frac{N}{2}} w_{r,s}], \tag{3.153}$$

and

$$\begin{aligned} a_{r,s|p,q} = \Big[&\langle \psi_{r,s}, C_{\xi\xi} \psi_{p,q} \rangle + w_N 2^{-\frac{N}{2}} \delta_{r,p} \delta_{s,q} \\ &+ \sum_{n=0}^{N} \sum_{m=0}^{2^{N-n}-1} w_{n,m} \langle \psi_{r,s}, \psi_{n,m} \psi_{p,q} \rangle \Big]. \end{aligned} \tag{3.154}$$

Special case of stationary signal and non-stationary noise. At the time being no assumptions about stationarity have been made, neither for the signal ξ nor for the noise n . The situation is simplified substantially, if the signal ξ is stationary and only the noise n is non-stationary. In this case one obtains

$$\langle \varphi_{N,0}, C_{\xi\xi} \varphi_{N,0} \rangle = 2^{-N} \int_0^{2^N} \int_0^{2^N} C_{\xi\xi}(s - t)\, ds\, dt, \tag{3.155}$$

$$
\begin{aligned}
\langle \varphi_{N,0}, C_{\xi\xi}\psi_{p,q}\rangle &= 2^{-\frac{N+p}{2}} \int_0^{2^N} \int_{-\infty}^{\infty} C_{\xi\xi}(s-t)\psi_{p,q}(2^{-p}t-q)\,dt\,ds \\
&= 2^{-\frac{N+p}{2}} \int_0^{2^N} \int_{2^p q}^{2^p q + 2^{p-1}} C_{\xi\xi}(s-t)\,dt\,ds \\
&\quad - 2^{-\frac{N+p}{2}} \int_0^{2^N} \int_{2^p q+2^{p-1}}^{2^p q + 2^{p}} C_{\xi\xi}(s-t)\,dt\,ds,
\end{aligned}
\tag{3.156}
$$

$$
\begin{aligned}
\langle \psi_{r,s}, C_{\xi\xi}\psi_{p,q}\rangle &= 2^{-\frac{r+p}{2}} \int_{-\infty}^{\infty} \int_{-\infty}^{\infty} C_{\xi\xi}(t'-t)\psi(2^{-r}t'-s)\psi(2^{-p}t-q)dt'\,dt \\
&= 2^{-\frac{r+p}{2}} [\int_{2^r s}^{2^r s+2^{r-1}} \int_{2^p q}^{2^p q+2^{p-1}} C_{\xi\xi}(t'-t)\,dt'\,dt \\
&\quad - \int_{2^r s}^{2^r s+2^{r-1}} \int_{2^p q+2^{p-1}}^{2^p q+2^{p}} C_{\xi\xi}(t'-t)\,dt'\,dt \\
&\quad - \int_{2^r s+2^{r-1}}^{2^r s+2^{r}} \int_{2^p q}^{2^p q+2^{p-1}} C_{\xi\xi}(t'-t)\,dt'\,dt \\
&\quad + \int_{2^r s+2^{r-1}}^{2^r s+2^{r}} \int_{2^p q+2^{p-1}}^{2^p q+2^{p}} C_{\xi\xi}(t'-t)\,dt'\,dt].
\end{aligned}
\tag{3.157}
$$

Introducing the abbreviation

$$
I(a,b,c,d) := \int_a^b \int_c^d C_{\xi\xi}(t'-t)\,dt'\,dt,
\tag{3.158}
$$

the following expressions for the scalar products are obtained:

$$
\langle \varphi_{N,0}, C_{\xi\xi}\varphi_{N,0}\rangle = 2^{-N} I(0, 2^N, 0, 2^N),
\tag{3.159}
$$

$$
\begin{aligned}
\langle \varphi_{N,0}, C_{\xi\xi}\psi_{p,q}\rangle = 2^{-\frac{N+p}{2}} [&I(0, 2^N, 2^p q, 2^p q + 2^{p-1}) \\
&- I(0, 2^N, 2^p q + 2^{p-1}, 2^p q + 2^p)],
\end{aligned}
\tag{3.160}
$$

and

$$
\begin{aligned}
(\psi_{r,s}, C_{\xi\xi}\psi_{p,q}) = 2^{-\frac{r+p}{2}} [&I(2^r s, 2^r s + 2^{r-1}, 2^p q, 2^p q + 2^{p-1}) \\
&- I(2^r s + 2^{r-1}, 2^r s + 2^r, 2^p q, 2^p q + 2^{p-1}) \\
&- I(2^r s, 2^r s + 2^{r-1}, 2^p q + 2^{p-1}, 2^p q + 2^p) \\
&+ I(2^r s + 2^{r-1}, 2^r s + 2^r, 2^p q + 2^{p-1}, 2^p q + 2^p)].
\end{aligned}
\tag{3.161}
$$

Since for the most common types of covariance functions $C_{\xi\xi}$ closed expressions can be given for (3.158), the computation of the *Galerkin* matrix can be done very efficiently.

Numerical example. For a numerical test a stationary signal ξ is chosen having a covariance function

$$C_{\xi\xi}(\tau) = \frac{1}{1+(\alpha\tau)^2}. \tag{3.162}$$

This signal is investigated on the interval $[0, T]$, $T = 512$. For the noise n two types of white noise are used:

- On the interval $[0, \frac{T}{2}]$ white noise with the variance $\sigma_1^2 = 4.0$.
- On the interval $[\frac{T}{2}, T]$ white noise with the variance $\sigma_2^2 = 1.0$.

The first step is the generation of a sample of the stationary signal ξ . For this purpose the following algorithm is applied:

- Generation of $N = 512$ independent random variables $\eta_i, i = 1, \dots 512$, which are equally distributed on the interval $[-\frac{1}{2}, \frac{1}{2}]$ and which have unit variance.
- Computation of the covariance matrix
$$C = (c_{ij}) = C_{\xi\xi}(i-j), \quad i, j = 0, \dots, 511. \tag{3.163}$$
- *Cholesky* decomposition of the covariance matrix
$$C = LL^\top. \tag{3.164}$$
- Construction of the signal
$$\xi_i := \sum_{j=1}^{i} l_{ij}\eta_j, \quad i = 1, \dots 512. \tag{3.165}$$

With the help of the rules of covariance propagation it can be shown that ξ has the covariance function $C_{\xi\xi}$:

$$E\{\xi_i\xi_j\} = E\Big\{\sum_{k=1}^{i} l_{ik}\eta_k \sum_{l=1}^{j} l_{lj}\eta_l\Big\} = E\{\eta^\top L^\top L\eta\} = LL^\top = C_{\xi\xi}.$$

Figure 3.50 shows a sample of the stationary signal ξ without and with superimposed noise. For the functional f two instances are chosen:

$$f_1 := \varphi_{0,128} \quad \text{and} \quad f_2 := \varphi_{0,384}. \tag{3.166}$$

Since ξ is a rather long-scale featured process, with a reasonable degree of approximation

$$\xi(128) \approx \langle f_1, \xi\rangle, \quad \xi(384) \approx \langle f_2, \xi\rangle \tag{3.167}$$

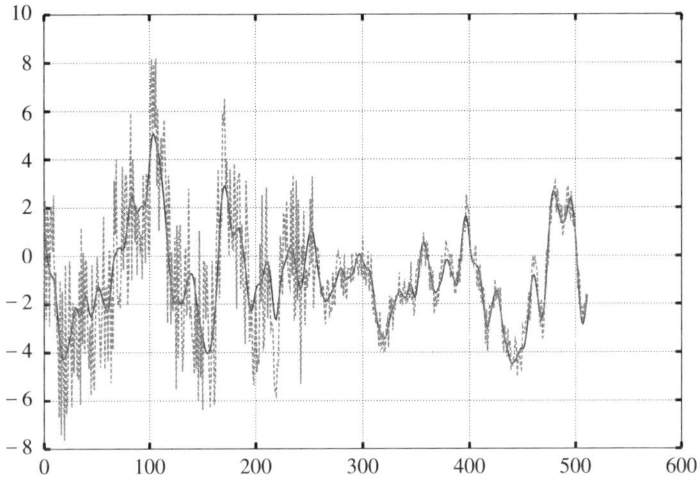

Figure 3.50. Noisy signal

holds. The first estimator f_1 works in a signal region with a high noise variance the second f_2 in a signal region with a low noise variance. Due to the fact that ξ and n are uncorrelated, $C_{\xi\eta} = C_{\xi\xi}$ is true. Now all previously compiled results can be used here:

$$
\begin{aligned}
I(a,b,c,d) &= \int_a^b \int_c^d \frac{1}{1+\alpha^2(t'-t)^2}\, dt'\, dt \\
&\quad - \frac{1}{\alpha^2} \int_{\alpha a}^{\alpha b} \int_{\alpha c}^{\alpha d} \frac{1}{1+(x-y)^2}\, dx\, dy \\
&= \frac{1}{\alpha^2}[\alpha(b-c)\arctan(\alpha(b-c)) - \frac{1}{2}\ln(1+\alpha^2(b-c)^2) \\
&\quad - \alpha(b-d)\arctan(\alpha(b-d)) + \frac{1}{2}\ln(1+\alpha^2(b-d)^2) \\
&\quad - \alpha(a-c)\arctan(\alpha(a-c)) + \frac{1}{2}\ln(1+\alpha^2(a-c)^2) \\
&\quad + \alpha(a-d)\arctan(\alpha(a-d)) - \frac{1}{2}\ln(1+\alpha^2(a-d)^2)].
\end{aligned}
\tag{3.168}
$$

With the help of the formulae (3.159)–(3.161) and (3.168) the Galerkin matrix can be computed. Figure 3.51 shows the structure of the Galerkin matrix for the non-stationary noise, given by the wavelet coefficients of its variance:

$$w_N = 2.5, \quad w_{N,0} = 1.5, \quad w_{n,m} = 0 \text{ else.}$$

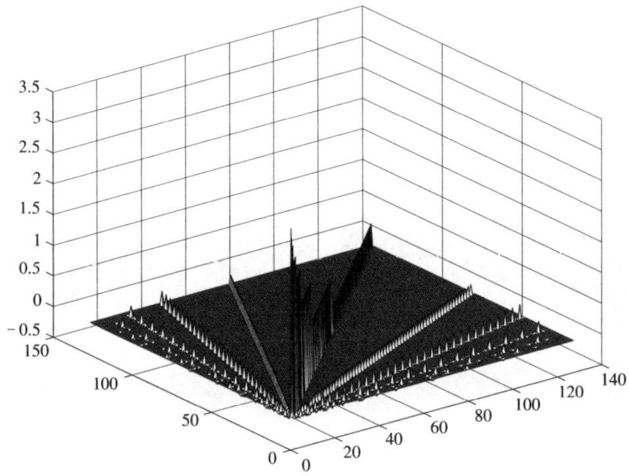

Figure 3.51. Galerkin matrix

Also the right-hand side can be computed easily

$$\begin{aligned}\langle \varphi_{N,0}, C_{\xi\eta} f_i\rangle &= (\varphi_{N,0}, C_{\xi\xi}\varphi_{0,i})\\ &= 2^{-\frac{N}{2}} \int_0^{2^N} \int_i^{i+1} C_{\xi\xi}(t'-t)dt'\,dt\\ &= 2^{-\frac{N}{2}} I(0, 2^N, i, i+1),\end{aligned}$$

and

$$\begin{aligned}\langle \psi_{r,s}, C_{\xi\eta} f_i\rangle &= (\psi_{r,s}, C_{\xi\xi}\varphi_{0,i})\\ &= 2^{-\frac{r}{2}} \Bigg(\int_{2^r s}^{2^r s+2^{r-1}} \int_i^{i+1} C_{\xi\xi}(t'-t)\,dt'\,dt\\ &\qquad - \int_{2^r s+2^{r-1}}^{2^r s+2^r} \int_i^{i+1} C_{\xi\xi}(t'-t)\,dt'\,dt \Bigg)\\ &= 2^{-\frac{r}{2}} \Big(I(2^r s, 2^r s + 2^{r-1}, i, i+1)\\ &\qquad - I(2^r s + 2^{r-1}, 2^r s + 2^r, i, i+1) \Big).\end{aligned}$$

The solutions a_i, $i \in \{128, 384\}$ are displayed in Figure 3.52.

Clearly, the prediction is basically a weighted mean of the data in the neighborhood of the value which has to be predicted. The stronger the data noise the more smoothing this weighted mean is. In Figure 3.53 the noise-free signal and its prediction are compared.

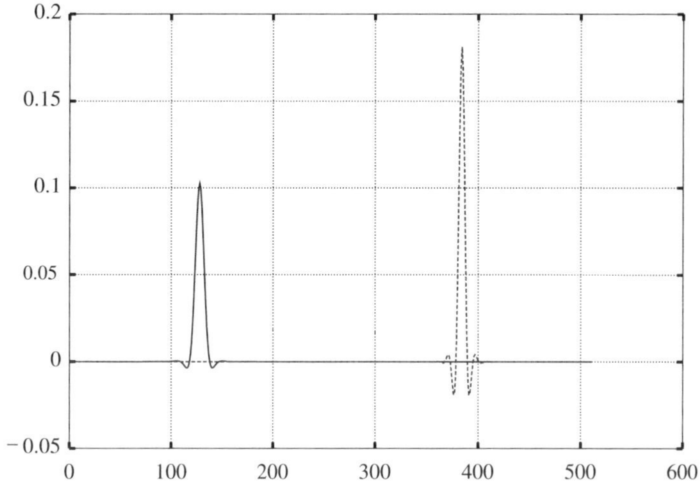

Figure 3.52. Predictors

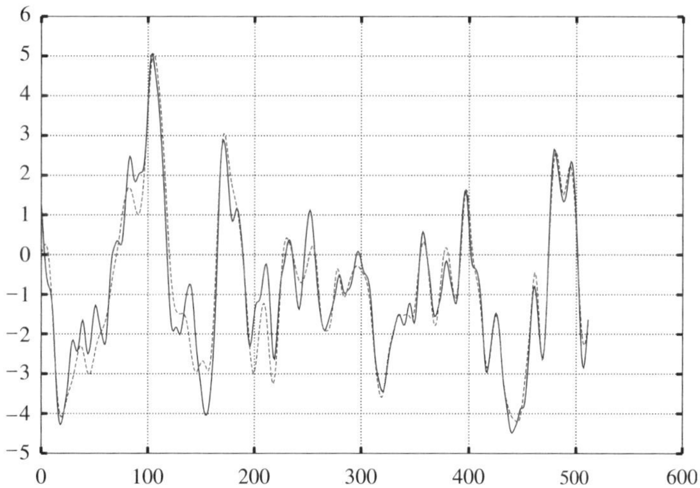

Figure 3.53. Non-stationary prediction

Obviously, the lower the noise is, the better is the prediction. In Figure 3.54 the result of the usual stationary prediction is displayed.

A comparison of the results shows that the non-stationary prediction yields better results, especially in the area of lower noise. The price which has to be paid is that in the non-stationary case the very efficient tool of FFT cannot be applied, which increases the computational load.

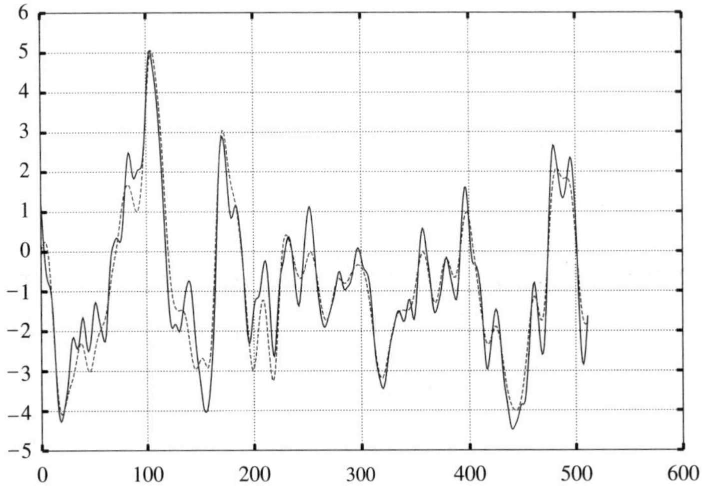

Figure 3.54. Conventional prediction

3.4 Operator approximation

3.4.1 Wavelet compression of operator equations

Many tasks have an integral or a differential equation as mathematical description. Since these infinite dimensional objects cannot be treated in a computer, they have to be approximated on finite dimensional subspaces. Depending on the choice of the approximating subspace and its base, the approximation of the corresponding operator can be more or less efficient. In many cases the approximation on spaces of an MRA is especially useful.

Consider operator equation

$$g(x) = \int_{\mathbb{R}} K(x, y) f(y)\, dy, \quad f \in H,\ g \in H. \tag{3.169}$$

This operator equation is to be solved by the Galerkin method. The basic idea of Galerkin's method is to look for an approximate solution f_n not in the whole space H but in a finite-dimensional subspace $H_n \subset H$. If $\{\varphi_1, \dots, \varphi_n\}$ is a base of H_n the approximate solution f_n has the structure

$$f_n(x) = \sum_{i=0}^{n} \alpha_i \varphi_i(x) \tag{3.170}$$

with unknown coefficients α_i. This leads to the Galerkin equations:

$$\int_{\mathbb{R}} \left(\int_{\mathbb{R}} (K(x,y) \cdot \sum_{i=0}^{N} \alpha_i \varphi_i(y))\, dy \right) \varphi_k(x) dx = \int_{\mathbb{R}} g(x) \cdot \varphi_k(x)\, dx, \quad k = 0, \ldots N. \tag{3.171}$$

Setting

$$a_{ki} := \int_{\mathbb{R}} \int_{\mathbb{R}} K(x,y) \varphi_i(y)\, dy \varphi_k(x)\, dx, \quad b_k := \int_{\mathbb{R}} g(x) \varphi_k(x)\, dx, \tag{3.172}$$

the linear system of equations

$$\sum_{i=0}^{N} a_{ki} \alpha_i = b_k, \quad k = 0, \ldots N \tag{3.173}$$

for the coefficients α_i is obtained. The representation of the Galerkin solution can be done with the help of those coefficients by

$$f_N(x) := \sum_{i=0}^{N} \alpha_i \cdot \varphi_i(x). \tag{3.174}$$

Usually, the dimension of the matrix (3.172) is very large. Therefore direct methods like Cholesky decomposition cannot be applied. On the other hand the matrix entries can be computed very easily. Thus the equations could be solved by iterative methods like Conjugate Gradients iteration. If CG is applied, the matrix has not to be stored in the computer's memory but can be computed on the fly. Unfortunately, the rate of convergence of the CG iteration is proportional to

$$\rho = \frac{\mu - \lambda}{\mu + \lambda}, \tag{3.175}$$

with μ being the largest and ν being the smallest eigenvalue of the matrix (3.172). This lead to convergence rates very close to 1, which means a very slow convergence.

The rate of convergence can be increased by pre-conditioning. Let C be a regular matrix and let us abbreviate $(C^\top)^{-1}$ by $C^{-\top}$. Then the two systems of linear equations

$$Ax = b \quad \text{and} \quad \tilde{A}\tilde{x} = \tilde{b}, \tag{3.176}$$

with

$$\tilde{A} = CAC^\top, \quad \tilde{x} = C^{-\top}x, \quad \tilde{b} = Cb \tag{3.177}$$

are equivalent. Since

$$C^\top CA = C^\top CA(C^\top C^{-\top}) = C^\top (CAC^\top) C^{-\top} = C^\top \tilde{A} C^{-\top}$$

holds, the matrices $\tilde{A}$ and $C^\top CA$ have the same spectrum. Hence the convergence rate of the CG iteration for the equation $\tilde{A}\tilde{x} = \tilde{b}$ can be increased if $C^\top C \approx A^{-1}$ holds.

The conditioning matrices C could be computed by a Cholesky decomposition of A. Unfortunately, for a fully occupied matrix A the numerical effort for the Cholesky decomposition is quite close to the total effort for the solution of $Ax = b$.

The situation changes, if the matrix A is sparse. In this case during the process of Cholesky decomposition the fill-in occurring in the computation of the Cholesky factors could be neglected, which leads to sparse pre-conditioning matrices C. In this case $CC^\top \approx A^{-1}$ holds only approximatively but this will already be sufficient to speed up the convergence. One method to generate a sparse Galerkin matrix is to use wavelets as base functions.

The structure of the matrix $A := (a_{ki})$ strongly depends on the choice of the base functions φ_i. The use of wavelets as base functions leads to a sparse matrix and provides a lot of numerical advantages.

Wavelets. Let $\{\emptyset\} \subset \cdots \subset V_2 \subset V_1 \subset V_0 \subset V_{-1} \subset \cdots \subset L_2(\mathbb{R})$ be a multi-resolution analysis of $L_2(\mathbb{R})$ and let φ be the scaling function of this MRA. One possibility for the choice of base function is to use the Riesz base of V_0 :

$$\varphi_k(x) := \varphi(x-k). \tag{3.178}$$

This leads to the following expression for the coefficients of the Galerkin matrix:

$$\begin{aligned} a_{ki} :&= \int_{\mathbb{R}} \int_{\mathbb{R}} K(x,y)\varphi(y-i)\,dy\,\varphi(x-k)\,dx \\ &= \int_{\mathbb{R}} \int_{\mathbb{R}} K(x+k,y+i)\varphi(y)\varphi(x)\,dy\,dx \end{aligned} \tag{3.179}$$

If the correlation length of the kernel K is large compared to the scale of φ, the scaling function φ acts approximatively like the delta distribution, i.e.

$$a_{ki} \approx K(k,i) \tag{3.180}$$

holds. This has the advantage that the elements of the Galerkin matrix can be computed very easily but it has also the disadvantage that the resulting Galerkin matrix is fully occupied. The scaling function φ is connected to the corresponding wavelet ψ and to the scaling function of the next coarser scaling space by the so-called scaling equations

$$\varphi(x) = \sqrt{2}\sum_{k\in Z} h_k\varphi(2x-k), \tag{3.181}$$

$$\psi(x) = \sqrt{2}\sum_{k\in Z} g_k\varphi(2x-k). \tag{3.182}$$

The space V_0 can be decomposed into $V_0 = V_1 \oplus W_1$. Hence the base of $V_1 \oplus W_1$ can also be used for the solution of the Galerkin equations.

$$\begin{aligned}
&\frac{1}{2}\int_{\mathbb{R}}\Big(\sum_{i\in\mathbb{Z}} K(x,y)c_i^{(1)}\cdot\varphi(2^{-1}y-i)\,dy\Big)\varphi(2^{-1}y-k)\,dx \\
&\quad+\frac{1}{2}\int_{\mathbb{R}}\Big(\sum_{i\in\mathbb{Z}} K(x,y)d_i^{(1)}\cdot\psi(2^{-1}y-i)\,dy\Big)\varphi(2^{-1}y-k)\,dx \qquad (3.183)\\
&=\frac{1}{\sqrt{2}}\int_{\mathbb{R}} g(x)\cdot\varphi(2^{-1}x-k)\,dx.
\end{aligned}$$

$$\begin{aligned}
&\frac{1}{2}\int_{\mathbb{R}}\Big(\sum_{i\in\mathbb{Z}} K(x,y)c_i^{(1)}\cdot\varphi(2^{-1}y-i)\,dy\Big)\psi(2^{-1}y-k)\,dx \\
&\quad+\frac{1}{2}\int_{\mathbb{R}}\Big(\sum_{i\in\mathbb{Z}} K(x,y)d_i^{(1)}\cdot\psi(2^{-1}y-i)\,dy\Big)\psi(2^{-1}y-k)\,dx \qquad (3.184)\\
&=\frac{1}{\sqrt{2}}\int_{\mathbb{R}} g(x)\cdot\psi(2^{-1}x-k)\,dx.
\end{aligned}$$

This again is a system of linear equations for the coefficients $c_i^{(1)}, d_i^{(1)}$ and with the help of this equation the solution of the operator equation can be approximated by

$$f(x) = \frac{1}{\sqrt{2}}\sum_{i\in\mathbb{Z}} c_i^{(1)}\varphi(2^{-1}x-i) + \frac{1}{\sqrt{2}}\sum_{i\in\mathbb{Z}} d_i^{(1)}\psi(2^{-1}x-i). \qquad (3.185)$$

The change to the alternative representation of the solution has the consequence that the Galerkin matrix is partitioned, i.e.

$$A^{(1)} = \begin{bmatrix} a_{ki}^1 & a_{ki}^2 \\ a_{ki}^3 & a_{ki}^4 \end{bmatrix} \qquad (3.186)$$

with

$$a_{ki}^1 = \frac{1}{2}\int_{\mathbb{R}}\int_{\mathbb{R}} K(x,y)\varphi(2^{-1}y-i)\varphi(2^{-1}x-k)\,dy\,dx, \qquad (3.187)$$

$$a_{ki}^2 = \frac{1}{2}\int_{\mathbb{R}}\int_{\mathbb{R}} K(x,y)\psi(2^{-1}y-i)\varphi(2^{-1}x-k)\,dy\,dx, \qquad (3.188)$$

$$a_{ki}^3 = \frac{1}{2}\int_{\mathbb{R}}\int_{\mathbb{R}} K(x,y)\varphi(2^{-1}y-i)\psi(2^{-1}x-k)\,dy\,dx, \qquad (3.189)$$

$$a_{ki}^4 = \frac{1}{2}\int_{\mathbb{R}}\int_{\mathbb{R}} K(x,y)\psi(2^{-1}y-i)\psi(2^{-1}x-k)\,dy\,dx. \qquad (3.190)$$

The advantage of this partition is that the coefficients a_{ki}^4 are very small, since they represent the shortest wave-length component of the kernel K and this kernel was supposed to be of a very long wavelength nature. Hence the change from a base of V_0 to a base of $V_1 \oplus W_1$ changes the character of the Galerkin matrix to a sparser structure. On first sight, this has to be paid by a more complicated computation of the matrix elements, since the support of $\varphi(2^{-1}x)$ and $\psi(2^{-1}x)$ is twice as large as the support of $\varphi(x)$. But due to the scaling equations the coefficients a_{ki}^j can be computed recursively from the coefficients a_{ki}:

$$\begin{aligned} a_{ki}^1 &= \frac{1}{2}\int_{\mathbb{R}}\int_{\mathbb{R}} K(x,y)\varphi(2^{-1}y-i)\varphi(2^{-1}x-k)\,dy\,dx \\ &= \int_{\mathbb{R}}\int_{\mathbb{R}} K(x,y)\sum_{l\in\mathbb{Z}} h_l\varphi(y-2i-l)\sum_{m\in\mathbb{Z}} h_m\varphi(x-2k-m)\,dy\,dx \\ &= \sum_{l\in\mathbb{Z}}\sum_{m\in\mathbb{Z}}\int_{\mathbb{R}}\int_{\mathbb{R}} K(x,y)\varphi(y-l)\varphi(x-m)\,dy\,dx h_{l-2i}h_{m-2k} \\ &= \sum_{l\in\mathbb{Z}}\sum_{m\in\mathbb{Z}} a_{lm}h_{l-2i}h_{m-2k}. \end{aligned} \tag{3.191}$$

In a quite similar way the recursions for the remaining parts of the Galerkin matrix can be found

$$a_{ki}^2 = \sum_{l\in Z}\sum_{m\in Z} a_{lm}g_{l-2i}h_{m-2k}, \tag{3.192}$$

$$a_{ki}^3 = \sum_{l\in Z}\sum_{m\in Z} a_{lm}h_{l-2i}g_{m-2k}, \tag{3.193}$$

$$a_{ki}^4 = \sum_{l\in Z}\sum_{m\in Z} a_{lm}g_{l-2i}g_{m-2k}. \tag{3.194}$$

A closer inspection of these recursion equations shows that the matrix $A^{(1)}$ is the two-dimensional wavelet transformation of the matrix $A = (a_{ki})$. In a quite analog way the right-hand side of the Galerkin equation can be considered as the wavelet transform of the inhomogeneity g. The advantage of this procedure is that it transforms the Galerkin equation into a sparse form.

Example 3.4.1. Let us consider the integral equation

$$\int_{\mathbb{R}} K(x,y)f(y)\,dy = g(x) \tag{3.195}$$

with the kernel

$$K(x,y) = \frac{1}{1+\alpha(x-y)^2}, \quad \alpha = 10^{-4} \tag{3.196}$$

which is displayed in Figure 3.55. The scaling space V_0 spanned by the scaling function of the Daubechies wavelet of order $N = 2$ is chosen as approximation space. The scaling function is displayed in Figure 2.39, p. 94. The entries a_{ik} of the Galerkin matrix are computed by (3.172) and they are displayed in Figure 3.56.

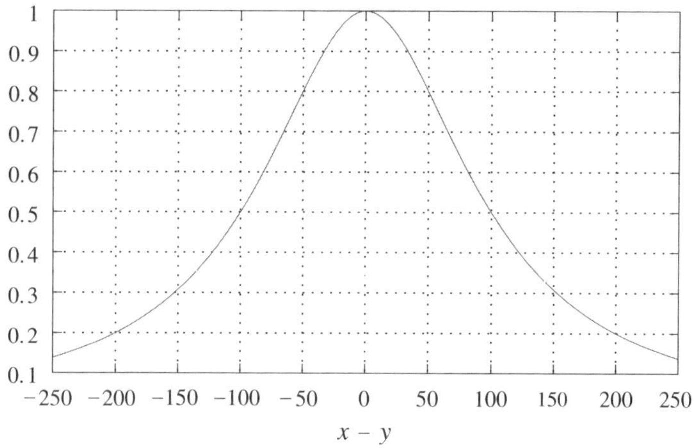

Figure 3.55. Kernel of the integral equation

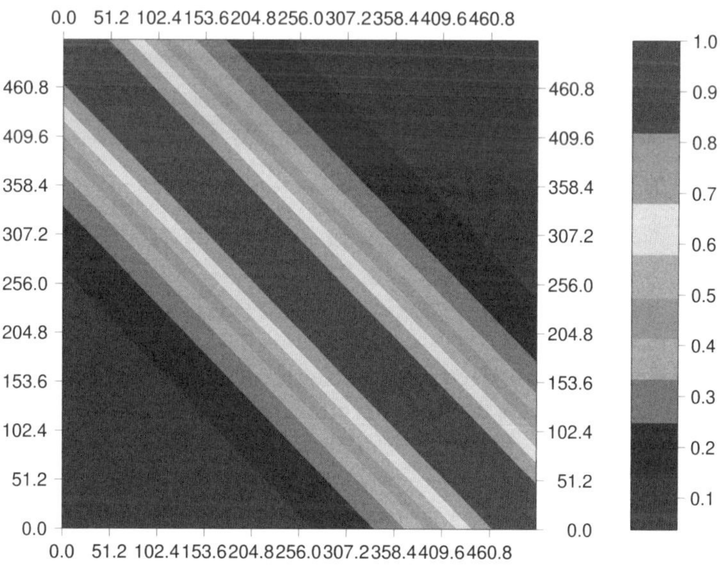

Figure 3.56. Galerkin matrix with respect to the base of V_0

The Galerkin matrix has a ridge along the main diagonal and decreases slowly towards the edges of the matrix. All in all, the matrix is far from being sparse and all matrix entries have to be taken into account for the solution of the resulting system of Galerkin equations.

The maximal and minimal eigenvalues of A are

$$\mu = 216.79, \quad \lambda = 1.2375 \cdot 10^{-9},$$

which leads to a convergence factor of

$$\rho = \frac{\mu - \lambda}{\mu + \lambda} = 0.99999999988538.$$

It can be expected that the iteration error $\|x - x_n\|$ is reduced in each iteration step of the CG iteration by the factor ρ only, and the convergence is very slow. Possibly, the convergence can even be prevented by the rounding errors. Figure 3.57 shows the iteration errors of the CG iteration in a logarithmic scale.

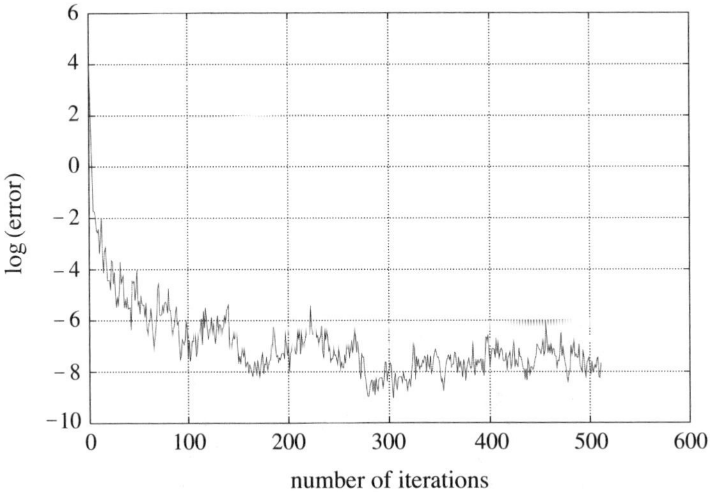

Figure 3.57. Norm of the iteration errors in logarithmic scale

In the beginning the error decreases quite fast, but in the end the interaction of the low rate of convergence and of the rounding errors prevents the computation of an exact solution. An error of about 10^{-8} remains independent of the number of iterations.

Since $V_0 = W_1 \oplus W_2 \oplus \cdots \oplus W_N \oplus V_N$ holds, the base of V_0 can be replaced by the base of $W_1 \oplus W_2 \oplus \cdots \oplus W_N \oplus V_N$. The transformation of the Galerkin matrix from the base of V_0 to the base of $W_1 \oplus W_2 \oplus \cdots \oplus W_N \oplus V_N$ can be accomplished by two-dimensional wavelet transformation with respect to the Daubechies 4 wavelet.

The result of this transformation is shown in Figure 3.58. It is clearly visible that the transformed matrix has only a very small number of dominant matrix entries.

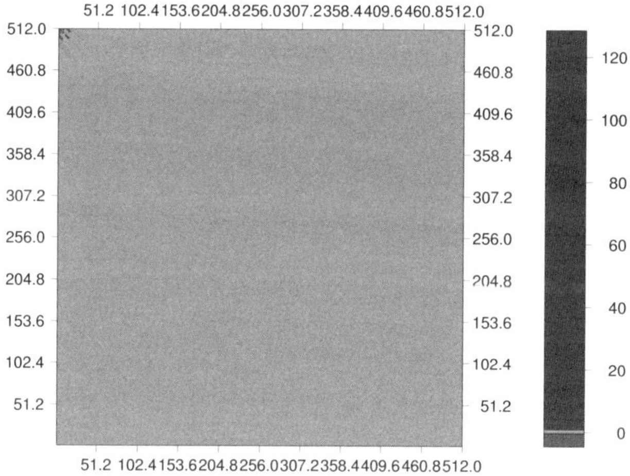

Figure 3.58. Wavelet transformed Galerkin matrix

All the other entries are small but not exactly equal to zero. The next step will be a thresholding. All elements with a modulus smaller than $10^{-7} \cdot \|A\|$ will be replaced by zero. The skeleton of the thresholded matrix, i.e. all elements which are different from zero are displayed in Figure 3.59.

Figure 3.59. Skeleton of the wavelet compressed matrix

Now the conditioning matrix C can be computed by incomplete Cholesky decomposition and can be used for pre-conditioning of the Galerkin equation. When preconditioned, the convergence is much faster and the results are much more precise than for the usual CG iteration. In Figure 3.60 the iteration errors of the preconditioned CG iteration are again displayed in a logarithmic scale.

Figure 3.60. Norm of the iteration errors of the PCG iteration

It is clearly visible that the iteration is much faster; instead of 100 steps and a final accuracy of 10^{-8} after only 30 steps an accuracy of 10^{-16} is obtained. Hence the use of a wavelet base for the Galerkin method is proved as a very efficient tool to increase booth speed and accuracy.

3.4.2 Multi-grid solvers for wavelet discretized operators

We will restrict our considerations to situations where the connection between the given data and the unknown solution can be formulated as a differential or an integral equation. The concept of a pseudo-differential operator (PDO) is the generalization both of the concept of a differential and an integral operator. It is defined as an operator of the following structure

$$pu(x) := \int_{-\infty}^{\infty} \hat{u}(\omega) \cdot a(\omega) e^{\imath x\omega} \, d\omega. \tag{3.197}$$

In the equation (3.197) the quantity $\hat{u}$ is the Fourier transform of the signal u. This means, the application of a PDO on a signal is a three-step procedure:

1. Computation of the spectrum $\hat{u}$ of the signal.
2. Filtering of the spectrum by the so-called symbol $a(\omega)$ of the PDO.
3. Back-transformation of the filtered signal into the space or time domain.

In general, a PDO equation

$$pu = f \tag{3.198}$$

cannot be solved analytically. A frequently applied technique for the generation of an approximate solution is Galerkin's method.

The idea of Galerkin's method is not to look for a solution in the whole definition space of the operator p, but in a finite-dimensional subspace. Let $\{\varphi_1, \dots, \varphi_n\}$ be a base of such an n-dimensional subspace. Then every candidate for an approximate solution is of the following structure:

$$u_n(x) = \sum_{i=1}^{n} u_i \cdot \varphi_i(x). \tag{3.199}$$

This leads to the Galerkin equations for the unknown weights u_i:

$$\sum_{j=1}^{n} a_{ij} \cdot u_j = b_i, \quad i = 1, \dots, n, \tag{3.200}$$

where the coefficients a_{ij}, b_i are given by

$$a_{ij} := \int_{-\infty}^{\infty} p\varphi_j \cdot \varphi_i \, dx, \quad b_i := \int_{-\infty}^{\infty} f \cdot \varphi_i \, dx. \tag{3.201}$$

Equation (3.201) is valid for any kind of an operator p. If the operator p is a PDO, the coefficients a_{ij} can also be determined by

$$a_{ij} = \sqrt{2\pi} \int_{-\infty}^{\infty} a(\omega)\hat{\varphi}_j(\omega)\hat{\varphi}_i(\omega) \, d\omega. \tag{3.202}$$

This means that for the establishment of the Galerkin equations it is not necessary to know the PDO exactly, it is already sufficient to know its symbol $a(\omega)$.

Multi-grid solvers. Usually the dimension of the Galerkin equations is so high that it has to be solved iteratively. Generally, it can be observed that an iterative solver reduces the short-wavelength error constituents in the initial guess for the solution but leaves the long-wavelength error constituents almost unchanged. For the elimination of the remaining long-wavelength error constituents a much coarser grid is sufficient. On such a coarse grid the dimension of the remaining problem is so small that the error elimination can be performed by a direct solver. This alternating treatment of the same problem on a fine grid by an iterative solver and on a coarse grid by a direct solver is the basic principle of a multi-grid method. Multi-grid solvers discretize a given problem on two hierarchical grids, having the mesh-widths h and H. The discretized problems are

$$A_h u_h = b_h, \quad A_H u_H = b_H. \tag{3.203}$$

For the restriction to a coarser and for the uplift to a finer grid two operators are necessary

$$I_h^H : U_h \to U_H, \quad I_H^h : U_H \to U_h. \tag{3.204}$$

On the fine grid the problem is solved iteratively:

$$u_h^{l+1} = T_h u_h^l + t_h, \quad l = 0, \dots, n-1, \tag{3.205}$$

where T_h denotes the iteration matrix and t_h the constant part of the iterative solver on the fine grid. The iteration will leave a long-wavelength residual

$$d_h = b_h - A_h u_h^n. \tag{3.206}$$

The cause of the residual can be determined on the coarse grid by

$$v_H = A_H^{-1}(I_h^H d_h). \tag{3.207}$$

After the uplift of the long-wavelength error v_H to a finer grid the solution can be corrected for the long-wavelength errors

$$u_h = u_h^n + I_H^h v_H. \tag{3.208}$$

This process has to be repeated several times. Compressed into a single formula a multi-grid solver is of the following structure (c.f. [44]):

$$u_h^{k+1} = T_h^n (I - I_H^h A_H^{-1} I_h^H A_h) T_h^n u_h^k + s_h. \tag{3.209}$$

For suitable operators p the multi-grid strategy provides a sufficiently high and dimension-independent convergence speed.

Multi-grid solvers on consecutive wavelet spaces. The solution spaces U_h and U_H on two hierarchical grids strongly remind of the hierarchical scaling spaces of a wavelet multi-resolution analysis (MRA). Therefore it is naturally to identify these solution spaces U_h and U_H with two spaces V_0 and V_1 of an MRA. The advantage of this choice is that the restriction operator I_h^H and the uplift operator I_H^h can be identified with Mallat's smoothing operator H and its adjoint H^*. From this setting two important conclusions can be drawn. First, the discretization of the problem on a coarser grid can be computed efficiently from the given discretization on the fine grid by a wavelet transformation

$$A_H = H A_h H^*. \tag{3.210}$$

Instead of a new discretization according to (3.202), a single step of Mallat's algorithm is sufficient to compute the matrix A_H from the given matrix A_h.

Secondly, it holds

$$\begin{aligned} H d_h^H H^* &:= H(I - I_H^h A_H^{-1} I_h^H A_h) H^* \\ &= HH^* - HH^* A_H^{-1} H A_h H^* \\ &= HH^* - HH^* A_H^{-1} A_H \\ &= 0. \end{aligned} \tag{3.211}$$

The interpretation of (3.211) is that the error d_h^H has no constituent on the coarse grid. The error d_h^H is produced by inverting A_H instead of A_h. This means that the direct solver deals only with the long-wavelength features. The short-wavelength features have do be dealt with by the iteration operator T_h. This is only possible if the iteration operator is a low-pass filter, which means in turn that the operator p has to be a high-pass filter. In this respect, equations (3.211) characterizes operators which are suitable for a multi-grid strategy.

Planar Stokes problem. The use of wavelet techniques for the construction of a multi-grid solver will be tested for two examples: First the Stokes problem in planar approximation, second the satellite gravity gradiometry boundary value problem.

The planar Stokes problem looks for a function u, which is harmonic in the upper half-space. The normal derivatives of this function have to coincide with given values δg at the $x_3 = 0$ plane.

$$\Delta u(x) = 0, \quad x_3 > 0, \tag{3.212}$$

$$\frac{\partial u}{\partial n}\Big|_{x_3=0} = \delta g. \tag{3.213}$$

Despite the fact that in spherical approximation the Stokes problem is a Robin-type boundary value problem, in planar approximation the Stokes problem simplifies to a Neumann problem.

The PDO formulation of this problem is

$$\delta g = pu = \frac{1}{2\pi}\int_{-\infty}^{\infty}\int_{-\infty}^{\infty} a(\omega)\hat{u}(\omega)e^{\imath x\omega}\, d\omega, \tag{3.214}$$

$$a(\omega) = |\omega|. \tag{3.215}$$

The plot of the symbol $a(\omega)$ up to the Nyquist frequency π shows that the Stokes PDO is a high-pass filter (see Figure 3.61). Hence the construction of a multi-grid solver for the problem might be useful.

For the MRA, generating a multi-grid solver for the planar Stokes problem, the simplest possible choice was made: The two-dimensional Haar tensor wavelet. This wavelet has the following scaling function

$$\varphi(\mathbf{x}) := \varphi(x_1) \cdot \varphi(x_2), \tag{3.216}$$

$$\varphi(x) := \begin{cases} 1, & 0 \leq x < 1 \\ 0, & \text{else}, \end{cases} \tag{3.217}$$

$$\hat{\varphi}(\omega) = \frac{1}{\imath\omega\sqrt{2\pi}}[1 - e^{-\imath\omega}]. \tag{3.218}$$

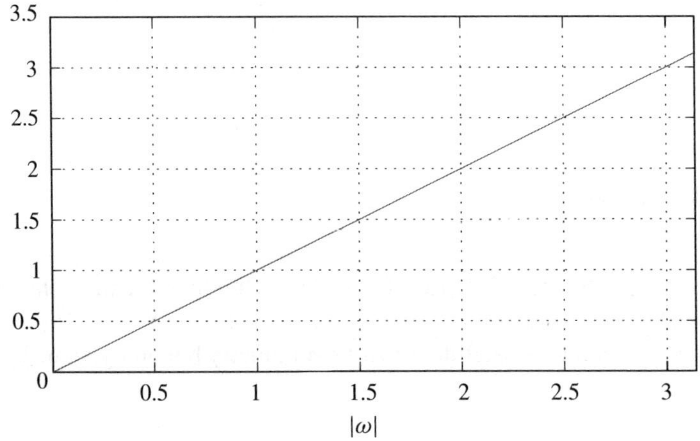

Figure 3.61. Symbol of the planar Stokes PDO

For this choice Mallat's smoothing operator H becomes

$$(Hu)_{ij} := \sum_{k=0}^{1}\sum_{l=0}^{1} u_{kl} h_{k-2i} h_{l-2j}, \tag{3.219}$$

$$h_0 = h_1 = \frac{1}{\sqrt{2}}. \tag{3.220}$$

According to (3.202), the coefficients of the Galerkin matrix can be computed by

$$\begin{aligned} a_{ijkl}^h &= 2\pi \int_{-\infty}^{\infty}\int_{-\infty}^{\infty} a(\omega)\hat{\varphi}_{ij}\hat{\varphi}_{kl}\, d\omega \\ &= 2\pi \int_{-\infty}^{\infty}\int_{-\infty}^{\infty} |\omega||\hat{\varphi}|^4 e^{-\imath(\omega_1(i-k)+\omega_2(j-l))}\, d\omega. \end{aligned} \tag{3.221}$$

Obviously, the coefficients a_{ijkl}^h of the Galerkin matrix are the Fourier transform of $2\pi|\omega||\hat{\varphi}|^4$ sampled at integer locations. Therefore these coefficients can be efficiently computed by FFT. For an $N \times N$ grid the Galerkin matrix contains N^4 coefficients. Fortunately it is sufficient to compute only the N^2 coefficients of a core block. Figure 3.62 shows the core block for the planar Stokes problem

Only this core block has to be stored in the computer's memory. The remaining coefficients can be determined on-the-fly by scanning this core block row wise.

In the examples to be discussed here N always is equal to 64, which means a 4096×4096 system has to be solved. In order to test the performance of the resulting multi-grid solver the Galerkin equations are solved twice: With the usual Gauß–Seidel

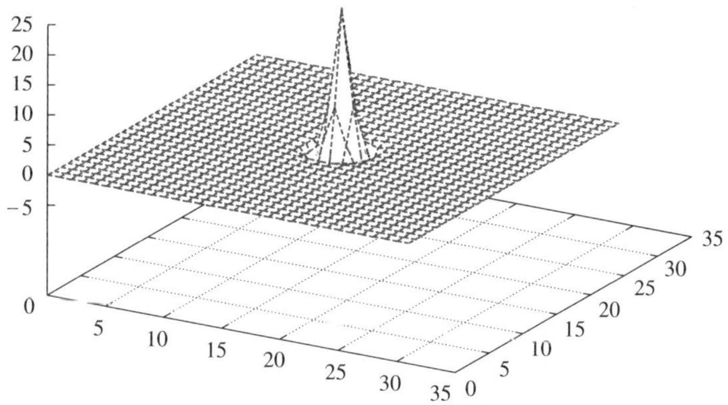

Figure 3.62. Core block for the planar Stokes Problem

iteration and by the multi-grid solver. (Of course, also other iterative solvers e.g. CG-solvers could be used. Independently from the iterative solver the multi-grid iteration will have a dimension-independent rate of convergence.) Both iterations start with an identical initial error distribution, which is displayed in Figure 3.63.

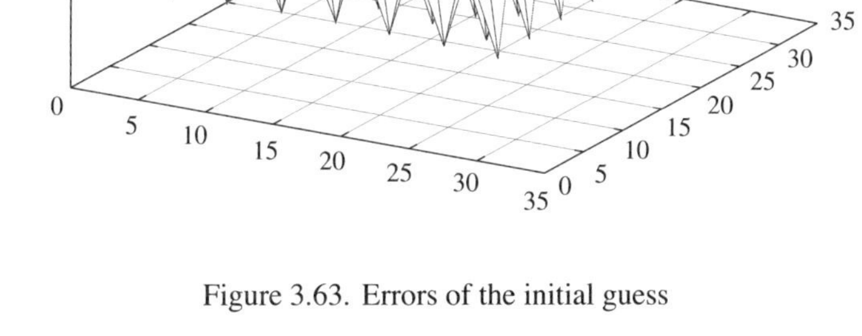

Figure 3.63. Errors of the initial guess

These initial errors consist of a long-wavelength and a short-wavelength part. The remaining errors after 20 steps of Gauß–Seidel iteration are displayed in Figure 5.64. The short-wavelength error constituents are completely eliminated, but the long-wavelength errors are almost unchanged. Only a slight change in its magnitude can be observed.

The situation is quite different for the multi-grid solver. The remaining errors after only 4 steps of multi-grid iteration are shown in Figure 3.65. Here, the long-wavelength errors are reduced by a factor of about 1000. The short-wavelength errors are so small that they are not visible in the plot.

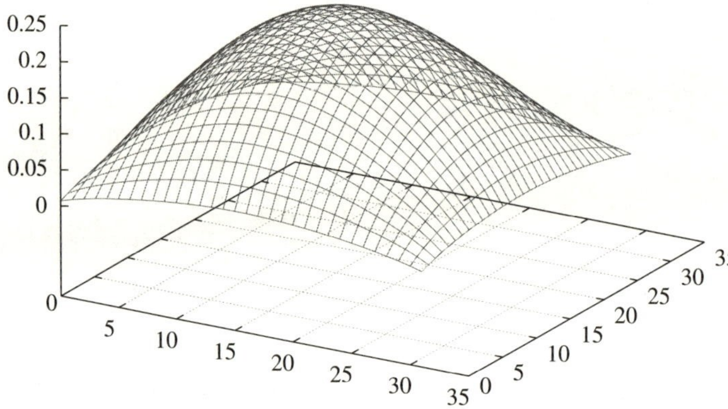

Figure 3.64. Remaining error after 20 Gauss–Seidel steps

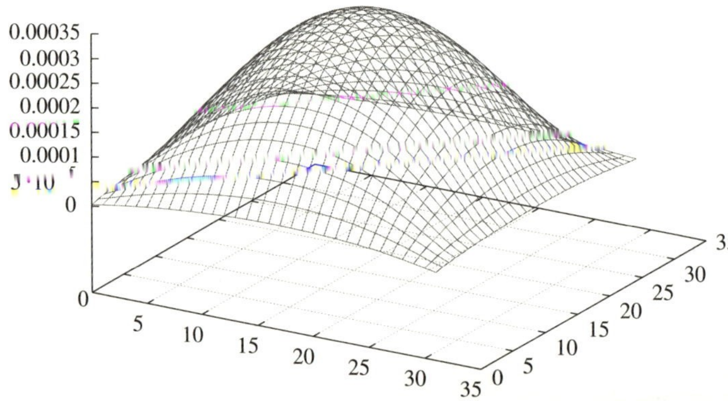

Figure 3.65. Remaining errors after 4 multi-grid steps.

Satellite gravity gradiometry. Solving the satellite gravity gradiometry boundary value problem means to look for a function u, which is harmonic in the upper half-space and which, at a certain height $x_3 = h$, has given second order normal derivatives.

$$\Delta u(x) = 0, \quad x_3 > 0, \tag{3.222}$$

$$\frac{\partial^2 u}{\partial n^2}\Big|_{x_3=h} = \Gamma. \tag{3.223}$$

The PDO formulation of this problem is

$$\Gamma = pu = \frac{1}{2\pi}\int_{-\infty}^{\infty}\int_{-\infty}^{\infty} a(\omega)\hat{u}(\omega)e^{\imath x\omega}\,d\omega \tag{3.224}$$

$$a(\omega) = |\omega|^2 e^{-h|\omega|}. \tag{3.225}$$

In contrast to the planar Stokes problem, here the symbol is the combination of a high-pass $|\omega|^2$ and a low-pass filter $e^{-h|\omega|}$. Which of the two components prevails depends on the orbital height h of the satellite. In Figure 3.66 the symbols for two flight-height scenarios are plotted.

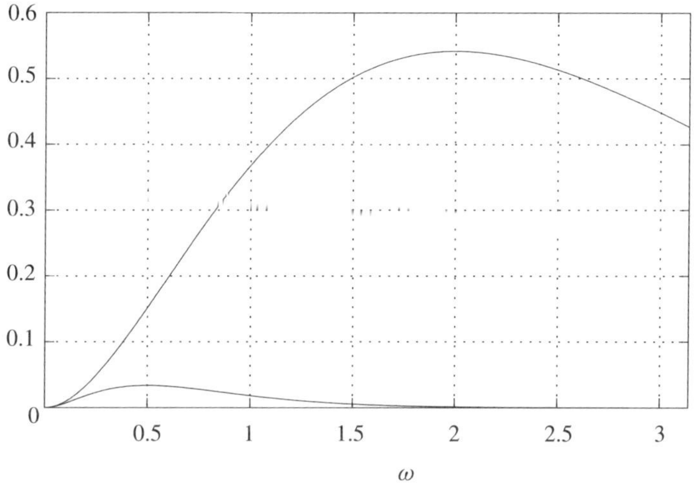

Figure 3.66. Symbols of gravity-gradient PDOs for two different orbital heights

The upper curve represents the symbol for an orbital height, which approximately equals the desired resolution on the ground; for the lower curve the orbital altitude is twice the desired resolution on the ground. In the first case the corresponding PDO is almost a high-pass filter and the use of a multi-grid solver seems promising. For the case of the higher satellite the PDO is a band-pass filter and therefore not suitable for multi-grid procedures. Hence a multi-grid solver was only constructed for the case of the low-flying satellite. Again the problem was solved twice: With the usual Gauß–Seidel iteration and with a multi-grid solver. In both cases 12 iteration steps were performed. Figure 3.67 shows the development of the error norms for both solvers.

For the Gauß–Seidel iteration the error norm is increasing, which means that the iteration is not convergent. The reason for this divergence might be the non-perfect high-pass filter of the underlying PDO.

The multi-grid solver is converging rapidly. After 11 steps 9 valid decimal places in the solution are obtained. This is more than for the most gravity field recovery purposes is necessary. For PDOs with a high-pass symbol multi-grid solvers provide an efficient

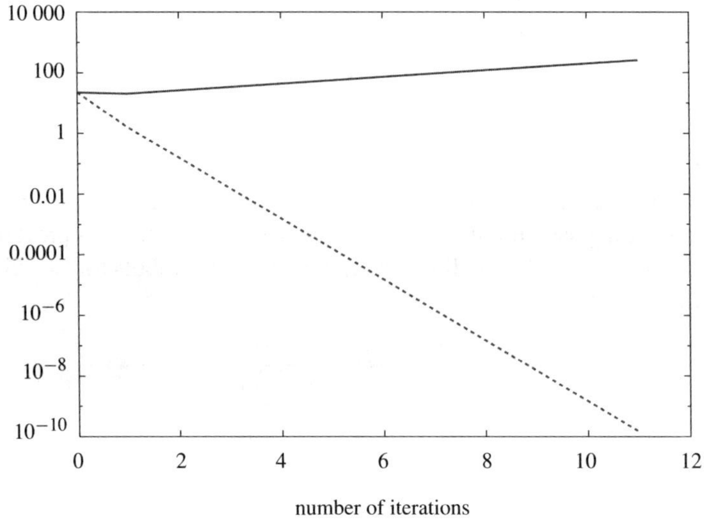

Figure 3.67. Error norms for Gauss–Seidel and multi-grid

tool to speed up the convergence of the iterative solution of the corresponding Galerkin equations. In principal, multi-grid solvers are possible on any kind of hierarchical grids. On regular grids the solution spaces can be identified with consecutive scaling spaces of an MRA. This enables the use of wavelet techniques for

- the discretization on different grids,
- the restriction from a finer to a coarser grid,
- the uplift from a coarser to a finer grid.

This technique is not restricted to planar cases. But if the underlying problem is a planar convolution, all matrices involved can be computed on the fly from much smaller core matrices. In this case really large problems can be tackled even on a PC.

3.5 Gravity field modelling

Usually, gravity field modeling is understood as the determination of the parameters c_j of a linear combination of known base functions φ_j such that the linear combination in some sense is an optimal approximation of the gravitational potential V of the Earth:

$$V(\boldsymbol{x}) \approx \sum_{j=0}^{M} c_j \varphi_j(\boldsymbol{x}). \tag{3.226}$$

Traditionally, spherical harmonics are used as base functions

$$\varphi_j(\boldsymbol{x}) = H_{n,m}(\boldsymbol{x}), \quad j = \frac{(n+1)(n+2)}{2} + n - m. \tag{3.227}$$

This spherical harmonics approximation technique has been extensively developed over the last decades and it is still the most frequently applied method. Nevertheless, it suffers from the conceptional disadvantage that the support of the base function covers the whole unit sphere σ. As a consequence the change of a single coefficient c_j changes the whole gravity field model. Since the coefficients c_j are derived from observations, also the change of a single observation changes all coefficients. In order to overcome this disadvantage, localized base functions such as harmonic wavelets on the sphere can be used for gravity field modeling purposes. In general not the values of the gravitational potential V but the values of several functionals applied to V are available for the determination of the coefficients c_j. In the general context of Physical Geodesy such functionals can be

- the gravity disturbancy functional $L_x^{\delta g} V = \frac{\boldsymbol{x}}{|\boldsymbol{x}|} \nabla V$,
- the gravity anomaly functional $L_x^{\Delta g} V = -\left(\frac{\boldsymbol{x}}{|\boldsymbol{x}|} \nabla + \frac{2}{|\boldsymbol{x}|} \mathbf{I}\right) V$ or
- the satellite-gravity-gradiometry functional $L_x^{\delta\delta g} V = \left(\frac{\boldsymbol{x}}{|\boldsymbol{x}|} \nabla\right)^2 V$.

Gravity field modeling by harmonic wavelets aims at the determination of the wavelet spectrum

$$(WT)V(j, y) = \langle \Psi_{j,y}, V \rangle \tag{3.228}$$

from the observed values $m_j = L_j V$ of certain functionals L_j of V. In order to accomplish this goal, the functionals cannot be arbitrarily chosen but have to form a so-called fundamental system on σ.

Definition 3.5.1. A set of $M = \sum_{l=p}^{q} (2l+1)$ functionals L_j, $j = 1, \dots, M$ is called a *fundamental system* of $\mathrm{Harm}_{p,\dots,q}$, if the matrix

$$\begin{bmatrix} L_1 H_{p,-p} & \dots & L_M H_{p,-p} \\ \vdots & & \vdots \\ L_1 H_{q,q} & \dots & L_M H_{q,q} \end{bmatrix} \tag{3.229}$$

is regular.

In [35] it is shown that for pairwise distinct $\boldsymbol{x}_j \in \sigma$ the gravity disturbance functional $L_{x_j}^{\delta g}$, the gravity anomaly functionals $L_{x_j}^{\Delta g}$ and also the satellite-gravity-gradiometry functionals $L_{x_j}^{\delta\delta g}$ form fundamental systems.

For a fundamental system a numerical quadrature formula can be developed, which computes the scalar product (3.228) from the observed values m_j of the functionals.

Lemma 3.5.1. *Let*

$$\cdots V_j \subset V_{j+1} \subset \cdots \subset \mathcal{H}(\{A_n\},\sigma) \tag{3.230}$$

be an MRA of $\mathcal{H}(\{A_n\},\sigma)$. *Let* $f \in V_{j+1}$, *let*

$$\Psi_j(\boldsymbol{x},\boldsymbol{y}) = \sum_{n=2^j+1}^{2^{j+1}} \sum_{m=-n}^{n} A_n^{-2} H_{n,m}(\boldsymbol{x}) H_{nm}(\boldsymbol{y}), \tag{3.231}$$

and let L_j *be a fundamental system of* $\mathrm{Harm}_{0,\dots,j+1}$. *If the weights* a_{nm} *are defined as the solution of the linear equations*

$$\sum_{n=0}^{2^{j+1}} \sum_{m=-n}^{n} (L_j H_{n,m}) a_{nm} = m_k\,,\ k = 1,\dots,M, \tag{3.232}$$

then

$$\langle \Psi_{j,y}, f\rangle = \sum_{n=2^j+1}^{2^{j+1}} \sum_{m=-n}^{n} a_{nm} A_n^{-2} H_{n,m}(\boldsymbol{x}) \tag{3.233}$$

holds.

Proof. Since f is supposed to be an element of V_{j+1}, it has the series expansion

$$f = \sum_{n=0}^{2^{j+1}} \sum_{m=-n}^{n} a_{nm} H_{n,m}.$$

For $k = 0,\dots,M$ holds

$$m_k = L_k f = \sum_{n=0}^{2^{j+1}} \sum_{m=-n}^{n} a_{nm} L_k H_{n,m}.$$

This is a linear system of equations for the coefficients a_{nm}, which has a unique solution since the functionals L_k form a fundamental system. Consequently

$$\begin{aligned}
\langle \Psi_{j,y}, f\rangle &= \Big\langle \sum_{n=2^j+1}^{2^{j+1}} \sum_{m=-n}^{n} A_n^{-2} H_{n,m}(\boldsymbol{x}) H_{n,m}(\bullet), f(\bullet)\Big\rangle \\
&= \sum_{n=2^j+1}^{2^{j+1}} \sum_{m=-n}^{n} A_n^{-2} H_{n,m}(\boldsymbol{x}) \sum_{p=0}^{j+1} \sum_{q=-p}^{p} a_{pq} \langle H_{n,m}, H_{p,q}\rangle \\
&= \sum_{n=2^j+1}^{2^{j+1}} \sum_{m=-n}^{n} a_{nm} A_n^{-2} H_{n,m}(\boldsymbol{x}).
\end{aligned}$$

□

The lemma shows how the wavelet decomposition of the gravitational potential can be obtained from the observed values $m_k = L_k$ of certain linear bounded functionals of this potential. In this way an almost arbitrary data combination can be used for a harmonic wavelet representation of the gravitational potential.

Nevertheless, some particularities of this approach have to me mentioned:

1. The approach uses the assumption $f \in V_{j+1} \subset \mathcal{H}(\{A_n\}, \sigma)$, which is not fulfilled in reality. But if the index j is chosen large enough, this is a sufficiently accurate approximation to reality.

2. Even if only the wavelet approximation $(WT)f(i, y)$ on a single scale i is to be computed, the total number of $M = (j+1)^2$ linear functionals L_k are necessary, which additionally have to form a fundamental system.

3. For the construction of exact quadrature formulas, the solution of a linear system of equations of the dimension M is necessary for the determination of the weights a_{nm} of these quadrature formulas.

4. Using the fact that the wavelets are rapidly decreasing, approximate quadrature formulas have been developed, which do not require the solution of a system of linear equations.

A vast amount of contributions about wavelet analysis comes from the Geomathematical Group of the Kaiserslautern University. For years this group, led and inspired by W. Freeden, develops the theory and application of harmonic wavelets on the sphere. In the period from 1999 to 2003 the application of the now mature theory to the analysis of the Earth's gravitational and magnetic field was in the focus of the activities of this group. Papers dealing with the spherical wavelet analysis of the Earth's magnetic field are [6], [5] and [4]. General questions related to the representation of scalar and vectorial functions on a sphere by spherical wavelets are discussed in [7], [26], [25], [28] and [32]. Since practically no data is noise free also the question of the denoising of the given signal using spherical wavelets has been discussed in the papers [29], [30], [31] and [33]. A great impact on geodetic research was made by the launches of the dedicated gravity field satellite missions CHAMP and GRACE. Naturally, these new data sources also have been reflected in techniques for the analysis of these data. Paper which are related to data collected from CHAMP or GRACE are [26], [34], [23], [24], [69], [70] and [74].

A Hilbert spaces

A.1 Definition of Hilbert spaces

The concept of a Hilbert space is the generalization of the familiar three-dimensional vector space into infinite dimensions. The typical properties of the three-dimensional vector space are:

- A scalar product $\langle \vec{x}, \vec{y} \rangle$ between vectors $\vec{x}, \vec{y}$ is defined.
- The length $|\vec{x}|$ of a vector $\vec{x}$ can be deduced from the scalar product: $|\vec{x}| = \sqrt{\langle \vec{x}, \vec{x} \rangle}$.
- There are three mutual orthogonal vectors $\vec{e}_1, \vec{e}_2, \vec{e}_3$ having unit length.
- Every vector $\vec{x}$ can be expressed as linear combination of these three base vectors $\vec{x} = x_1\vec{e}_1 + x_2\vec{e}_2 + x_3\vec{e}_3$.
- The weights x_1, x_2, x_3 in this linear combination are the projections of $\vec{x}$ to the corresponding base-vector: $x_i = \langle \vec{x}, \vec{e}_i \rangle$.

  The weights are frequently called coordinates of $\vec{x}$ with respect to the base $\vec{e}_1, \vec{e}_2, \vec{e}_3$.
- The square of length of a vector is equal to the sum of the squares of its coordinates

  $|\vec{x}|^2 = x_1^2 + x_2^2 + x_3^3$.

The theory of Hilbert spaces generalizes this concept with respect to two directions:

- First, the elements of a Hilbert space do not have to be vectors. They can also be different mathematical objects.
- Secondly, the dimension can be infinite.

In the following, a short sketch of the theory of Hilbert spaces will be given. The basic building-block of a Hilbert space is a so-called real linear space.

Definition A.1.1. A set V is called a *linear space* over the real numbers, if two operations

$$\oplus : V \times V \to V \quad \text{and} \quad \odot : \mathbb{R} \times V \to V, \tag{A.1}$$

called addition and scaling, are defined having the following properties:

$$x \oplus y = y \oplus x, \quad x, y \in V, \tag{A.2}$$

$$x \oplus (y \oplus z) = (x \oplus y) \oplus z, \quad x, y, z \in V. \tag{A.3}$$

There is a neutral element $\mathcal{O}$ with

$$x \oplus \mathcal{O} = x, \quad x \in V. \tag{A.4}$$

For each element x there is an inverse element $-x$, with

$$x \oplus (-x) = \mathcal{O}. \tag{A.5}$$

The following distributive laws are valid:

$$\alpha \odot (\beta \odot x) = (\alpha\beta) \odot x, \qquad x \in V,\ \alpha, \beta \in \mathbb{R}, \tag{A.6}$$
$$(\alpha + \beta) \odot x = (\alpha \odot x) \oplus (\beta \odot x), \quad x \in V,\ \alpha, \beta \in \mathbb{R}, \tag{A.7}$$
$$\alpha \odot (x \oplus y) = (\alpha \odot x) \oplus (\alpha \odot y), \quad x, y \in V,\ \alpha \in \mathbb{R}. \tag{A.8}$$

Obviously, the familiar vectors in the three-dimensional space with the operations "$\oplus$" = vector addition and "$\odot$" = prolongation are a linear space in the sense of the definition given above. But also other objects, like polynomials of maximum degree n, form a linear space with respect to the operations

$$(P \oplus Q)(t) := P(t) + Q(t), \tag{A.9}$$
$$(\alpha \odot P)(t) := \alpha P(t). \tag{A.10}$$

Besides the operations $\oplus$, $\odot$, the scalar product is the next structure in a linear space.

Definition A.1.2. A mapping $\langle \bullet, \bullet \rangle : V \times V \to \mathbb{R}$ is called a *scalar product* in V, if the following relations hold:

$$\langle x, y \rangle = \langle y, x \rangle, \qquad x, y \in V, \tag{A.11}$$
$$\langle x \oplus y, z \rangle = \langle x, z \rangle + \langle y, z \rangle, \quad x, y, z \in V, \tag{A.12}$$
$$\langle \alpha \odot x, y \rangle = \alpha \langle x, y \rangle, \qquad x, y \in V,\ \alpha \in \mathbb{R}, \tag{A.13}$$
$$\langle x, x \rangle \geq 0, \qquad \langle x, x \rangle = 0 \Leftrightarrow x = \mathcal{O}. \tag{A.14}$$

Again, it is easy to see that the traditional scalar product of three-dimensional vectors

$$\langle \vec{x}, \vec{y} \rangle := |\vec{x}||\vec{y}| \cos \varphi$$

fulfills all of the above mentioned conditions. Hence it is also a scalar product in the sense of the Hilbert space theory.

For each linear space with a scalar product an important result, the so-called Cauchy–Schwartz inequality, holds.

Theorem A.1.1. *Let V be a linear space with the scalar product $\langle \bullet, \bullet \rangle$ then the inequality*

$$|\langle x, y \rangle| \leq \sqrt{\langle x, x \rangle} \cdot \sqrt{\langle y, y \rangle} \tag{A.15}$$

holds.

Proof. [13]. □

Every Hilbert space has a mapping, which measures the length of its elements. This mapping is called a norm.

Definition A.1.3. A mapping $\|\bullet\| : V \to \mathbb{R}$ is called a *norm* in V, if the following conditions are fulfilled:

$$\|\alpha x\| = |\alpha| \cdot \|x\|, \qquad \alpha \in \mathbb{R},\ x \in V, \tag{A.16}$$

$$\|x\| \geq 0, \qquad x \in V, \tag{A.17}$$

$$\|x\| = 0 \Leftrightarrow x = \mathcal{O}, \tag{A.18}$$

$$\|x \oplus y\| \leq \|x\| + \|y\|. \tag{A.19}$$

Remark. There is an instructive interpretation for these three conditions (A.16)–(A.19): The first condition means that the length of a scaled element equals the length of the original element multiplied by the scaling factor. The second condition indicates that the length of an element of a linear space always has to be larger than or equal to zero. The length zero is only possible if the element itself is the neutral element. The last condition is also known under the name *triangle inequality*. It states that the sum of the lengths of two sides of a triangle is always larger than the length of the third side.

Once a scalar product is given, there is a simple way to derive a norm from this scalar product.

Lemma A.1.1. *The mapping* $\|\bullet\| : V \to \mathbb{R}^+$ *defined by*

$$\|x\| := \sqrt{\langle x, x\rangle} \tag{A.20}$$

is a norm in V.

Proof. (A.17) and (A.18) are direct consequences of (A.14).

Moreover,

$$\|\alpha x\| = \sqrt{\langle \alpha x, \alpha x\rangle} = \sqrt{\alpha^2 \langle x, x\rangle} = |\alpha| \sqrt{\langle x, x\rangle} = |\alpha| \cdot \|x\|,$$

which shows (A.16).

The triangle inequality (A.19) can by proved by using (A.15):

$$\begin{aligned}
\|x \oplus y\| &= \sqrt{\langle x \oplus y, x \oplus y\rangle} = \sqrt{\langle x, x\rangle + 2\langle x, y\rangle + \langle y, y\rangle} \\
&\leq \sqrt{\langle x, x\rangle + 2\sqrt{\langle x, x\rangle}\sqrt{\langle y, y\rangle} + \langle y, y\rangle} \\
&= \sqrt{\left(\sqrt{\langle x, x\rangle} + \sqrt{\langle y, y\rangle}\right)^2} \\
&= \|x\| + \|y\|.
\end{aligned}$$

□

Besides the norm, derived from the scalar product, there are many other possibilities to define a norm. Whether a given norm is compatible with the scalar product can be tested by the so-called parallelogram equation.

Lemma A.1.2. *A norm $\|\bullet\|$ is compatible with the scalar product $\langle\bullet,\bullet\rangle$, if and only if the equation*

$$\|x \oplus y\|^2 + \|x \ominus y\|^2 = 2(\|x\|^2 + \|y\|^2) \tag{A.21}$$

holds.

Proof. [50]. □

With the help of the norm, the concept of convergence can be introduced.

Definition A.1.4. A sequence $\{x_n\} \subset V$ is called *convergent* to $x \in V$, if for every $\varepsilon > 0$ there is an index $n_0 = n_0(\varepsilon)$, with

$$\|x_n \ominus x\| \leq \varepsilon, \quad \text{for all } n > n_0. \tag{A.22}$$

Every convergent sequence has an interesting property: It is a so-called Cauchy sequence

Definition A.1.5. A sequence $\{x_n\}$ is called *Cauchy sequence*, if for every $\varepsilon > 0$ there is an index $n_0 = n_0(\varepsilon)$ with

$$\|x_m \ominus x_n\| < \varepsilon, \quad n, m \geq n_0. \tag{A.23}$$

Lemma A.1.3. *Every convergent sequence is a Cauchy sequence.*

Proof. Let $\{x_n\}$ converge to x and let $\varepsilon > 0$. Then there exists an n_0 with

$$\|x \ominus x_n\| < \frac{\varepsilon}{2}, \quad n \geq n_0.$$

Using the triangle inequality (A.16) one obtains for $n, m > n_0$

$$\|x_n \ominus x_m\| \leq \|x \ominus x_n\| + \|x \ominus x_m\| \leq \frac{\varepsilon}{2} + \frac{\varepsilon}{2} = \varepsilon.$$ □

It has to be mentioned that the reverse statement is not true. There are Cauchy sequences which have no limit, as will be shown in the following example.

Example A.1.1. Let V be the set of rational numbers. With respect to the operations $\oplus = +$ and $\odot = \cdot$, the set V is a linear space over $\mathbb{Q}$. The scalar product in V is the usual multiplication of rational numbers and the norm of a rational number equals its absolute value.

Consider the sequence $\{x_n\} = \{\frac{a_n+b_n}{2}; n = 0, 1, 2 \ldots\}$ given as follows. Let

$$a_0 = 1, \quad b_0 = 2$$

and set

$$a_{n+1} = \begin{cases} a_n, & (a_n^2 - 2)(x_n^2 - 2) < 0, \\ x_n, & \text{else}; \end{cases}$$

$$b_{n+1} = \begin{cases} b_n, & (b_n^2 - 2)(x_n^2 - 2) < 0, \\ x_n, & \text{else} \end{cases}$$

for $n = 0, 1, 2, \ldots$

Obviously, $\|x_n - x_{n-1}\| = 2^{-n-1}$ holds. Hence, for $m > n$ the following relation ist true:

$$\begin{aligned} \|x_m - x_n\| &= \|x_m - x_{m-1} + x_{m-1} - \cdots - x_n\| \\ &\le \|x_m - x_{m-1}\| + \|x_{m-1} - x_{m-2}\| + \cdots + \|x_{n+1} - x_n\| \\ &= 2^{-n-2}(1 + 2 + \cdots + 2^{-m+n+1}) \\ &= 2^{-n-2}(2 - 2^{-m+n+1}) \\ &\le 2^{-n-1} \end{aligned}$$

On the other hand from the definition of the sequence $\{x_n\}$ follows

$$|a_n - b_n| = 2^{-n}, \quad a_n < \sqrt{2} \quad \text{and} \quad \sqrt{2} < b_n.$$

This leads to the conclusion

$$\begin{aligned} |x_n - \sqrt{2}| &= |b_n - a_n - \sqrt{2} + a_n - b_n + x_n| \le |b_n - a_n| + |a_n - \sqrt{2}| + |x_n - b_n| \\ &\le |b_n - a_n| + |a_n - b_n| + |a_n - b_n| = 3|b_n - a_n| \\ &= 3 \cdot 2^{-n} \to 0 \quad \text{for } n \to \infty. \end{aligned}$$

Since $\sqrt{2} \notin V = \mathbb{Q}$, and since a sequence cannot have more than one limit, the sequence $\{x_n\}$ is not convergent.

The deeper reason for the fact that in the previous example a Cauchy sequence is not convergent is that the rational numbers are not complete. The space of rational numbers has – as one might say – holes: The irrational numbers. Therefore an indication of the completeness of a linear space is that all Cauchy sequences are convergent.

Definition A.1.6. A linear space V with scalar product is called *complete*, if all Cauchy sequences in V are convergent.

Now all preparations for the definition of a Hilbert space have been made:

Definition A.1.7. A linear space H with scalar product $\langle \bullet, \bullet \rangle$ which is complete, is called a *Hilbert space*.

A.2 Complete orthonormal systems in Hilbert spaces

So far the concept of a Hilbert space has been defined. We now introduce the dimension of a Hilbert space. The concept of the dimension of a linear space is closely connected to the concept of a base of a linear space.

Definition A.2.1. The elements $x_1, x_2 \ldots, x_n \in V$ of a linear space V are called *linear independent*, if

$$\mathcal{O} = \alpha_1 \odot x_1 \oplus \alpha_2 \odot x_2 \oplus \cdots \oplus \alpha_n \odot x_n \Leftrightarrow \alpha_1 = \alpha_2 = \cdots = \alpha_n = 0 \quad (A.24)$$

holds. Otherwise they are called *linear dependent.*

Remark. The concept of linear dependence is motivated by the fact that for linear dependent elements it is always possible to express one element by the remaining elements: Let for instance $x_1, \ldots, x_n$ be linear dependent. Then there are real numbers $\alpha_1, \ldots, \alpha_n$, such that

$$\mathcal{O} = \alpha_1 \odot x_1 \oplus \cdots \oplus \alpha_n \odot x_n$$

holds and at least one α_i is not zero. Without any restriction of generality let $\alpha_1 \neq 0$. Then

$$x_1 = \frac{\alpha_2}{-\alpha_1} \odot x_2 \oplus \cdots \oplus \frac{\alpha_n}{-\alpha_1} \odot x_n$$

follows, i.e. the element x_1 is dependent on $x_2, \ldots, x_n$.

Definition A.2.2. The maximal number of linear independent elements of a linear space V is called the *dimension* of V.

The idea behind the introduction of linear independent elements in a linear space is to represent every element of the linear space as a linear combination of the independent elements. This representation is particularly simple, if the linear space is a Hilbert space.

Definition A.2.3. Let H be a Hilbert space. Two elements $x, y \in H$ are called *orthogonal*, if

$$\langle x, y \rangle = 0 \quad (A.25)$$

holds. A subset $D \subset H$ is called an *orthogonal system* in H, if any two different elements of D are orthogonal.

If for all elements $z \in D$ of an orthogonal system $\|z\| = 1$ holds, D is called an *orthonormal system.*

Remark. The definition of orthogonality is again motivated from the prototype of the three-dimensional vectors. For two orthogonal vectors $\vec{x}, \vec{y}$ the relation

$$\langle \vec{x}, \vec{y} \rangle = |\vec{x}| \cdot |\vec{y}| \cos\left(\frac{\pi}{2}\right) = 0$$

holds. This motivates the usage of a vanishing scalar product as an indicator of orthogonality.

An important property of an orthogonal system is the linear independence of its elements. Let $D = \{e_1, e_2, \dots, e_n\}$ be an orthogonal system in H and let hold

$$\mathcal{O} = \alpha_1 \odot e_1 \oplus \cdots \oplus \alpha_n \odot e_n.$$

Then it follows that

$$0 = \langle e_i, \mathcal{O} \rangle = \alpha_i \langle e_i, e_i \rangle, \quad i = 1, \dots, n.$$

This is only possible for $\alpha_i = 0, i = 1, \dots, n$, i.e. the elements of D are linear independent. Since with the help of a linear independent system the representation of all other elements of a Hilbert space as a linear combination of them is possible, the question remains, how to choose the weights in this linear combination. The answer is particularly simple, if the independent elements form an orthonormal system.

Definition A.2.4. Let $D = \{e_n\}$ be an orthonormal system in a Hilbert space H. For an arbitrary element $x \in H$ the numbers

$$\langle x, e_k \rangle, \quad k = 1, 2, \dots \tag{A.26}$$

are called the *Fourier coefficients* of x with respect to the orthonormal system D.

The name Fourier coefficient for the scalar product $\langle x, e_k \rangle$ can be motivated by a simple example.

Example A.2.1. Let H be the set of all functions x, defined on $[-\pi, \pi]$ which fulfill the condition

$$\int_{-\pi}^{\pi} x^2(t)\, dt < \infty.$$

Then H is a linear space with respect to the operations $(x \oplus y)(t) := x(t) + y(t)$ and $(\alpha \odot x)(t) := \alpha \cdot x(t)$. By

$$\langle x, y \rangle := \frac{1}{2\pi} \int_{-\pi}^{\pi} x(t) \cdot y(t)\, dx$$

a scalar product is introduced in H. The elements

$$e_0 = 1, \; e_{2n} = \sqrt{2}\cos(nt), \; e_{2n-1} = \sqrt{2}\sin(nt), \quad n = 1, 2, \dots$$

form an orthonormal system in H. If now, according to (A.26), the abstract Fourier coefficients are computed, one obtains

$$\langle x, e_0 \rangle = \frac{1}{2\pi} \int_{-\pi}^{\pi} x(t)\, dt =: a_0,$$

$$\langle x, e_{2n}\rangle = \frac{\sqrt{2}}{2\pi}\int_{-\pi}^{\pi} x(t)\cos(nt)\,dt =: a_n,$$

$$\langle x, e_{2n-1}\rangle = \frac{\sqrt{2}}{2\pi}\int_{-\pi}^{\pi} x(t)\sin(nt)\,dt =: b_n.$$

For the complex numbers $a_n + \imath b_n$ holds

$$\begin{aligned} a_n + \imath b_n &= \frac{\sqrt{2}}{2\pi}\int_{-\pi}^{\pi} x(t)(\cos(nt) + \imath\sin(nt))\,dt \\ &= \sqrt{2}\frac{1}{2\pi}\int_{-\pi}^{\pi} x(t)e^{\imath nt}\,dt. \end{aligned}$$

If this is compared to the classical definition (1.1) of the Fourier coefficients, one obtains $\langle x, e_n\rangle = \sqrt{2}c_{-n}$, i.e. the abstract definition (A.26) of Fourier coefficients is a generalization of its classical definition.

The Fourier coefficients $\langle x, e_n\rangle$ of an element x with respect to an orthonormal system $\{e_k\}$ define the best approximation of x by the orthonormal system: If one considers the approximation error $r := x - \sum_{k=1}^{n}\langle x, e_k\rangle e_k$, the relation

$$\langle r, e_n\rangle = \langle x, e_n\rangle - \sum_{k=1}^{n}\langle x, e_k\rangle\langle e_k, e_n\rangle = \langle x, e_n\rangle - \langle x, e_n\rangle = 0$$

holds, i.e. the approximation cannot be improved further by the orthonormal system. An improvement would only be possible if the orthonormal system was enhanced by additional independent elements. If such an enhancement is not necessary, the orthonormal system is called complete.

Definition A.2.5. Let $\{e_n\}$ be an orthonormal system in H. The orthonormal system is called *a complete orthonormal system* (CONS), if

$$0 = \langle z, e_n\rangle \Leftrightarrow z = 0, \quad n = 1, 2, \ldots \tag{A.27}$$

holds.

For complete orthonormal systems the so-called *orthonormal expansion theorem* is valid.

Theorem A.2.1. *Let H be a Hilbert space and $\{e_n\}$ be an orthonormal system in H. Then the following statements are equivalent:*

- *The orthonormal system $\{e_n\}$ is complete.*
- *Every $x \in H$ has a Fourier expansion*

$$x = \sum_{n=1}^{\infty}\langle x, e_n\rangle e_n. \tag{A.28}$$

- *For every $x \in H$ Parseval's identity*

$$\|x\|^2 = \sum_{n=1}^{\infty} \langle x, e_n \rangle^2 \tag{A.29}$$

holds.

Proof. [18]. □

A.3 Linear functionals – dual space

The theory of Hilbert spaces is the core of a mathematical discipline, called *Functional Analysis*. In this context the concept of a functional is the generalization of the concept of a function. A function is an injective mapping from the real numbers into the real numbers. A functional is more general, it is a mapping from a linear space into the real numbers. The linear functionals are especially important.

Definition A.3.1. Let V be a linear space. A mapping $f : V \to \mathbb{R}$ is called a *linear functional*, if the following conditions are fulfilled

$$f(x \oplus y) = f(x) + f(y), \quad x, y \in V, \tag{A.30}$$

$$f(\alpha \odot x) = \alpha \cdot f(x), \quad \alpha \in \mathbb{R},\ x \in V. \tag{A.31}$$

If the linear space is equipped with a scalar product the concepts of continuity and boundedness can also be defined for linear functionals.

Definition A.3.2. Let V be a linear space with the scalar product $\langle \bullet, \bullet \rangle$. A linear functional $f : V \to \mathbb{R}$ is called *continuous*, if

$$\lim_{n\to\infty} x_n = x \quad \Rightarrow \lim_{n\to\infty} f(x_n) = f(x) \tag{A.32}$$

holds.

Definition A.3.3. Let V be a linear space with the scalar product $\langle \bullet, \bullet \rangle$. A linear functional $f : V \to \mathbb{R}$ is called *bounded*, if there is a constant $M > 0$ with

$$|f(x)| \le M \cdot \|x\|, \quad x \in V. \tag{A.33}$$

In contrast to the case of classical functions, where a continuous function is not automatically bounded, for linear functionals the concepts of continuity and boundedness coincide.

Theorem A.3.1. *A linear functional on a linear space with scalar product is continuous if and only if it is bounded.*

Definition A.3.4. The set V^* of all linear, bounded functionals on V is called the *dual space* of V.

If the linear space V is even a Hilbert space H, the elements of the dual space H^* can be characterized by the elements of H. This is the content of *Riesz's* representation theorem.

Theorem A.3.2 (Riesz's representation theorem). *Let H be a Hilbert space. For each $f \in H^*$ there is a uniquely determined $g \in H$, such that*

$$f(x) = \langle g, x \rangle, \quad x \in H \tag{A.34}$$

holds.

Proof. [60]. □

A.4 Examples of Hilbert spaces

After compiling the most important concepts and theorems of the theory of Hilbert spaces, now some examples will be given. The examples are all taken from geodetic applications.

Example A.4.1 (The space of real n-tuples). The space $\mathbb{R}^n$ of the real n-tuples $x = (x_1, \dots, x_n)$ form a linear space with respect to the operations

$$\begin{aligned} x \oplus y &:= (x_1 + y_1, \dots, x_n + x_n), \\ \alpha \odot x &:= (\alpha x_1, \dots, \alpha x_n). \end{aligned}$$

This can be proved by verifying that the conditions starting with (A.1) are fulfilled.

$$x \oplus y = (x_1 + y_1, \dots, x_n + y_n) = (y_1 + x_1, \dots, y_n + x_n) = y \oplus x.$$

$$\begin{aligned} x \oplus (y \oplus z) &= x \oplus (y_1 + z_1, \dots, y_n + z_n) = (x_1 + y_1 + z_1, \dots, x_n + y_n + z_n) \\ &= (x_1 + y_1, \dots, x_n + yn_n) \oplus z = (x \oplus y) \oplus z. \end{aligned}$$

$$\begin{aligned} x \oplus \mathcal{O} &= (x_1, \dots, x_n) \oplus (0, \dots, 0) = (x_1 + 0, \dots, x_n + 0) \\ &= (x_1, \dots, x_n) = x. \end{aligned}$$

$$x \oplus (-x) = (x_1, \dots, x_n) \oplus (x_1 + (-x_1), \dots, x_n + (-x_n)) = \mathcal{O}.$$

$$\alpha \odot (\beta \odot x) = \alpha \odot (\beta_1 x_1, \dots, \beta_n x_n) = (\alpha_1 \beta_1 x_1, \dots, \alpha_n \beta_n x_n) = (\alpha\beta) \odot x.$$

$$\begin{aligned} \alpha \odot (x \oplus y) &= \alpha \odot (x_1 + y_1, \dots, x_n + y_n) = (\alpha x_1 + \alpha y_1, \dots, \alpha x_n + \alpha y_n) \\ &= \alpha \odot x \oplus \alpha \odot y. \end{aligned}$$

$$\begin{aligned} (\alpha + \beta) \odot x &= ((\alpha + \beta) x_1, \dots, (\alpha + \beta) x_n) = (\alpha x_1, \dots, \alpha x_n) \oplus (\beta x_1, \dots, \beta x_n) \\ &= \alpha \odot x \oplus \beta \odot x. \end{aligned}$$

A scalar product in $\mathbb{R}^n$ can be defined by

$$\langle x, y\rangle := \sum_{i=1}^{n} x_i \cdot y_i \,. \tag{A.35}$$

Obviously, the conditions (A.11)–(A.14) are fulfilled:

$$\langle x, y\rangle = \sum_{i=1}^{n} x_i \cdot y_i = \sum_{i=1}^{n} y_i \cdot x_i = \langle y, x\rangle,$$

$$\langle x, y \oplus z\rangle = \sum_{i=1}^{n} x_i(y_i + z_i) = \sum_{i=1}^{n} x_i \cdot y_i + \sum_{i=1}^{n} x_i \cdot z_i = \langle x, y\rangle + \langle x, z\rangle,$$

$$\langle x, x\rangle = \sum_{i=1}^{n} x_i^2 \geq 0,$$

$$\langle \alpha \cdot x, y\rangle = \sum_{i=1}^{n} (\alpha x_i) y_i = \alpha \sum_{i=1}^{n} x_i \cdot y_i = \alpha\langle x, y\rangle.$$

Hence the norm in $\mathbb{R}^n$ is

$$\|x\| := \sqrt{\sum_{i=1}^{n} x_i^2}. \tag{A.36}$$

Let now $\{x_n\}$ be a Cauchy sequence in $\mathbb{R}^n$. Then for every $i \in \{1, \ldots, n\}$ one has

$$|x_{i,n} - x_{i,m}| = \sqrt{(x_{i,n} - x_{i,m})^2} \leq \sqrt{\sum_{i=1}^{n} (x_{i,n} - x_{i,m})^2} = \|x_n - x_m\|.$$

Since $\{x_n\}$ is a Cauchy sequence, each component $\{x_{i,n}\}$ is a Cauchy sequence in $\mathbb{R}$. Due to the completeness of $\mathbb{R}$ there are limits of all component sequences

$$\lim_{n\to\infty} x_{i,n} = x_i \,.$$

Therefore, for every $\varepsilon > 0$, there is an $n_0 = n_0(\varepsilon)$ with

$$|x_i - x_{i,n}| \leq \frac{\varepsilon}{\sqrt{n}}, \quad n \geq n_0.$$

Hence

$$\|x - x_n\| = \sqrt{\sum_{i=1}^{n} (x_i - x_{i,n})^2} \leq \sqrt{\sum_{i=1}^{n} \left(\frac{\varepsilon}{\sqrt{n}}\right)^2} = \varepsilon.$$

Consequently, $\{x_n\}$ converges to x and therefore the space $\mathbb{R}^n$ is complete, i.e. $\mathbb{R}^n$ is a Hilbert space.

A set of vectors $\{e_1, \ldots, e_k\} \subset \mathbb{R}^n$ is linear independent, if the matrix

$$E := \begin{bmatrix} e_{1,1} & e_{2,1} & \cdots & e_{k,1} \\ e_{1,2} & e_{2,2} & \cdots & e_{k,2} \\ & & \vdots & \\ e_{1,n} & e_{2,n} & \cdots & e_{k,n} \end{bmatrix}$$

formed from these vectors has the full rank. Since the rank of an $n \times k$ matrix cannot exceed n, the maximal number of linear independent vectors in $\mathbb{R}^n$, the maximal number of independent vectors is n. Hence the dimension of $\mathbb{R}^n$ equals n. It is easy to see that the vectors

$$e_1 = (1, 0, 0, \ldots, 0), \; e_2 = (0, 1, 0, \ldots, 0), \ldots, e_n = (0, \ldots, 0, 1)$$

form a complete orthonormal system in the Hilbert space $\mathbb{R}^n$: Obviously,

$$\langle e_i, e_j \rangle = \sum_{k=1}^{n} e_{i,k} e_{j,k} = \delta_{ij}$$

holds, which means that the e_i are orthonormal. Furthermore, if for an arbitrary $x \in \mathbb{R}^n$ the relation

$$0 = \langle x, e_i \rangle, \quad i = 1, \ldots, n \tag{A.37}$$

is valid, then

$$0 = \langle x, e_i \rangle = \sum_{k=1}^{n} x_k e_{i,k} = x_i \quad i = 1, \ldots, n,$$

follows, which implicates $x = 0$. Therefore the orthonormal system $\{e_1, \ldots, e_n\}$ is complete. This orthonormal system is frequently called the *canonical base* of the Hilbert space $\mathbb{R}^n$.

Let f be an arbitrary linear functional from the dual space $(\mathbb{R}^n)^*$, and let

$$g_i := f(e_i), \quad i = 1, \ldots, n.$$

For the vector

$$g := (g_1, \ldots, g_n) \in \mathbb{R}^n$$

holds

$$f(x) = f\Big(\sum_{k=1}^{n} x_k e_k\Big) = \sum_{k=1}^{n} x_k f(e_k) = \sum_{k=1}^{n} x_k g_k = \langle g, x \rangle.$$

Ergo, the vector g is the Riesz-representer of the linear bounded functional f.

Example A.4.2 (The space of complex $2N$-tuples). The space $\mathbb{C}^{2N}$ of the complex $2N$-tuples $x = (x_0, \dots, x_{2N-1})$ forms a linear space with respect to the operations

$$x \oplus y := (x_0 + y_0, \dots, x_{2N-1} + y_{2N-1}),$$
$$\alpha \odot x := (\alpha x_0, \dots, \alpha x_{2N-1}).$$

The scalar product in $\mathbb{C}^{2N}$ is defined in the following way:

$$\langle x, y \rangle := \sum_{i=0}^{2N-1} x_i \overline{y_i} \tag{A.38}$$

and the norm is again

$$\|x\| := \sqrt{\langle x, x \rangle}.$$

The linear space $\mathbb{C}^{2N}$ is a Hilbert space with respect to the scalar product (A.38).

Besides the canonical base there is a different complete orthogonal system in $\mathbb{C}^n$, the so-called discrete trigonometric base.

Definition A.4.1. The vectors

$$f_k := (e^{-\imath(-N)\cdot k\frac{\pi}{N}}, e^{-\imath(-N+1)\cdot k\frac{\pi}{N}}, \dots, e^{-\imath(N-1)\cdot k\frac{\pi}{N}})/2N, \qquad k = -N, \dots, N-1 \tag{A.39}$$

are called *discrete trigonometric base.*

It is easy to see that the discrete trigonometric base forms a complete orthogonal system in $\mathbb{C}^{2N}$. The Fourier coefficients of a given vector g with respect to the discrete trigonometric base have a special meaning.

Lemma A.4.1. *Let* $g_i := g\big((i-N)\frac{\pi}{N}\big)$, $i = 0, \dots, 2N-1$ *and* $\tilde{g} := (g_0, \dots, g_{2N-1})$. *Then the Fourier coefficients of* $\tilde{g}$ *are identical with the DFT* (1.35), *see p.* 11, *of* g:

$$\bar{c}_n = \langle \tilde{g}, f_n \rangle, \quad n = -N, \dots, N-1. \tag{A.40}$$

Proof. With the relation (1.35) the following is true:

$$\begin{aligned}\langle \tilde{g}, f_n \rangle &= \sum_{k=-N}^{N-1} g_{k+N} e^{-\imath kn\frac{\pi}{N}} \\ &= \sum_{k=-N}^{N-1} g(k\frac{\pi}{N}) e^{-\imath kn\frac{\pi}{N}} \\ &= \bar{c}_n.\end{aligned}$$

□

Let h be an arbitrary linear functional $h \in (\mathbb{C}^{2N})^*$, and let $h_k := h(f_k) = h((e^{-\imath(-N)\cdot k\frac{\pi}{N}}, e^{-\imath(-N+1)\cdot k\frac{\pi}{N}}, \ldots, e^{-\imath(N-1)\cdot k\frac{\pi}{N}})/2N)$, then for every $x \in \mathbb{C}^{2N}$ holds

$$
\begin{aligned}
h(x) &= h\Big(\sum_{k=-N}^{N-1} x_k f_k\Big) = \sum_{k=-N}^{N-1} x_k h(f_k) \\
&= \sum x_k \bar{h}_k \\
&= \Big\langle x, \sum_{k=-N}^{N-1} h_k f_k \Big\rangle.
\end{aligned}
$$

Hence the Riesz representer of h is $\sum_{k=-N}^{N-1} h_k f_k$.

Example A.4.3. The space of summable sequences of complex numbers
Let $l_2(\mathbb{C})$ be the set of all complex sequences $\{x_n\}$, fulfilling

$$\sum_{n\in\mathbb{Z}} |x_n|^2 < \infty.$$

The set $l_2(\mathbb{C})$ forms a linear space with respect to the operations

$$
\begin{aligned}
x \oplus y &:= \{x_n + y_n\}, \\
\alpha \odot x &:= \{\alpha x_n\}.
\end{aligned}
$$

The operation

$$\langle x, y\rangle := \sum_{n\in\mathbb{Z}} x_n \bar{y}_n$$

defines a scalar product in $\ell_2(\mathbb{C})$, and with respect to the norm

$$\|x\| := \sqrt{\langle x, x\rangle} = \sqrt{\sum_{n\in\mathbb{Z}} |x_n|^2}$$

the set $l_2(\mathbb{C})$ is a Hilbert space.

Example A.4.4. The square integrable functions on $[-\pi, \pi]$

Definition A.4.2. A function $f : [-\pi, \pi] \to \mathbb{R}$ is called *square integrable* on $[-\pi, \pi]$, if the relation

$$\int_{-\pi}^{\pi} f^2(x)\, dx < \infty \tag{A.41}$$

holds. The set of all square integrable functions on $[-\pi, \pi]$ is denoted by $L_2([-\pi, \pi])$.

With respect to the operations

$$(f \oplus g)(x) := f(x) + g(x) \quad f, g \in L_2([-\pi, \pi]) \tag{A.42}$$

$$(\alpha \odot f)(x) := \alpha \cdot f(x) \quad f \in L_2([-\pi, \pi]), \alpha \in \mathbb{R} \tag{A.43}$$

the set $L_2([-\pi, \pi])$ is a linear space. A scalar product can be introduced by

$$\langle f, g \rangle := \int_{-\pi}^{\pi} f(x) \cdot g(x)\, dx. \tag{A.44}$$

With respect to the scalar product (A.44) the linear space $L_2([-\pi, \pi])$ becomes a Hilbert space. The dimension of $L_2([-\pi, \pi])$ is infinite and the trigonometric functions

$$\left\{ e^{-\imath nx} \sqrt{\frac{1}{2\pi}} \right\}, \quad n \in \mathbb{Z} \tag{A.45}$$

form a complete orthonormal system. The Fourier coefficients with respect to this orthonormal system are proportional to the classical Fourier coefficients:

$$\left\langle e^{-\imath nx} \sqrt{\frac{1}{2\pi}}, f(x) \right\rangle = \sqrt{\frac{1}{2\pi}} \int_{-\pi}^{\pi} f(x) e^{\imath nx}\, dx = \sqrt{2\pi}\, c_n.$$

Let again h be an arbitrary linear bounded functional on $L_2[-\pi, \pi]$. Define

$$h_n := h\left(e^{-\imath nx} \sqrt{\frac{1}{2\pi}} \right), \quad n \in \mathbb{Z} \tag{A.46}$$

and then the Riesz representer of h is given by

$$\sum_{n \in \mathbb{Z}} h_n e^{-\imath nx} \sqrt{\frac{1}{2\pi}}.$$

Example A.4.5 (Square integrable functions on $[-1, 1]$). A function $f : [-1, 1] \to \mathbb{R}$ is called *square integrable on* $[-1, 1]$, if the relation

$$\int_{-1}^{1} f^2(x)\, dx < \infty \tag{A.47}$$

holds. The set of all square integrable functions on $[-1, 1]$ is denoted by $L_2([-1, 1])$.

With respect to the operations

$$(f \oplus g)(x) := f(x) + g(x) \quad f, g \in L_2([-1, 1]) \tag{A.48}$$

$$(\alpha \odot f)(x) := \alpha \cdot f(x) \quad f \in L_2([-1, 1]), \alpha \in \mathbb{R} \tag{A.49}$$

the set $L_2([-1, 1])$ is a linear space. A scalar product can be introduced by

$$\langle f, g\rangle := \int_{-1}^{1} f(x) \cdot g(x)\, dx. \tag{A.50}$$

With respect to the scalar product (A.50) the linear space $L_2([-1, 1])$ becomes a Hilbert space. There are several complete orthogonal systems in $L_2([-1, 1])$ which will be introduced now.

Lemma A.4.2. *Let $m \geq 0$. Then the so-called associated Legendre functions*

$$P_n^{(m)}(t) := (1 - t^2)^{m/2} \frac{1}{2^n n!} \frac{d^{n+m}}{dt^{n+m}} (t^2 - 1)^n, \quad n \geq m \tag{A.51}$$

fulfill

$$\langle P_n^{(m)}, P_p^{(m)}\rangle = \frac{2}{2n+1} \frac{(n+m)!}{(n-m)!} \delta_{np}, \quad n, p \geq m. \tag{A.52}$$

For an arbitrary but fixed m the Legendre functions (A.51) are orthogonal but not orthonormal in $L_2([-1, 1])$. Hence, by simply normalizing the Legendre functions, a complete orthonormal system in $L_2([-1, 1])$ is obtained.

Lemma A.4.3. *The functions*

$$\bar{P}_n^{(m)}(t) := \sqrt{\frac{2n+1}{2} \frac{(n-m)!}{(n+m)!}} P_n^{(m)}, \quad n \geq m \tag{A.53}$$

are called fully normalized Legendre functions. They form a complete orthonormal system on $L_2([-1, 1])$.

Example A.4.6 (Square integrable functions on the unit sphere σ). A function $f : \sigma \to \mathbb{R}$ is called *square integrable on σ*, if the relation

$$\int_\sigma f^2(x)\, dx < \infty \tag{A.54}$$

holds. The set of all square integrable functions on σ is denoted by $L_2(\sigma)$.

With respect to the operations

$$(f \oplus g)(x) := f(x) + g(x) \quad f, g \in L_2(\sigma), \tag{A.55}$$

$$(\alpha \odot f)(x) := \alpha \cdot f(x) \quad f \in L_2(\sigma), \alpha \in \mathbb{R} \tag{A.56}$$

the set $L_2(\sigma)$ is a linear space. A scalar product can be introduced by

$$\langle f, g\rangle := \int_\sigma f(x) \cdot g(x)\, dx. \tag{A.57}$$

With respect to the scalar product (A.57) the linear space $L_2(\sigma)$ becomes a Hilbert space.

Now local surface coordinates are introduced in σ:

$$x := \begin{bmatrix} \sin(\vartheta)\cos(\lambda) \\ \sin(\vartheta)\sin(\lambda) \\ \cos(\vartheta) \end{bmatrix} \in \sigma, \quad \vartheta \in [0, \pi],\ \lambda \in [-\pi, \pi].$$

In these local coordinates the scalar product is expressed as

$$\langle f, g \rangle = \int_0^\pi \int_{-\pi}^\pi f(\vartheta, \lambda) g(\vartheta, \lambda) \sin(\vartheta) d\lambda\, d\vartheta. \tag{A.58}$$

With the substitution $t := \cos(\vartheta)$ the representation

$$\langle f, g \rangle = \int_{-1}^1 \int_{-\pi}^\pi f(\arccos t, \lambda) g(\arccos t, \lambda)\, d\lambda\, dt \tag{A.59}$$

can be found. Recalling example A.4.5 and example A.4.4, a complete orthonormal system of $L_2(\sigma)$ can be found by $\bar{P}_n^{(m)}(\cos\vartheta)e^{\imath m\lambda}\sqrt{\frac{1}{2\pi}},\ n = 0, 1, \ldots,\ m = 0, \ldots, m$, since

$$\begin{aligned}
&\left\langle \bar{P}_n^{(m)}(\cos\vartheta)e^{\imath m\lambda}\sqrt{\frac{1}{2\pi}}, \bar{P}_n^{(q)}(\cos\vartheta)e^{\imath q\lambda}\sqrt{\frac{1}{2\pi}} \right\rangle \\
&= \int_0^\pi \int_{-\pi}^\pi \bar{P}_n^{(m)}(\cos\vartheta)e^{\imath m\lambda}\sqrt{\frac{1}{2\pi}} \cdot \bar{P}_p^{(q)}(\cos\vartheta)\overline{e^{\imath q\lambda}}\sqrt{\frac{1}{2\pi}} \sin\vartheta\, d\lambda\, d\vartheta \\
&= \int_0^\pi \bar{P}_n^{(m)}(\cos\vartheta)\bar{P}_p^{(q)}(\cos\vartheta)\sin\vartheta \underbrace{\int_{-\pi}^\pi e^{\imath m\lambda}\overline{e^{\imath q\lambda}}\, d\lambda\, d\vartheta \cdot \frac{1}{2\pi}}_{\delta_{mq}} \\
&= \int_0^\pi \bar{P}_n^{(m)}(\cos\vartheta)\bar{P}_p^{(m)}(\cos\vartheta)\sin\vartheta\ d\lambda\, d\vartheta \\
&= \delta_{np}\delta_{mq}.
\end{aligned}$$

Changing from the complex to the real notation of the orthonormal base in $L_2(\sigma)$ the so-called surface spherical harmonics are found.

Definition A.4.3. The functions

$$\bar{Y}_{nm}(\vartheta, \lambda) := \sqrt{\frac{1}{\pi(1+\delta_{0n})}} \begin{cases} \bar{P}_n^{(|m|)}(\cos\vartheta)\cos(m\lambda) &, m \geq 0 \\ \bar{P}_n^{(|m|)}(\cos\vartheta)\sin(|m|\lambda) &, m < 0 \end{cases} \tag{A.60}$$

are called *fully normalized surface spherical harmonics*.

Lemma A.4.4. *The fully normalized surface spherical harmonics $\bar{Y}_{nm}$ form a complete orthonormal system in $L_2(\sigma)$.*

For later use an important property of the fully normalized surface spherical harmonics will be given now:

Lemma A.4.5. *Let (ϑ, λ) and (ϑ', λ') be two points on the unit sphere σ. Let ψ with*

$$\cos\psi = \cos\vartheta\cos\vartheta' + \sin\vartheta\sin\vartheta'\cos(\lambda - \lambda') \tag{A.61}$$

be the spherical distance between these two points. Then

$$P_n^{(0)}(\cos\psi) = \frac{4\pi}{2n+1}\sum_{m=-n}^{n} \bar{Y}_{nm}(\vartheta, \lambda)\bar{Y}_{nm}(\vartheta', \lambda') \tag{A.62}$$

holds.

A.5 Linear operators – Galerkin method

Besides linear functionals, mapping a Hilbert space into the real numbers, also linear operators, mapping a Hilbert space to another Hilbert space, have to be considered.

Definition A.5.1. Let G, H be two Hilbert spaces. A mapping $T : G \to H$ is called a *linear operator* from G to H, if

$$T(x \oplus y) = Tx + Ty, \quad x, y \in G, \tag{A.63}$$

$$T(\alpha \odot x) = \alpha \cdot (Tx), \quad \alpha \in \mathbb{R},\ x \in G \tag{A.64}$$

holds.

Definition A.5.2. A linear operator $T : G \to H$ is called *bounded*, if an $M > 0$ exists such that

$$\|Tx\|_H \leq \|x\|_G, \quad \text{for all } x \in G \tag{A.65}$$

is true. The space of all linear bounded operators is denoted by $\mathcal{L}(G, H)$.

Definition A.5.3. Let H be a Hilbert space. An operator $T \in \mathcal{L}(H, H)$ fulfilling

$$\langle Tx, x\rangle \geq \gamma\|x\|^2, \quad \gamma > 0, \tag{A.66}$$

for all $x \in H$ is called *strictly positive*.

Definition A.5.4. Let H be a Hilbert space and let $T \in \mathcal{L}(H, H)$. An operator $T^* \in \mathcal{L}(H, H)$ is called the *adjoint operator* of T, if for all $x, y \in H$

$$\langle Tx, y\rangle = \langle x, T^*y\rangle \tag{A.67}$$

holds.

If an operator coincides with its adjoint operator, it is called a self-adjoint operator.

Definition A.5.5. Let H be a Hilbert space. An operator $T \in \mathcal{L}(H, H)$ is called *selfadjoint*, if

$$\langle Tx, y\rangle = \langle x, Ty\rangle \tag{A.68}$$

holds for all $x \in H$.

Let us consider the operator equation

$$Tx = y, \quad x, y \in H. \tag{A.69}$$

The operator equation (A.69) has a unique solution, if T is strictly positive. In this case numerical methods have to be constructed for the approximation of the unique solution. One frequently used technique is Galerkin's method. The basic idea of Galerkin's method is to search for a solution of (A.69) not in the complete Hilbert space H, but in a finite dimensional subspace $H_n \subset H$. Let $\{e_1, \dots, e_n\}$ be a base of H_n, then the approximate solution $x_n \in H_n$ has the following representation:

$$x_n = \sum_{j=1}^{n} \alpha_j \cdot e_j. \tag{A.70}$$

No matter how the weights α_j are chosen, it cannot be expected that x_n fulfills the operator equation (A.69) exactly. There always remains a residual

$$r := y - \sum_{j=1}^{n} \alpha_j \cdot e_j. \tag{A.71}$$

A reasonable choice for the weights α_j is to make this residual as small as possible. In the context of Galerkin's method a residual is considered small, if it is orthogonal to all base functions:

$$0 = \langle r, e_i\rangle = \langle y, e_i\rangle - \sum_{j=1}^{n} \langle Te_j, e_i\rangle \alpha_j, \quad i = 1, \dots, n. \tag{A.72}$$

The equations (A.72) are a linear system of equations for the determination of the unknown optimal weights α_j. They are called *Galerkin equations*. The convergence of Galerkin's method is described by Cea's lemma.

Theorem A.5.1 (Cea's lemma). *Let H be a Hilbert space and $T \in \mathcal{L}(H, H)$ a strictly positive operator. Then equations* (A.69) *and* (A.72) *have unique solutions x, x_n and*

$$\|x - x_n\| \le \frac{M}{\gamma} \inf_{v \in H_n} \|x - v\| \tag{A.73}$$

holds.

Proof. [44]. □

A.6 Hilbert space valued random variables

Let H be a Hilbert space with the scalar product $\langle \bullet, \bullet \rangle$ and let $[\Omega\, \mathcal{A}, P]$ be a probability space.

The mathematical model for a random event giving real numbers as output is a random variable. Random variables are mappings from a probability space to the real numbers. In many applications the result of a random event is not only a single number, but a time or space dependent function. If these functions are members of some Hilbert space, these random events can be described mathematically by a Hilbert space valued random variable.

Definition A.6.1. A mapping $\xi : [\Omega\, \mathcal{A}, P] \to H$ is called *Hilbert space valued random variable*.

For a Hilbert space valued random variable, moments of first and second order have to be defined.

Definition A.6.2. $m \in H$ is called the *mean value* of a Hilbert space valued random variable ξ, if

$$E\{\langle f, \xi \rangle\} = \langle f, m \rangle \quad \text{for all } f \in H \tag{A.74}$$

holds.

Definition A.6.3. An operator $C : H \to H$ is called the *covariance operator* of a Hilbert space valued random variable ξ, if

$$E\{\langle f, \xi - m \rangle \cdot \langle g, \xi - m \rangle\} = \langle f, Cg \rangle \quad \text{for all } f, g \in H \tag{A.75}$$

holds.

Completely analog the cross-covariance between two *Hilbert* space valued random variables ξ and η is defined.

Definition A.6.4. An operator $C_{\xi\eta} : H \to H$ is called the *cross-covariance operator* of the two random variables ξ and η, if

$$E\{\langle f, \xi - m_\xi \rangle \cdot \langle g, \eta - m_\eta \rangle\} = \langle f, C_{\xi\eta} g \rangle, \quad \text{for all } f, g \in H \tag{A.76}$$

holds.

Let ξ be a random variable with the mean value zero and the covariance operator $C_{\xi\xi}$. To the unknown signal ξ some random noise will be added. This random noise is modeled by a Hilbert space valued random variable n with the mean value zero and the covariance operator C_{nn}. The signal ξ itself cannot be observed but the sum $\eta = \xi + n$ of the signal and the noise is supposed to be observable. The goal is to find an optimal prediction of $z = \langle f, \xi \rangle$, $f \in H$, from the noisy data η. In this context the concept of an optimal prediction is given by three requirements:

- The prediction has to be linear

$$\widehat{z} = \langle a, \eta \rangle. \tag{A.77}$$

- The prediction has to be unbiased

$$E\{\widehat{z}\} = z. \tag{A.78}$$

- Among all linear unbiased estimations it must have the minimal error variance

$$E\{(\widehat{z} - z)^2\} = \min . \tag{A.79}$$

Theorem A.6.1. *The optimal prediction of $z = \langle f, \xi \rangle$ is*

$$\widehat{z} = \langle a, \eta \rangle \tag{A.80}$$

with

$$(C_{\xi\xi} + C_{nn})a = C_{\eta\xi} f. \tag{A.81}$$

Proof. Since both ξ and n have zero mean values, the unbiasedness condition is automatically fulfilled. The estimation $\hat{z}$ is linear per definitionem. The only condition which has to be verified is the condition of minimal error variance:

$$\begin{aligned} E\{(\hat{z} - z)^2\} &= E\{(\langle f, \xi \rangle - \langle a, \eta \rangle)^2\} \\ &= \langle f, C_{\xi\xi} f \rangle - 2\langle f, C_{\xi\eta} a \rangle + \langle a, C_{\eta\eta} a \rangle \\ &\to \min . \end{aligned}$$

The necessary extremal condition is

$$0 = -2C_{\xi\eta} f + 2C_{\eta\eta} a,$$

which leads to the following Wiener–Kolmogorov equations for the determination of a:

$$(C_{\xi\xi} + C_{nn})a = C_{\xi\eta} f. \qquad \square$$

B Distributions

Many mathematical operations like differentiation or Fourier transform are only defined for functions being sufficiently regular. On the other hand, functions lacking this regularity or which are only piecewise regular occur in applications frequently. Hence the mathematical operations have to be generalized to be applicable to these functions as well. This is done by the generalization of the concept of a function and by the definition of mathematical operations for these generalized functions. This twofold generalization is performed using the so-called *regular test functions.*

Definition B.1. The space $C_0^\infty(\mathbb{R})$ of all functions φ which have a bounded support and which are infinitely often differentiable is called the *space of the test functions.*

In the space of the test functions the concept of convergence is introduced:

Definition B.2. A function $\varphi \in C_0^\infty(\mathbb{R})$ is called the *limit of a sequence* $\{\varphi_k\} \subset C_0^\infty(\mathbb{R})$ if

$$\lim_{k\to\infty} \frac{d^n}{dx^n}\varphi_k(x) = \frac{d^n}{dx^n}\varphi(x) \tag{B.1}$$

holds for all n and for all x.

Definition B.3. A mapping $f : C_0^\infty(\mathbb{R}) \to \mathbb{R}$ is called *generalized function* or *distribution*, if

1. $(f, \alpha\varphi) = \alpha(f, \varphi), \quad \alpha \in \mathbb{R}, \quad \varphi \in C_0^\infty(\mathbb{R})$
2. $(f, \varphi + \psi) = (f, \varphi) + (f, \psi), \quad \varphi, \psi \in C_0^\infty(\mathbb{R})$
3. $\lim_{k\to\infty}(f, \varphi_k) = (f, \varphi)$ for $\lim_{k\to\infty}\varphi_k = \varphi$

holds.

Remark. The definition of a generalized function is rather abstract and has little to do with the usual understanding of a function. Nevertheless, each traditional function f can be understood as a special case of a generalized function.

With the help of the traditional function f a mapping $\bar{f} : C_0^\infty(\mathbb{R}) \to \mathbb{R}$ is defined by

$$(\bar{f}, \varphi) := \int_{\mathbb{R}} f(x)\varphi(x)\,dx. \tag{B.2}$$

Obviously

$$\begin{aligned}
(\bar{f}, \alpha\varphi) &= \alpha(\bar{f}, \varphi),\\
(\bar{f}, \varphi + \psi) &= (\bar{f}, \varphi) + (\bar{f}, \psi),\\
\lim_{k\to\infty}(\bar{f}, \varphi_k) &= (\bar{f}, \varphi)
\end{aligned}$$

holds. Hence f is a generalized function. Those generalized functions which are induced by classical functions are called *regular generalized functions* or *regular distributions*.

Nevertheless, the concept of generalized functions is indeed an extension of the classical concept of functions, since there are generalized functions which cannot be induced by classical functions. An example for such an irregular distribution is the famous δ-function, defined by

$$(\delta, \varphi) := \varphi(0). \tag{B.3}$$

The δ-function is the mathematical model of the unit-pulse, i.e. the limit of rectangle functions with decreasing widths and unit integral

$$\delta_\varepsilon(x) := \frac{1}{2\varepsilon}\Pi\left(\frac{x}{\varepsilon}\right). \tag{B.4}$$

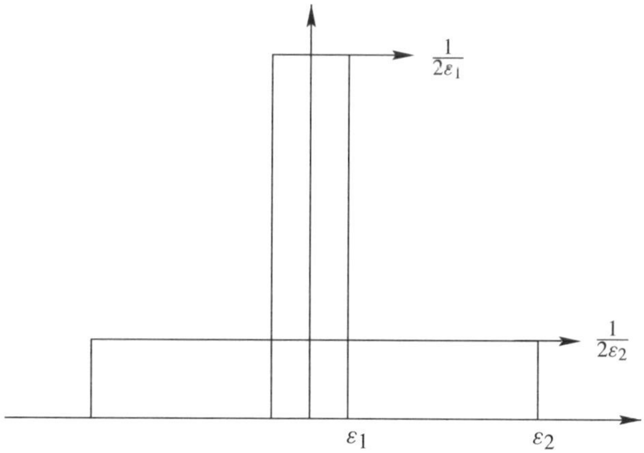

Figure B.1. Approximation of delta-distribution by rectangular functions

Obviously, for all $\varepsilon > 0$

$$\int_{\mathbb{R}} \delta_\varepsilon(x)\, dx = 1$$

holds, but the pointwise limit has no proper meaning:

$$\delta(x) := \lim_{\varepsilon \to 0} \delta_\varepsilon(x) = \begin{cases} \infty, & x = 0 \\ 0, & x \neq 0, \end{cases} \tag{B.5}$$

which contradicts the condition

$$\int_{\mathbb{R}} \delta(x)\, dx = 1.$$

This means that the δ-function cannot be the limit of rectangle functions in the classical but rather in a generalized sense:

$$\begin{aligned}
(\delta, \varphi) &:= \lim_{\varepsilon \to 0} \int_{\mathbb{R}} \delta_\varepsilon(x)\varphi(x)\,dx \\
&= \lim_{\varepsilon \to 0} \frac{1}{2\varepsilon} \int_{-\varepsilon}^{\varepsilon} \varphi(x)\,dx \\
&= \lim_{\varepsilon \to 0} \frac{1}{2\varepsilon} \int_{-\varepsilon}^{\varepsilon} \varphi(0) + \varphi'(\tau)x\,dx \\
&= \lim_{\varepsilon \to 0} \left(\varphi(0) + \frac{\varphi'(\tau)}{2\varepsilon} \left(\frac{\varepsilon^2}{2} - \frac{\varepsilon^2}{2} \right) \right) \\
&= \varphi(0).
\end{aligned}$$

Now mathematical operations have to be defined for generalized functions.

Definition B.4. A generalized function Df is called the *derivative of a generalized function* f, if

$$(Df, \varphi) = -(f, \varphi'), \quad \varphi \in C_0^\infty(\mathbb{R}) \tag{B.6}$$

holds.

First of all, it can be shown that if f is a regular distribution, the newly defined concept of a derivative coincides with the classical one:

$$\begin{aligned}
(Df, \varphi) = -(f, \varphi') &= -\int_{\mathbb{R}} f(x)\varphi'(x)\,dx \\
&= -\left(f(x)\varphi(x)\big|_{-\infty}^{\infty} - \int_{\mathbb{R}} f'(x)\varphi(x)\,dx \right) \\
&= (f', \varphi).
\end{aligned}$$

Additionally, the derivative in the distributional sense exists even for functions which do not have a classical derivative. Let us, for example, consider the ramp-function

$$r(x) := \begin{cases} 0, & x < -1 \\ \frac{1}{2}(x+1), & |x| \leq 1 \\ 1, & x > 1. \end{cases} \tag{B.7}$$

At the points $x = \pm 1$ the function r does not have a derivative in the classical sense.

For its distributional derivative we have

$$\begin{aligned}
(Dr, \varphi) &= -(r, \varphi') \\
&= -\int_{-1}^{1} \frac{1}{2}(x+1)\varphi'(x)\,dx - \int_{1}^{\infty} \varphi'(x)\,dx \\
&= -\frac{1}{2}(x+1)\varphi\Big|_{-1}^{1} + \int_{-1}^{1} \frac{1}{2}\varphi(x)\,dx - \varphi\Big|_{1}^{\infty} + \int_{1}^{\infty} 0 \cdot \varphi\,dx \\
&= \int_{-1}^{1} \frac{1}{2}\varphi(x)\,dx \\
&= \int_{-\infty}^{\infty} \frac{1}{2}\Pi(x)\varphi(x)\,dx.
\end{aligned}$$

This means that in the generalized sense the derivative of r exists and is equal to $\frac{1}{2}\Pi$.

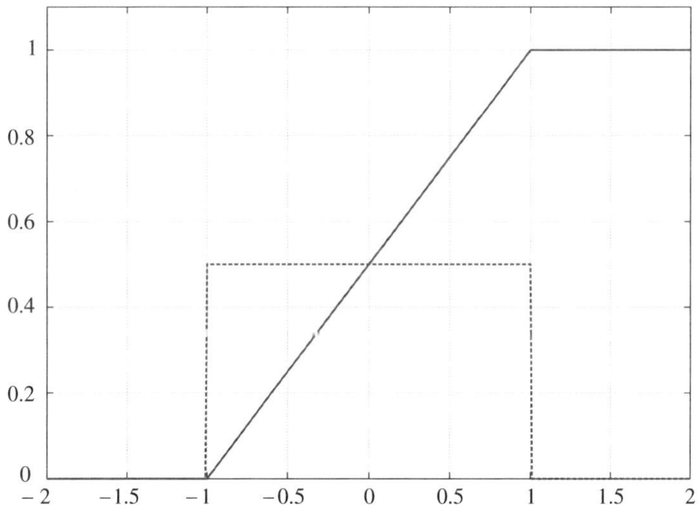

Figure B.2. Generalized derivative of ramp function

Another example is the generalized derivative of the Heaviside function

$$H(x) := \begin{cases} 0, & x < 0 \\ 1, & x \geq 0, \end{cases} \tag{B.8}$$

which represents a unit-jump. For its generalized derivative the following relation holds:

$$\begin{aligned}(DH, \varphi) &= -\int_{\mathbb{R}} H(x)\varphi'(x)\,dx \\ &= -\int_0^{\infty} \varphi'(x)\,dx \\ &= -\varphi(\infty) + \varphi(0) = \varphi(0) \\ &= (\delta, \varphi).\end{aligned}$$

This means that the δ -distribution is the generalized derivative of the Heaviside function.

The next important generalized operation is the generalized Fourier transform.

Definition B.5. A generalized function $\hat{f}$ is called *generalized Fourier transform of a distribution* f, if

$$(\hat{f}, \varphi) := (f, \mathcal{F}\{\varphi\}), \quad \varphi \in C_0^{\infty}(\mathbb{R}). \tag{B.9}$$

Again, it can be shown that if $\mathcal{F}\{f\}$ exists in the classical sense the generalized Fourier transform coincides with the classical one.

Lemma B.1. For the generalized Fourier transform the following rules apply:

$$\widehat{D^n f} = \imath^n \omega^n \hat{f}, \tag{B.10}$$

$$\widehat{(f(a\bullet))}(\omega) = \frac{1}{|a|}\hat{f}\left(\frac{\omega}{|a|}\right), \tag{B.11}$$

$$\widehat{f(\bullet + b)}(\omega) = e^{-\imath\omega b}\hat{f}(\omega), \tag{B.12}$$

$$\widehat{f * g}(\omega) = \sqrt{2\pi}\,\hat{f}(\omega) \cdot \hat{g}(\omega). \tag{B.13}$$

Here the operations convolution and differentiation have to be understood in the generalized sense.

Proof. See [103]. □

Example B.1. If $\delta(x - x_0)$ is defined by $(\delta(x - x_0), \varphi) = \varphi(x_0)$, the following relations hold:

- $\widehat{\delta(x - x_0)} = e^{-\imath x_0 \omega}\frac{1}{\sqrt{2\pi}}$,
- $\hat{1} = \sqrt{2\pi}\delta(x)$,
- $\hat{\delta} = \frac{1}{\sqrt{2\pi}}$.

Proof.

$$
\begin{aligned}
(\widehat{\delta(x-x_0)}, \varphi) &= (\delta(x-x_0), \hat{\varphi}) = \hat{\varphi}(x_0) \\
&= \frac{1}{\sqrt{2\pi}} \int_{\mathbb{R}} \varphi(\xi) e^{-\imath x_0 \xi}\, d\xi \\
&= \frac{1}{\sqrt{2\pi}} (e^{-\imath x_0 \xi}, \varphi(\xi)).
\end{aligned}
$$

$$
\begin{aligned}
(\hat{\delta}, \varphi) &= (\delta, \hat{\varphi}) = \hat{\varphi}(0) \\
&= \frac{1}{\sqrt{2\pi}} \int_{\mathbb{R}} e^{-\imath 0 x} \varphi(x)\, dx = \int_{\mathbb{R}} \left(\frac{1}{\sqrt{2\pi}} \cdot 1 \right) \varphi(x)\, dx \\
&= \left(\frac{1}{\sqrt{2\pi}} \cdot 1, \varphi \right).
\end{aligned}
$$

$$
\begin{aligned}
(\hat{1}, \varphi) &= (1, \hat{\varphi}) = \int_{\mathbb{R}} \hat{\varphi}(\omega)\, d\omega \\
&= \frac{\sqrt{2\pi}}{\sqrt{2\pi}} \int_{\mathbb{R}} e^{\imath 0 \omega} \hat{\varphi}(\omega)\, d\omega = \sqrt{2\pi}\, \varphi(0) \\
&= \left(\sqrt{2\pi}\, \delta, \varphi\right).
\end{aligned}
$$

□

Naturally, the theory of distributions is much more extended than the material presented here. This appendix serves as only the compilation of material, which is used in the main part of this book.

Exercises

Fourier analysis

Exercise 1. Prove the relation

$$ {}^n\mathcal{F}\{f(\bullet - \boldsymbol{b})\} = e^{-\imath \boldsymbol{b}^\top \omega} \cdot {}^n\mathcal{F}\{f\}(\omega). $$

Exercise 2. Show that

$$ {}^n\mathcal{F}\{f(a\cdot\bullet)\}(\omega) = \frac{1}{a}\,{}^n\mathcal{F}\{f\}(\frac{\omega}{a}) $$

is valid.

Exercise 3. Prove that

$$ \mathcal{F}\{f'\}(\omega) = \imath\omega\mathcal{F}\{f\}(\omega) $$

holds.

Exercise 4. Compute the Fourier transform of the following functions

$$ \varphi(x) = \begin{cases} 1, & 0 \le x \le 1 \\ 0, & \text{else,} \end{cases} $$

$$ \psi(x) = \begin{cases} -1, & 0 \le x < 0.5 \\ 1, & 0.5 \le x < 1 \\ 0, & \text{else} \end{cases} $$

and plot $|\hat{\varphi}|$, $|\hat{\psi}|$.

Exercise 5. With the help of the differentiation theorem compute the Fourier transform of

$$ \psi(x) = (1 - x^2)e^{-\frac{x^2}{2}} $$

and plot $|\hat{\psi}(\omega)|$.

Exercise 6. Compute the Fourier transformation of

$$ \psi(x) = \pi^{-1/4} e^{\imath\omega_0 x} e^{-\frac{x^2}{2}} $$

Exercise 7. Let

$$B_0(x) := \begin{cases} 1, & -0.5 \le x \le 0.5 \\ 0, & \text{else} \end{cases}$$

and $B_{n+1}(x) := B_0 * B_n(x)$. With the help of the convolution theorem compute the Fourier transform of B_n. For $n = 0, 1, 2, 3$ plot $\hat{B}_n$.

Exercise 8. Compute $B_1(x)$. Additionally, compute $B_i(x), i = 1, 2, 3, 4$ on the interval $[-3, 3]$ numerically with the help of a FFT algorithm. Compare the numerically computed B_1 with the exact B_1.

Exercise 9. Let $g_{t,\omega_0}(x) := \pi^{-1/4} e^{-\frac{(x-t)^2}{2}} e^{-\imath x \omega_0}$. Compute

- $\hat{g}_{t,\omega_0}(\omega)$
- $\int_{-\infty}^{\infty} |g_{t,\omega_0}|^2 \, dx$
- $\int_{-\infty}^{\infty} |\hat{g}_{t,\omega_0}|^2 \, d\omega$.

Exercise 10. Compute, for the same g as in Exercise 9,

- $\bar{x} = \int_{-\infty}^{\infty} x |g_{t,\omega_0}|^2 \, dx$
- $\bar{\omega} = \int_{-\infty}^{\infty} \omega |\hat{g}_{t,\omega_0}|^2 \, d\omega$
- $\sigma_x^2 := \int_{-\infty}^{\infty} (x - \bar{x})^2 |g_{t,\omega_0}|^2 \, dx$
- $\sigma_\omega^2 := \int_{-\infty}^{\infty} (\omega - \bar{\omega})^2 |\hat{g}_{t,\omega_0}|^2 \, d\omega$.

Linear Filters

Exercise 1. Compute the z-transform $H(z)$ of the convolution filters with the following coefficients:

- high pass filter: $h_0 = 1, \; h_1 = -1$,
- low -pss filters: $h_n = \frac{1}{2N+1}, \; n = -N, \dots, N$,

and plot $|H(e^{\imath\omega})|^2, \omega \in [-\pi, \pi], N = 1, 2, 3$.

Exercise 2. Show that the filters H, G, H^*, G^*, defined by

$$(Hu)_n := \sum_{k\in\mathbb{Z}} h_k u_{n-k}$$
$$(H^*u)_n := \sum_{k\in\mathbb{Z}} h_k^* u_{n-k}$$
$$(Gu)_n := \sum_{k\in\mathbb{Z}} g_k u_{n-k}$$
$$(G^*u)_n := \sum_{k\in\mathbb{Z}} g_k^* u_{n-k}$$

with coefficients given by

$$h_0 = \frac{1}{4\sqrt{2}}(1+\sqrt{3}), \quad h_{-1} = \frac{1}{4\sqrt{2}}(3+\sqrt{3}),$$
$$h_{-2} = \frac{1}{4\sqrt{2}}(3-\sqrt{3}), \quad h_{-3} = \frac{1}{4\sqrt{2}}(1-\sqrt{3})$$
$$g_k = (-1)^k h_{-k-1}, \quad h_k^* = h_{-k}, \quad g_k^* = g_{\ k}$$

form a PR filter bank.

Exercise 3. Determine all filters h of length 2, which fulfill

$$\sum_{k\in\mathbb{Z}} h_k h_{k-2n} = \delta_{0n} \quad \text{and} \quad \sum_{k\in\mathbb{Z}} h_k = \sqrt{2}.$$

Exercise 4. To the low-pass filter H of Exercise 3 compute the corresponding high-pass filter G according to

$$G(-z) = z^l H\left(\frac{1}{z}\right).$$

Garbor transform

Exercise 1. Compute the Garbor transform $\mathcal{G}\{f\}(\omega, t)$ of the signal

$$f(t) = e^{\imath \nu t}$$

and discuss the result.

Exercise 2. Besides the traditional Garbor transform a modified Garbor transform, including an additional parameter a, is frequently considered:

$$G_a\{f\}(\omega, t) := \frac{1}{\sqrt{2\pi}} \int_{-\infty}^{\infty} f(u) \cdot g_{t,\omega,a}(u)\, du$$

with

$$g_{t,\omega,a}(u) = \pi^{-1/4} a^{-1/4} e^{-\frac{(u-t)^2}{2a}} e^{-\imath\omega u}.$$

Compute $\mu_g, \mu_{\hat{g}}, \sigma_g^2, \sigma_{\hat{g}}^2$ as functions of a.

Exercise 3. For any choice of the window parameter a, determine grid-points (t_i, ω_j), $i, j \in \mathbb{Z}$ in such a way that the resolution windows

$$[t_i - \sigma_g, t_i + \sigma_g] \times [\omega_j - \sigma_{\hat{g}}, \omega_j + \sigma_{\hat{g}}]$$

covers the whole (t, ω) phase-space.

Exercise 4.

Compute the parameter dependent Garbor transform

$$\mathcal{G}_a\{f\}(\omega, t)$$

of the signal

$$f(t) = e^{\imath\nu t}$$

and discuss the influence of the parameter a on the result.

Continuous wavelet transformation

Exercise 1. Show that the function

$$\psi_1(x) = (1 - x^2) e^{-\frac{x^2}{2}}$$

is a wavelet and determine the corresponding constants c_ψ.

Exercise 2. The continuous wavelet transform

$$W\{f\}(\lambda, t) = \frac{1}{\sqrt{c_\psi}} |\lambda|^{-1/2} \int_{-\infty}^{\infty} f(u) \psi\left(\frac{u-t}{\lambda}\right) du$$

can be considered as a linear filter applied to the signal f. Compute the filter characteristic of these filters for the wavelets

$$\psi(x) = \begin{cases} -1, & 0 \le x < 0.5 \\ 1, & 0.5 \le x < 1 \\ 0, & \text{else,} \end{cases}$$

$$\psi(x) = (1 - x^2)e^{-\frac{x^2}{2}},$$

$$\psi(x) = \pi^{-1/4}e^{\iota\omega_0 x}e^{-\frac{x^2}{2}}.$$

Exercise 3. For $\omega_0 = 5.0$ and for $\lambda \in \{1.0, 2.0, 4.0, 8.0\}$ make plots of the filter characteristics of the three wavelets from Exercise 2.

Exercise 4. For the Haar wavelet, the Mexican hat wavelet and the Morlet wavelet compute the moments

$$\bar{x} := \int_{-\infty}^{\infty} x|\psi^2(x)|\,dx,$$

$$\sigma_x^2 = \int_{-\infty}^{\infty} (x - \bar{x})^2|\psi(x)|^2\,dx,$$

$$\bar{\omega} = \int_{-\infty}^{\infty} \omega|\hat{\psi}(\omega)|^2\,d\omega,$$

$$\sigma_\omega^2 = \int_{-\infty}^{\infty} (\omega - \bar{\omega})^2|\hat{\psi}(\omega)|^2\,d\omega.$$

Exercise 5. Let $\bar{x}, \sigma_{x^2}, \bar{\omega}, \sigma_\omega^2$ be the moments for a wavelet ψ. Compute the moments for the shifted and dilated wavelet $\lambda^{-1/2}\psi(\frac{x-t}{b})$.

Exercise 6. The resolution window of the wavelet $\lambda^{-1/2}\psi(\frac{x-t}{\lambda})$ is

$$[\bar{x}_{\lambda,t} - \sigma_{x,\lambda,t}, \bar{x}_{\lambda t} + \sigma_{x,\lambda,t}] \times [\bar{\omega}_{\lambda t} - \sigma_{\omega\,\lambda,t}, \bar{\omega}_{\lambda t} + \sigma_{\omega,\lambda,t}].$$

Find grid points t_i, λ_j in such a way that the corresponding resolution windows

$$[\bar{x}_{\lambda_j,t_i} - \sigma_{x,\lambda_j,t_i}, \bar{x}_{\lambda_j\,t_i} + \sigma_{x,\lambda_j,t_i}] \times [\bar{\omega}_{\lambda_j,t_i} - \sigma_{\omega\,\lambda_j,t_i}, \bar{\omega}_{\lambda_j,t_i} + \sigma_{\omega,\lambda_j,t_i}]$$

cover the whole (x, ω) phase plane.

Exercise 7. Let the signal f be sampled with a spacing Δx. Define the dilatation parameter $\lambda_{\min}$ such that the $\frac{\bar{\omega}}{\lambda_{\min}} = \nu$, with ν being the Nyquist frequency.

Let T be the sampling interval. Define the dilatation parameter $\lambda_{\max}$ such that $\frac{\bar{\omega}}{\lambda_{\max}} = \frac{2\pi}{T}$ holds.

Mallat algorithm

Exercise 1. Let ψ be the Haar wavelet. Show that the functions

$$\psi_{m,n}(x) = 2^{-m/2}\psi(2^{-m}x - n)$$

form an orthonormal system.

Exercise 2. For $\varphi \in L^2(\mathbb{R})$ let hold

$$\langle \varphi(\bullet - k), \varphi(\bullet - n) \rangle = \delta_{kn}.$$

Show that this property is equivalent to

$$\sum_{n \in \mathbb{Z}} |\hat{\varphi}(\omega + 2\pi n)|^2 = \frac{1}{2\pi}.$$

Hilbert spaces

Exercise 1. Prove that all square matrices of the dimension n form a vector space $L(\mathbb{R}^n)$with respect to the operations $\oplus$ (= matrix addition) and $\odot$ (= multiplication with a real number).

Exercise 2. Prove that

$$\|A\| := \max_i \sum_{j=1}^{n} |a_{ij}|$$

is a norm in $L(\mathbb{R}^n)$.

Exercise 3. Let $V := \left\{ f : [a, b] \to \mathbb{R}^1 | \int_a^b f^2(x)\,dx < \infty \right\}$. Show that V forms a linear space with respect to the operations

$$\oplus : V \times V \to V : (f \oplus g)(x) = f(x) + g(x)$$
$$\odot : \mathbb{R}^1 \times V \to V : (\alpha \odot f)(x) := \alpha \cdot f(x)$$

Prove that the product,

$$\langle f, g \rangle := \int_a^b f(x) g(x)\,dx$$

is always defined and that $\langle f, g \rangle$ is a scalar product in V.

Exercise 4. Let $V := \left\{ f : [0, 2\pi] \to \mathbb{R}^1 \mid \int_0^{2\pi} f^2(x)\,dx \right\}$. Show that the functions $\left\{ \cos(mx), \sin(mx), m = 0, 1, \ldots \right\}$ are orthogonal with respect to the scalar product

$$\langle f, g \rangle := \int_0^{2\pi} f(x) \cdot g(x)\,dx$$

Exercise 5. Let $\mathbb{C}^n := \{\boldsymbol{c} = (c_1, \dots c_n) \mid c_i \in \mathbb{C}\}$ equipped with the scalar product

$$\langle \boldsymbol{c}, \boldsymbol{d} \rangle := \frac{2}{n} \sum_{i=1}^{n} c_i \bar{d}_i .$$

Let

$$\boldsymbol{c}_k = \left(\cos\left(\frac{2\pi k 0}{n}\right), \dots, \cos\left(\frac{2\pi k(n-1)}{n}\right) \right)$$
$$\boldsymbol{d}_k = \left(\sin\left(\frac{2\pi k 0}{n}\right), \dots, \sin\left(\frac{2\pi k(n-1)}{n}\right) \right).$$

Show that $\boldsymbol{c}_k, \boldsymbol{d}_k$ are orthogonal elements of $\mathbb{C}^n$.

Exercise 6. Let $C[a, b]$ be the set of all functions which are continuous on $[a, b]$. Show that

$$\|f\| = \max_{x \in [a,b]} |f(x)|$$

is a norm on $C[a, b]$.

Exercise 7. Show that $f : C[a, b] \to \mathbb{R}^1$, $f(x) := \int_a^b x(t)\, dt$ is a linear bounded functional on $C[a, b]$. Compute the norm of f.

Exercise 8. Show that the evaluation functional $ev_{t_0}(x) := x(t_0)$ is a linear bounded functional on $C[a, b]$.

Exercise 9. Let

$$x_n(t) := \begin{cases} 0, & -1 \le t \le -\frac{1}{n} \\ \frac{n}{2}(t + \frac{1}{n}), & -\frac{1}{n} \le t \le \frac{1}{n} \\ 1, & \frac{1}{n} \le t \le 1 \end{cases}$$

and

$$x(t) = \begin{cases} 0, & -1 \le t < 0 \\ 1, & 0 \le t \le 1. \end{cases}$$

Show that $\lim_{n\to\infty} x_n = x$ holds in $L^2([-1, 1])$.

Exercise 10. Show that the evaluation functional $ev_0(x) := x(0)$ is *not* a linear bounded functional on $L^2([-1, 1])$.

Solutions

Fourier analysis

Exercise 1.

$$\begin{aligned}
{}^n\mathcal{F}\{f(\bullet - \boldsymbol{b})\}(\omega) &= (2\pi)^{-n/2}\int_{\mathbb{R}^n} f(\mathbf{x}-\boldsymbol{b})e^{-\imath\mathbf{x}^\top\omega}d\mathbf{x}\\
&= (2\pi)^{-n/2}\int_{\mathbb{R}^n} f(\mathbf{x})e^{-\imath(\mathbf{x}+\boldsymbol{b})^\top\omega}d\mathbf{x}\\
&= e^{-\imath\boldsymbol{b}^\top\omega}(2\pi)^{-n/2}\int_{\mathbb{R}^n} f(\mathbf{x})e^{-\imath\mathbf{x}^\top\omega}d\mathbf{x}\\
&= e^{-\imath\boldsymbol{b}^\top\omega}\cdot{}^n\,\mathcal{F}\{f(\bullet)\}(\omega).
\end{aligned}$$

Exercise 2.

$$\begin{aligned}
{}^n\mathcal{F}\{f(a\cdot\bullet)\}(\omega) &= (2\pi)^{-n/2}\int_{\mathbb{R}^n} f(a\mathbf{x})e^{-\imath\mathbf{x}^\top\omega}d\mathbf{x}\\
&= \frac{1}{a}(2\pi)^{-n/2}\int_{\mathbb{R}^n} f(\mathbf{x})e^{-\imath\mathbf{x}^\top\frac{\omega}{a}}d\mathbf{x}\\
&= \frac{1}{a}{}^n\,\mathcal{F}\{f(\bullet)\}\left(\frac{\omega}{a}\right).
\end{aligned}$$

Exercise 3. Integration by parts yields

$$\begin{aligned}
\mathcal{F}\{f'(\cdot)\}(\omega) &= \frac{1}{\sqrt{2\pi}}\int_{\mathbb{R}} f'(x)e^{-\imath x\omega}dx\\
&= \frac{1}{\sqrt{2\pi}}\left(f(x)\cdot e^{-\imath x\omega}\right)_{-\infty}^{\infty} - \frac{1}{\sqrt{2\pi}}\int_{\mathbb{R}} f(x)(-\imath\omega)e^{-\imath x\omega}dx\\
&= \frac{\imath\omega}{\sqrt{2\pi}}\int_{\mathbb{R}} f(x)e^{-\imath x\omega}d\mathbf{x} = \imath\omega\mathcal{F}\{f(\bullet)\}(\omega).
\end{aligned}$$

Exercise 4.

$$\begin{aligned}
\hat{\varphi}(\omega) &= \frac{1}{\sqrt{2\pi}}\int_{-\infty}^{\infty}\varphi(x)e^{-\imath x\omega}dx = \frac{1}{\sqrt{2\pi}}\int_0^1 e^{-\imath x\omega}dx\\
&= \frac{1}{\sqrt{2\pi}}\left[-\frac{1}{\imath\omega}e^{-\imath x\omega}\right]_0^1 = \frac{1}{\sqrt{2\pi}}\left(-\frac{1}{\imath\omega}(e^{-\imath\omega}-1)\right)\\
&= \frac{1}{\imath\omega\sqrt{2\pi}}(1-e^{-\imath\omega}) = \frac{e^{-\imath\omega/2}}{\sqrt{2\pi}}\frac{e^{\imath\omega/2}-e^{-\imath\omega/2}}{\imath\omega}\\
&= \frac{e^{-\imath\omega/2}}{\sqrt{2\pi}}\frac{\sin(\omega/2)}{\omega/2}.
\end{aligned}$$

$$
\begin{aligned}
\hat{\psi}(\omega) &= \frac{1}{\sqrt{2\pi}} \int_{-\infty}^{\infty} \psi(x) e^{-\imath x\omega} dx \\
&= \frac{-1}{\sqrt{2\pi}} \int_{0}^{1/2} e^{-\imath x\omega} dx + \frac{1}{\sqrt{2\pi}} \int_{1/2}^{1} e^{-\imath x\omega} dx \\
&= \frac{1}{\sqrt{2\pi}} \frac{-1}{\imath\omega} \left(e^{-\imath\omega} - 2 \cdot e^{-\imath\omega/2} + 1 \right) \\
&= \frac{e^{-\imath\omega/2}}{\sqrt{2\pi}} \frac{-1}{\imath\omega} \left(e^{-\imath\omega/2} - 2 + e^{\imath\omega/2} \right) \\
&= \frac{e^{-\imath\omega/2}}{\sqrt{2\pi}} \frac{-1}{\imath\omega} \left(e^{\imath\omega/4} - e^{-\imath\omega/4} \right)^2 \\
&= \frac{\imath e^{-\imath\omega/2}}{\sqrt{2\pi}} \frac{\sin(\omega/4)\sin(\omega/4)}{(\omega/4)}.
\end{aligned}
$$

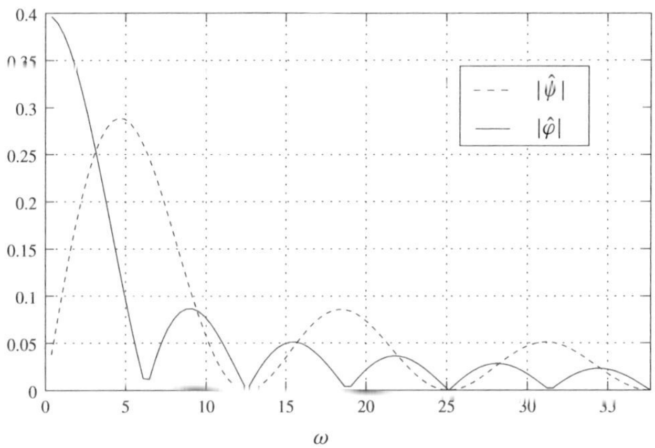

Spectra $|\hat{\varphi}|$, $|\hat{\psi}|$ of the Haar scaling function φ and the Haar wavelet ψ

Exercise 5. Obviously

$$
\psi(x) = -\frac{d^2}{dx^2} e^{-\frac{x^2}{2}}
$$

is true. Consequently

$$
\hat{\psi}(\omega) = \omega^2 \cdot \mathcal{F}\left\{ e^{-\frac{x^2}{2}} \right\} = \omega^2 e^{-\frac{\omega^2}{2}}.
$$

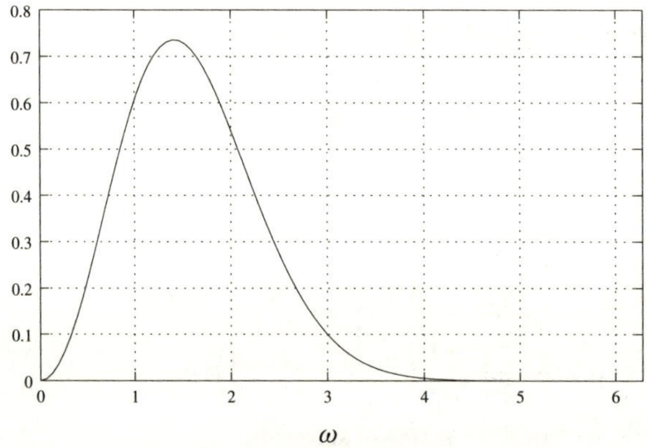

Spectrum of the Mexican hat wavelet

Exercise 6.

$$\begin{aligned}\hat{\psi}(\omega) &= \frac{1}{\sqrt{2\pi}} \int_{-\infty}^{\infty} \pi^{-1/4} e^{-\frac{x^2}{2}} e^{-\iota(\omega-\omega_0)x} dx \\ &= \pi^{-1/4} \mathcal{F}\left\{e^{-\frac{x^2}{2}}\right\}(\omega - \omega_0) \\ &= \pi^{-1/4} e^{-(\omega-\omega_0)^2/2}.\end{aligned}$$

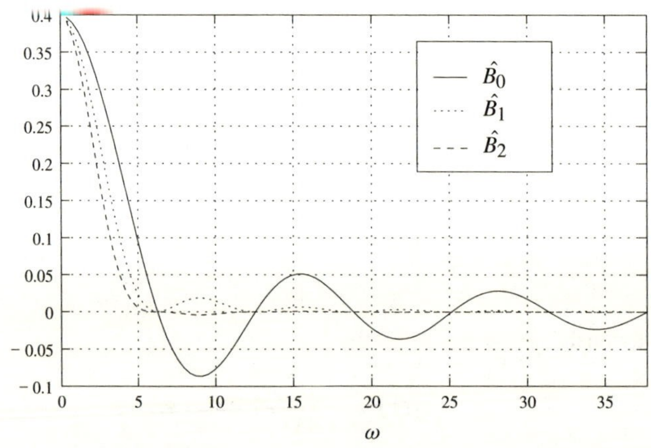

Spectra of spline wavelets B_0, B_1, B_2

Exercise 7. Obviously $B_0(x) = \varphi(x + \frac{1}{2})$. Using the shifting theorem, one obtains

$$\hat{B}_0(\omega) = e^{\imath\omega/2}\hat{\varphi}(\omega) = \frac{1}{\sqrt{2\pi}} \operatorname{sinc}(\omega/2).$$

Hence

$$\hat{B}_n = \sqrt{2\pi}^n (\hat{B}_0)^{n+1} = \frac{1}{\sqrt{2\pi}} \operatorname{sinc}^{n+1}(\omega/2).$$

Exercise 8.

$$\begin{aligned}
B_1(x) &= \int_{-\infty}^{\infty} B_0(y)B_0(x-y)dy = \int_{-\frac{1}{2}}^{\frac{1}{2}} B_0(x-y)dy \\
&= -\int_{x+1/2}^{x-1/2} B_0(z)dz = \int_{x-1/2}^{x+1/2} B_0(z)dz \\
&= \begin{cases} 0, & |x| > 1 \\ \int_{-1/2}^{x+1/2} dz, & -1 \le x \le 0 \\ \int_{x-1/2}^{1/2} dz, & 0 \le x < 1 \end{cases} \\
&= \begin{cases} 0, & |x| > 1 \\ x+1, & -1 \le x \le 0 \\ 1-x, & 0 \le x \le 1. \end{cases}
\end{aligned}$$

Exercise 9.

$$\begin{aligned}
\hat{g}_{t,\omega_0}(\omega) &= \frac{1}{\sqrt{2\pi}} \int_{-\infty}^{\infty} \pi^{-1/4} e^{-\frac{(x-t)^2}{2}} e^{-\imath(\omega+\omega_0)x} dx \\
&= \pi^{-1/4} \mathcal{F}\left\{e^{-\frac{(x-t)^2}{2}}\right\}(\omega+\omega_0) = \pi^{-1/4} e^{-\imath\omega t} \mathcal{F}\left\{e^{-\frac{x^2}{2}}\right\}(\omega+\omega_0) \\
&= \pi^{-1/4} e^{-\imath\omega t} e^{-\frac{(\omega+\omega_0)^2}{2}}.
\end{aligned}$$

$$\int_{-\infty}^{\infty} |g_{t,\omega_0}(x)|^2 dx = \pi^{-1/2} \int_{-\infty}^{\infty} e^{-(x-t)^2} dx = \frac{1}{\sqrt{2\pi\frac{1}{2}}} \int_{-\infty}^{\infty} e^{-\frac{z^2}{2\frac{1}{2}}} dz = 1.$$

From Parseval's theorem follows

$$\int_{-\infty}^{\infty} |\hat{g}_{t,\omega_0}|^2 d\omega = \int_{-\infty}^{\infty} |g_{t,\omega_0}|^2 dx = 1.$$

Exercise 10.

$$\int_{-\infty}^{\infty} x|g_{t,\omega_0}|^2 dx = \pi^{-1/2} \int_{-\infty}^{\infty} xe^{-(x-t)^2} dx = \frac{1}{\sqrt{2\pi\frac{1}{2}}} \int_{-\infty}^{\infty} xe^{-\frac{(x-t)^2}{2\frac{1}{2}}} dx = t.$$

$$\int_{-\infty}^{\infty} \omega|\hat{g}_{t,\omega_0}|^2 d\omega = \pi^{-1/2} \int_{-\infty}^{\infty} \omega e^{-(\omega+\omega_0)^2} d\omega = -\omega_0.$$

$$\sigma_x^2 = \pi^{-1/2} \int_{-\infty}^{\infty} (x-t)^2 e^{-(x-t)^2} dx = \frac{1}{\sqrt{2\pi \frac{1}{2}}} \int_{-\infty}^{\infty} z^2 e^{-\frac{z^2}{2\frac{1}{2}}} dz = \frac{1}{2}.$$

$$\sigma_\omega^2 = \pi^{-1/2} \int_{-\infty}^{\infty} (\omega + \omega_0) e^{-(\omega+\omega_0)^2} d\omega = 1/2.$$

Linear Filters

Exercise 1.

$$H(z) = 1 - z^{-1} = \frac{z-1}{z}$$

$$|H(e^{\imath\omega})|^2 = (1 - e^{-\imath\omega})(1 - e^{\imath\omega}) = 2 - 2\cos(\omega)$$

$$H(z) = \frac{1}{2N+1} \sum_{n=-N}^{N} z^{-n}$$

$$|H(z)| = \Big(\frac{1}{2N+1} \sum_{n=-N}^{N} e^{-\imath n\omega}\Big)^2 = \Big(\frac{1}{2N+1}[1 + 2\sum_{n=1}^{N} \cos(n\omega)]\Big)^2$$

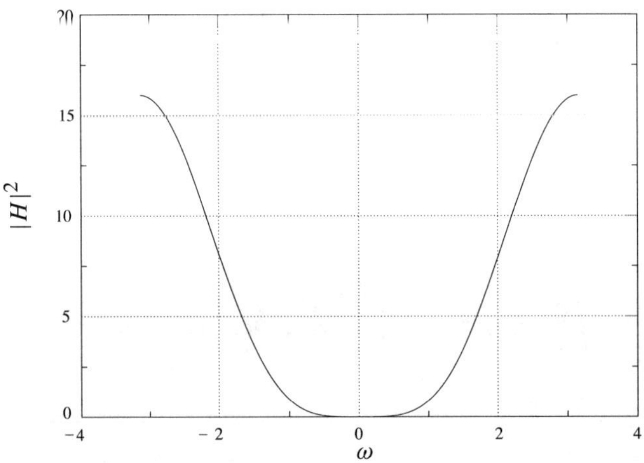

High pass filter characteristic $|H(e^{\imath\omega})|^2$

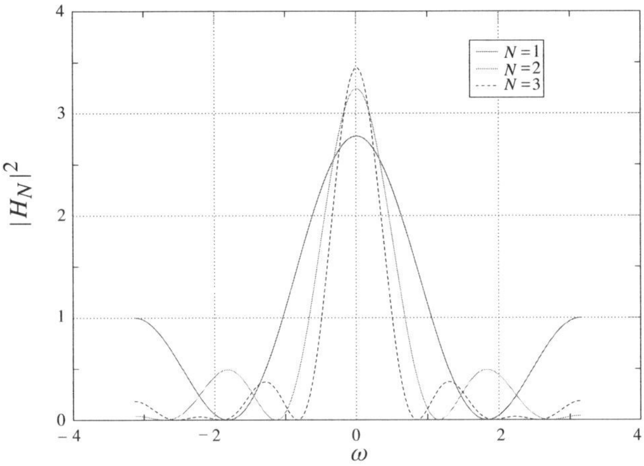

Low pass filter characteristics $|H_N(e^{\imath\omega})|^2$ for $N = 1, 2, 3$

Exercise 2. The z -transform of the filter H, H^* is

$$H(z) = h_0 + h_{-1}z + h_{-2}z^2 + h_{-3}z^3,$$

$$H^*(z) = h_0 + h_{-1}z^{-1} + h_{-2}z^{-2} + h_{-3}z^{-3},$$

$$G(z) = g_{-1}z + g_0 + g_1 z^{-1} + g_2 z^{-2} = h_0 z - h_{-1} + h_{-2}z^{-1} - h_{-3}z^{-2},$$

$$G^*(z) = g_1^* z^{-1} + g_0^* + g_{-1}^* z + g_{-2}^* z^2 = h_0 z^{-1} - h_{-1} + h_{-2}z - h_{-3}z^2.$$

Hence

$$\begin{aligned}
H(-z)H^*(z) + G(-z)G^*(z) &= (h_0 - h_{-1}z + h_{-2}z^2 - h_{-3}z^3) \\
&\quad \times (h_0 + h_{-1}z^{-1} + h_{-2}z^{-2} + h_{-3}z^{-3}) \\
&\quad + (-h_0 z - h_{-1} - h_{-2}z^{-1} - h_{-3}z^{-2}) \\
&\quad \times (h_0 z^{-1} - h_{-1} + h_{-2}z - h_{-3}z^2) \\
&= (h_0^2 - h_{-1}^2 + h_{-2}^2 - h_{-3}^2) \\
&\quad + (h_0 h_{-1} - h_{-1}h_{-2} + h_{-2}h_{-3})(z^{-1} - z) \\
&\quad + (h_0 h_{-2} - h_{-1}h_{-3})(z^{-2} + z^2) + h_0 h_{-3}(z^{-3} - z^3) \\
&\quad + (-h_0^2 + h_{-1}^2 - h_{-2}^2 + h_{-3}^2) \\
&\quad + (-h_0 h_{-1} + h_{-1}h_{-2} - h_{-2}h_{-3})(z^{-1} - z) \\
&\quad + (-h_0 h_{-2} + h_{-1}h_{-3})(z^{-2} + z^2) \\
&\quad - h_0 h_{-3}(z^{-3} - z^3) \\
&= 0
\end{aligned}$$

and

$$
\begin{aligned}
H(z)H^*(z) + G(z)G^*(z) = {} & (h_0 + h_{-1}z + h_{-2}z^2 + h_{-3}z^3)\times \\
& \times (h_0 + h_{-1}z^{-1} + h_{-2}z^{-2} + h_{-3}z^{-3}) \\
& + (h_0 z - h_{-1} + h_{-2}z^{-1} - h_{-3}z^{-2}) \\
& \times (h_0 z^{-1} - h_{-1} + h_{-2}z - h_{-3}z^2) \\
= {} & (h_0^2 + h_{-1}^2 + h_{-2}^2 + h_{-3}^2) \\
& + (h_0 h_{-1} + h_{-1}h_{-2} + h_{-2}h_{-3})(z + z^{-1}) \\
& + (h_0 h_{-2} + h_{-1}h_{-3})(z^2 + z^{-2}) \\
& + h_0 h_{-3}(z^3 + z^{-3}) \\
& + (h_0^2 + h_{-1}^2 + h_{-2}^2 + h_{-3}^2) \\
& + (-h_0 h_{-1} - h_{-1}h_{-2} - h_{-2}h_{-3})(z + z^{-1}) \\
& + (h_0 h_{-2} + h_{-1}h_{-3})(z^2 + z^{-2}) \\
& + (-h_0 h_{-3})(z^3 + z^{-3}) \\
= {} & 2((h_0^2 + h_{-1}^2 + h_{-2}^2 + h_{-3}^2) + (h_0 h_{-2} + h_{-1}h_{-3})(z^2 + z^{-2})) \\
= {} & \frac{1}{16}((1+\sqrt{3})^2 + (3+\sqrt{3})^2 + (3-\sqrt{3})^2 + (1-\sqrt{3})^2) \\
& + \frac{1}{32}((1+\sqrt{3})((3-\sqrt{3}) + (3+\sqrt{3})(1-\sqrt{3}))(z^2 + z^{-2}) \\
= {} & \frac{1}{16}(8+24) + \frac{1}{32}(2\sqrt{3} - 2\sqrt{3})(z^2 + z^{-2}) \\
= {} & ?
\end{aligned}
$$

Exercise 3. The two conditions form a system of two nonlinear equations for four unknowns:

$$
\begin{aligned}
h_0^2 + h_1^2 &= 1 \\
h_0 + h_1 &= \sqrt{2}
\end{aligned}
$$

The geometric interpretation of these two equations is the intersection of a straight line with the unit circle. The straight line has the normal vector $\boldsymbol{n} = (1, 1)^\top$ and the distance of the straight line form the origin equals 1.0, which is also the radius of the circle. The two intersection points therefore coincide. They form the so-called contact point p. The contact point is the intersection of the prolongation of the normal vector $\boldsymbol{n}$ with the unit circle. This yields

$$
p = \begin{bmatrix} h_0 \\ h_1 \end{bmatrix} = \begin{bmatrix} \cos \pi/4 \\ \sin \pi/4 \end{bmatrix} = \begin{bmatrix} \frac{1}{2}\sqrt{2} \\ \frac{1}{2}\sqrt{2} \end{bmatrix}
$$

Exercise 4. Obviously,

$$
H(z) = \frac{1}{2}\sqrt{2}(1 + z^{-1})
$$

holds. This leads to

$$\begin{aligned} g_0 - g_1 z^{-1} &= G(-z) \\ &= z^l H(\frac{1}{z}) \\ &= z^l \frac{1}{2}\sqrt{2}(1+z) \end{aligned}$$

This is only possible if $l = -1$ holds and leads to

$$g_0 = \frac{1}{2}\sqrt{2}, \; g_1 = -\frac{1}{2}\sqrt{2}.$$

Garbor transform

Exercise 1.

$$\begin{aligned} \mathcal{G}\{f\}(\omega, t) &= \frac{1}{\sqrt{2\pi}} \int_{\mathbb{R}} e^{\imath \nu u} \cdot \pi^{-1/4} e^{-(u-t)^2/2} e^{-\imath u \omega} du \\ &= \frac{1}{\sqrt{2\pi}} \int_{\mathbb{R}} \pi^{-1/4} e^{-(u-t)^2/2} e^{-\imath u(\omega-\nu)} du = \pi^{-1/4} \mathcal{F}\{e^{-(u-t)^2/2}\}(\omega - \nu) \\ &= \pi^{-1/4} e^{-\imath t \omega} \mathcal{F}\{e^{-u^2/2}\}(\omega - \nu) = \pi^{-1/4} e^{-\imath t \omega} e^{-\frac{(\omega-\nu)^2}{2}}. \end{aligned}$$

For a fixed frequency ω the Garbor transform of $e^{\imath \nu t}$ is a harmonic oscillation with this frequency ω. The amplitude of this oscillation is the smaller the more the frequency ω differs from the frequency ν of the transformed signal.

Exercise 2.

$$\int_{-\infty}^{\infty} x |g_{t,\omega_0,a}|^2 dx = \pi^{-1/2} a^{-1/2} \int_{-\infty}^{\infty} x e^{-\frac{(x-t)^2}{a}} dx = \frac{1}{\sqrt{2\pi \frac{1}{2} a}} \int_{-\infty}^{\infty} x e^{-\frac{(x-t)^2}{2\frac{1}{2}a}} dx = t.$$

$$\sigma_x^2 = \pi^{-1/2} a^{-1/2} \int_{-\infty}^{\infty} (x-t)^2 e^{-\frac{(x-t)^2}{a}} dx = \frac{1}{\sqrt{2\pi \frac{1}{2} a}} \int_{-\infty}^{\infty} z^2 e^{-\frac{z^2}{2\frac{1}{2}a}} dz = \frac{1}{2} a.$$

$$\begin{aligned} \hat{g}_{t,\omega_0,a}(\omega) &= \frac{1}{\sqrt{2\pi}} \int_{-\infty}^{\infty} \pi^{-1/4} a^{-1/4} e^{-\frac{(x-t)^2}{2a}} e^{-\imath(\omega+\omega_0)x} dx \\ &= \pi^{-1/4} \mathcal{F}\left\{a^{-1/4} e^{-\frac{(x-t)^2}{2a}}\right\}(\omega + \omega_0) \\ &= \pi^{-1/4} a^{1/4} e^{-\imath \omega t} \mathcal{F}\left\{e^{-\frac{x^2}{2}}\right\}(a^{1/4}(\omega + \omega_0)) \\ &= \pi^{-1/4} a^{1/4} e^{-\imath \omega t} e^{-\frac{a^{1/2}(\omega+\omega_0)^2}{2}}. \end{aligned}$$

Hence $\bar{\omega} = -\omega_0$ and $\sigma_\omega^2 = \frac{1}{2a}$ holds.

Exercise 3.

$$t_i := i\sqrt{2a}, \quad \omega_j := j\sqrt{\frac{2}{a}}.$$

Exercise 4.

$$\begin{aligned}
\mathcal{G}_a\{f\}(\omega,t) &= \frac{1}{\sqrt{2\pi}}\int_{\mathbb{R}} e^{\imath\nu u}\cdot\pi^{-1/4}a^{-1/4}e^{-\frac{(u-t)^2}{2a}}e^{-\imath u\omega}du\\
&= \frac{1}{\sqrt{2\pi}}\int_{\mathbb{R}} \pi^{-1/4}a^{-1/4}e^{-\frac{(u-t)^2}{2a}}e^{-\imath u(\omega-\nu)}du\\
&= \pi^{-1/4}a^{-1/4}\mathcal{F}\left\{e^{-\frac{(u-t)^2}{2a}}\right\}(\omega-\nu)\\
&= \pi^{-1/4}a^{1/4}e^{-\imath t\omega}\mathcal{F}\{e^{-u^2/2}\}(a^{1/2}(\omega-\nu))\\
&= \pi^{-1/4}a^{1/4}e^{-\imath t\omega}e^{-\frac{a^{1/4}(\omega-\nu)^2}{2}}.
\end{aligned}$$

The parameter a controls the speed of amplitude attenuation. The larger a the faster the amplitude decreases for ω departing from the signal frequency ν.

Continuous wavelet transformation

Exercise 1.

$$c_\psi = 2\pi\int_{-\infty}^{\infty}\frac{|\hat{\psi}_1(\omega)|^2}{|\omega|}d\omega = 2\pi\int_{-\infty}^{\infty}\frac{\omega^4e^{-\omega^2}}{|\omega|}d\omega = 4\pi\int_0^{\infty}\omega^3e^{-\omega^2}d\omega = 2\pi.$$

Exercise 2.

$$\begin{aligned}
\mathcal{F}\left\{|\lambda|^{-1/2}\psi\left(\frac{(x-t)}{\lambda}\right)\right\}(\omega) &= |\lambda|^{1/2}\mathcal{F}\{\psi(x-t)\}(|\lambda|\omega)\\
&= |\lambda|^{1/2}e^{-\imath t\omega}\hat{\psi}(|\lambda|\omega).
\end{aligned}$$

Hence

$$U(\omega) = |\mathcal{F}\left\{|\lambda|^{-1/2}\psi\left(\frac{(x-t)}{\lambda}\right)\right\}|^2 = \lambda\hat{\psi}^2(\lambda\omega).$$

This leads to

$$U_1(\omega) = \frac{\lambda}{2\pi}\frac{\sin^4(\lambda\omega/4)}{(\lambda\omega/4)^2}$$

for the Haar Wavelet, to

$$U_2(\omega) = \lambda(\lambda\omega)^2e^{-(\lambda\omega)^2}$$

for the Mexican hat wavelet and to

$$U_3(\omega) = \frac{\lambda}{\sqrt{\pi}}e^{-(\lambda\omega-\omega_0)^2}$$

for the Morlet wavelet.

Exercise 3.

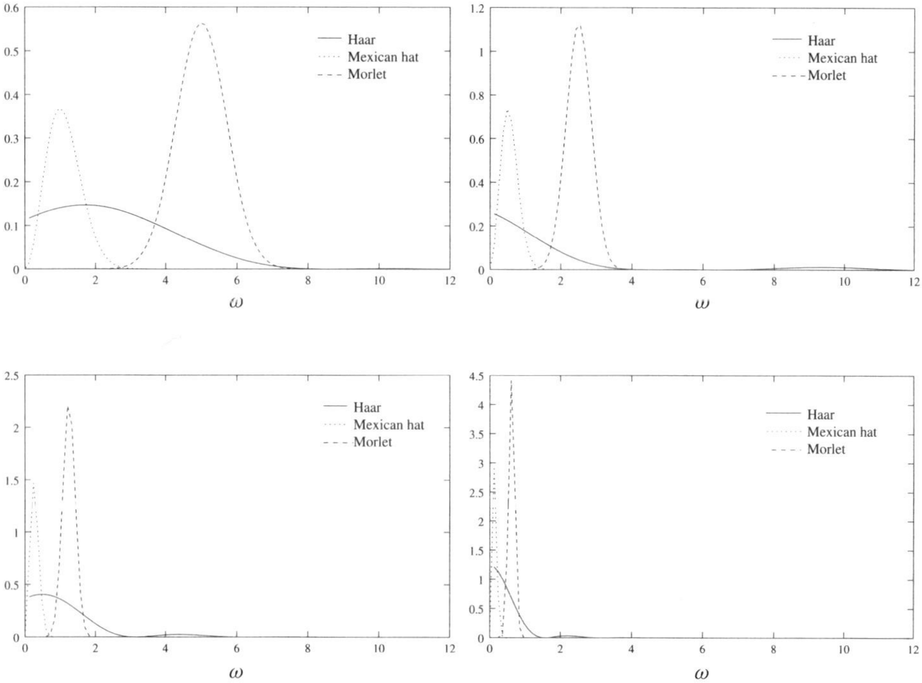

Filter characteristic of the Haar, Mexican hat and Morlet wavelet for scale parameters $\lambda = 1$ (top left), $\lambda = 2$ (top right), $\lambda = 4$ (bottom left) and $\lambda = 8$ (bottom right)

Exercise 4. Haar wavelet:

$$\bar{x} = \int_0^1 x\,dx = \frac{1}{2}.$$

$$\sigma_x^2 = \int_0^1 (x - \frac{1}{2})^2 dx = \int_0^1 x^2 - x + \frac{1}{4} dx = \frac{1}{3} - \frac{1}{2} + \frac{1}{4} = \frac{1}{12}.$$

$$\bar{\omega} = \frac{1}{2\pi} \int_0^\infty \omega \frac{\sin^4(\omega/4)}{(\omega/4)^2} d\omega = \frac{16}{2\pi} \int_0^\infty \frac{\sin^4 z}{z} dz = 8.3848.$$

Mexican hat wavelet:

$$\bar{x} = \int_{-\infty}^{\infty} x(1 - x^2)^2 e^{-x^2} dx = 0.$$

$$\sigma_x^2 = \int_{-\infty}^{\infty} x^2(1 - x^2)^2 e^{-x^2} dx = \int_{-\infty}^{\infty} (x^2 - 2x^4 + x^6) e^{-x^2} dx = 2\sqrt{\pi} \left(\frac{1}{4} - \frac{3}{4} + \frac{15}{16} \right)$$
$$= \sqrt{\pi} \frac{7}{8}.$$

$$\bar{\omega} = \int_0^\infty \omega \cdot \omega^4 e^{-\omega^2} d\omega = 1.$$

$$\sigma_{\omega}^2 = \int_0^\infty (\omega - 1)^2 \omega^4 e^{-\omega^2} d\omega = \int_0^\infty (\omega^6 - 2\omega^5 + \omega^4) e^{-\omega^2} d\omega$$
$$= \sqrt{\pi}\frac{15}{16} - 2 + \sqrt{\pi}\frac{3}{4} = \sqrt{\pi}\frac{3}{2} - 2.$$

Morlet wavelet:

$$\bar{x} = \pi^{-1/2} \int_{-\infty}^\infty x e^{-x^2} dx = 0.$$

$$\sigma_x^2 = \pi^{-1/2} \int_{-\infty}^\infty x^2 e^{-x^2} dx = \frac{1}{\sqrt{2\pi\frac{1}{2}}} \int_{-\infty}^\infty x^2 e^{-\frac{x^2}{2\frac{1}{2}}} dx = \frac{1}{2}.$$

$$\bar{\omega} = \pi^{-1/2} \int_{-\infty}^\infty \omega e^{-(\omega-\omega_0)^2} d\omega = \frac{1}{\sqrt{2\pi\frac{1}{2}}} \int_{-\infty}^\infty \omega e^{-\frac{(\omega-\omega_0)^2}{2\frac{1}{2}}} d\omega = \omega_0.$$

$$\sigma_{\omega}^2 = \pi^{-1/2} \int_{-\infty}^\infty (\omega - \omega_0)^2 e^{-(\omega-\omega_0)^2} d\omega = \frac{1}{2}.$$

Exercise 5.

$$\bar{x}_{\lambda,t} = \lambda^{-1} \int_{\infty}^\infty x\psi^2\left(\frac{x-t}{\lambda}\right) dx = \int_{-\infty}^\infty (\lambda z + t)\psi^2(z) dz = \lambda\bar{x} + t.$$

$$\sigma_{x,\lambda t}^2 = \lambda^{-1} \int_{-\infty}^\infty (x - \lambda\bar{x} - t)^2 \psi^2(\frac{x-t}{\lambda}) dx = \int_{-\infty}^\infty \lambda^2 z^2 \psi^2(z) dz = \lambda^2 \sigma_x^2.$$

$$\bar{\omega}_{\lambda,t} = \lambda \int_{-\infty}^\infty \omega\hat{\psi}^2(\lambda\omega) d\omega = \frac{1}{\lambda} \int_{-\infty}^\infty z\hat{\psi}^2(z) dz = \frac{\bar{\omega}}{\lambda}.$$

$$\sigma_{\omega\lambda t}^2 = \lambda \int_{-\infty}^\infty (\omega - \frac{\bar{\omega}}{\lambda})^2 \hat{\psi}^2(\lambda\omega) d\omega = \frac{1}{\lambda^2} \int_{-\infty}^\infty (z - \bar{\omega})^2 \hat{\psi}^2(z) dz = \frac{\sigma_{\omega^2}}{\lambda^2}.$$

Exercise 6. Centers in x-direction: $\lambda_j \bar{x} + t_i$. Interval length in x direction: $\lambda_j \sigma_x$. Hence

$$t_{i+1} - t_i = 2\lambda_j \sigma_x \Leftrightarrow t_i = 2\lambda_j \sigma_x.$$

Centers in ω direction $\frac{\bar{\omega}}{\lambda_j}$. Interval length in ω direction: $2\frac{\sigma_\omega}{\lambda_j}$. Therefore

$$\frac{\bar{\omega}}{\lambda_{j+1}} = \frac{\bar{\omega}}{\lambda_j} - \frac{\sigma_\omega}{\lambda_j} - \frac{\sigma_\omega}{\lambda_{j+1}} \Leftrightarrow \frac{\bar{\omega} + \sigma_\omega}{\lambda_{j+1}} = \frac{\bar{\omega} - \sigma_\omega}{\lambda_j}$$
$$\Leftrightarrow \lambda_{j+1} = \frac{\bar{\omega} + \sigma_\omega}{\bar{\omega} - \sigma_\omega} \lambda_j.$$

Hence

$$\lambda_j = \left(\frac{\omega + \sigma_\omega}{\bar{\omega} - \sigma_\omega}\right)^j \text{ and } t_i = 2\lambda_j \sigma_x.$$

Exercise 7. Haar wavelet:

$$\lambda_{\min} = \frac{8.3848\Delta x}{\pi}, \quad \lambda_{\max} = \frac{8.3848T}{2\pi}.$$

Mexican hat wavelet:

$$\lambda_{\min} = \frac{\Delta x}{\pi}, \quad \lambda_{\max} = \frac{T}{2\pi}.$$

Morlet wavelet:

$$\lambda_{\min} = \frac{\omega_0 \Delta x}{\pi}, \quad \lambda_{\max} = \frac{\omega_0 T}{2\pi}.$$

Mallat algorithm

Exercise 1. Obviously $\operatorname{supp} \psi_{m,n} = [2^m n, 2^m(n+1)]$. Without any restriction of generality let $m \geq p$. Thus $\operatorname{mes}(\operatorname{supp} \psi_{m,n}) \geq \operatorname{mes}(\operatorname{supp} \psi_{p,q})$, where $\operatorname{mes}(M)$ stands for the Lebesgue measure of the set M. Therefore

$$\langle \psi_{m,n}, \psi_{p,q} \rangle = 0 \Leftrightarrow m \neq p \vee n \neq q.$$

On the other hand

$$\begin{aligned}
\langle \psi_{m,n}, \psi_{m,n} \rangle &= \int_{-\infty}^{\infty} \psi_{m,n}^2(x)dx = 2^{-m} \int_{-\infty}^{\infty} \psi^2(2^{-m}x - n)dx \\
&= 2^{-m} \int_{2^m n}^{2^m(n+1)} dx = 1.
\end{aligned}$$

Exercise 2.

$$\begin{aligned}
\delta_{0k} = \langle \varphi(\bullet), \varphi(\bullet - k) \rangle = \langle \hat{\varphi}, \hat{\varphi} e^{-\imath k\omega} \rangle &= \int_{-\infty}^{\infty} |\hat{\varphi}|^2 e^{\imath k\omega} d\omega \\
&= \sum_{n \in Z} \int_{2\pi n}^{2\pi(n+1)} |\hat{\varphi}|^2 e^{\imath k\omega} d\omega \\
&= \int_0^{2\pi} \sum_{n \in \mathbb{Z}} |\varphi(\omega + 2\pi n)|^2 e^{\imath k\omega} d\omega.
\end{aligned}$$

Obviously, $f(\omega) := \sum_n |\hat{\varphi}(\omega + 2\pi n)|^2$ is a 2π periodic function. Hence its Fourier series is

$$f(\omega) = \sum_{k \in \mathbb{Z}} c_k e^{-\imath k\omega}, \quad c_k := \frac{1}{2\pi} \int_0^{2\pi} f(\omega) e^{\imath k\omega} d\omega.$$

This means that all Fourier coefficients of f except of $c_0 = 1/2\pi$ vanish. Therefore

$$\sum_{n \in \mathbb{Z}} |\hat{\varphi}(\omega + 2\pi n)|^2 = f(\omega) = c_0 e^{-\imath 0\omega} = \frac{1}{2\pi}.$$

Hilbert spaces

Exercise 1.

$$A \oplus B = (A \oplus B)_{ij} = a_{ij} + b_{ij} = b_{ij} + a_{ij} = (B \oplus A)_{ij} = B \oplus A.$$

$$\begin{aligned} A \oplus (B \oplus C) &= (A \oplus (B \oplus C))_{ij} = a_{ij} + (B \oplus C)_{ij} = a_{ij} + (b_{ij} + c_{ij}) \\ &= (a_{ij} + b_{ij}) + c_{ij} = (A \oplus B)_{ij} + c_{ij} = (A \oplus B) \oplus C. \end{aligned}$$

Let $O = (o_{ij})$, $o_{ij} = 0$ for all i, j, then

$$A \oplus O = (A \oplus O)_{ij} = a_{ij} + o_{ij} = a_{ij} = A$$

holds.

Let $A = (a_{ij})$. Define $-A$ by $-A = (-a_{ij})$. Then

$$A \oplus -A = a_{ij} + (-a_{ij}) = 0 = o_{ij} = O,$$

$$\alpha \odot (\beta \odot A) = \alpha \odot (\beta \cdot a_{ij}) = (\alpha \cdot \beta \cdot a_{ij}) = ((\alpha \cdot \beta) \cdot a_{ij}) = (\alpha \cdot \beta) \odot A,$$

and so on, and so on.

Exercise 2.

$$\|A\| = \max_i \sum_{j=1}^{n} |a_{ij}| > \sum_{i}^{n} |a_{1j}| > |a_{11}| \geq 0.$$

Let $0 = \|A\|$ and suppose that there is at least one i and one j with $a_{ij} \neq 0$. Then

$$0 = \|A\| > |a_{ij}| > 0$$

follows. Hence $\|A\| = 0 \Leftrightarrow A = 0$.

$$\|\alpha A\| = \max_i \sum_{j=1}^{n} |\alpha \cdot a_{ij}| = \max_i |\alpha| \sum_{j=1}^{n} |a_{ij}| = |\alpha| \cdot \|A\|.$$

$$\begin{aligned} \|A \oplus B\| &= \max_i \sum_{j=1}^{n} |a_{ij} + b_{ij}| \leq \max_i \sum_{j=1}^{n} (|a_{ij}| + |b_{ij}|) \\ &= \max_i \left(\sum_{j=1}^{n} |a_{ij}| + \sum_{j=1}^{n} |b_{ij}| \right) \leq \max_i \sum_{j=1}^{n} |a_{ij}| + \max_i \sum_{j=1}^{n} |b_{ij}| \\ &= \|A\| + \|B\|. \end{aligned}$$

Exercise 3. Let $f, g \in V$. Then

$$\begin{aligned}\int_a^b (f \oplus g)^2 dx &= \int_a^b (f(x) + g(x))^2 dx \\ &\le \int_a^b (f(x) + g(x))^2 dx + \int_a^b (f(x) - g(x))^2 dx \\ &= 2\Big(\int_a^b f^2(x)dx + \int_a^b g^2(x)dx\Big) < \infty.\end{aligned}$$

Consequently, $f \oplus g \in V$.

$$\begin{aligned}|\langle f, g\rangle| &= \Big|\int_a^b f \cdot g dx\Big| \le 2 \cdot \int_a^b |f \cdot g| dx \le \int_a^b (f+g)^2 + f^2 + g^2 dx \\ &= \int_a^b (f+g)^2 dx + \int_a^b f^2 dx + \int_a^b g^2 dx < \infty.\end{aligned}$$

$$\langle f, g\rangle = \int_a^b f \cdot g dx = \int_a^b g \cdot f dx = \langle g, f\rangle.$$

$$\begin{aligned}\langle f \oplus g, h\rangle &= \int_a^b (f \oplus g)(x) \cdot h(x) dx = \int_a^b (f(x) + g(x)) \cdot h(x) dx \\ &= \int_a^b f(x) \cdot h(x) + g(x) \cdot h(x) dx = \int_a^b f \cdot h dx + \int_a^b g \cdot h dx \\ &= \langle f, h\rangle + \langle g, h\rangle.\end{aligned}$$

$$\langle \alpha \odot f, g\rangle = \int_a^b \alpha \cdot f \cdot g dx = \alpha \cdot \int_a^b f \cdot g dx = \alpha\langle f, g\rangle.$$

$$0 = \langle f, f\rangle = \int_a^b f^2 dx \Leftrightarrow f^2 = 0 \Leftrightarrow f = 0.$$

Exercise 4. Obviously

$$\begin{aligned}\int_0^{2\pi} \sin^2(mx) dx &\le \int_0^{2\pi} 1 dx = 2\pi < \infty, \\ \int_0^{2\pi} \cos^2(mx) dx &\le \int_0^{2\pi} 1 dx = 2\pi < \infty.\end{aligned}$$

holds. Hence the functions $\sin mx$, $\cos mx$ indeed belong to V.

$$
\begin{aligned}
\langle \cos mx, \cos nx\rangle &= \int_0^{2\pi} \frac{1}{2}(e^{\imath mx} + e^{-\imath mx})\frac{1}{2}(e^{\imath nx} + e^{-\imath nx})dx \\
&= \frac{1}{4}\int_0^{2\pi} e^{\imath(m+n)x} + e^{-\imath(m-n)x} + e^{\imath(m-n)x} + e^{-\imath(m+n)x}dx \\
&= \frac{1}{2}\int_0^{2\pi} \cos(m+n)x dx + \frac{1}{2}\int_0^{2\pi} \cos(m-n)x dx \\
&= \frac{1}{2(m+n)}(\sin(m+n)2\pi - \sin(m+n)0) \\
&\quad + \frac{1}{2(m-n)}(\sin(m-n)2\pi - \sin(m-n)0) \\
&= \begin{cases} \pi, & m = n \\ 0, & m \neq n. \end{cases}
\end{aligned}
$$

The orthogonality of $\langle \cos mx, \sin nx\rangle$, $\langle \sin mx, \cos nx\rangle$ is proved completely analog.

Exercise 5. Let $\omega_k^j := e^{\imath j \frac{2\pi k}{n}}$. Obviously

$$0 = \omega_k^n - 1 = (\omega_k - 1)(\omega_k^{n-1} + \omega_k^{n-2} + \cdots + 1)$$

holds. Since $\omega_k \neq 1$, the following relations follow:

$$\sum_{i=0}^{n-1} \omega_i^j \bar{\omega}_i^k = \sum_{i=0}^{n-1} \omega_i^{j-k} = \sum_{i=0}^{n-1} \omega_{j-k}^i = (\omega_{j-k}^{n-1} + \cdots + 1) = \begin{cases} n, & j = h \\ 0, & j \neq h. \end{cases}$$

$$
\begin{aligned}
\langle \mathbf{c}_k, \mathbf{c}_l\rangle &= \frac{2}{n}\sum_{i=0}^{n-1} \cos\left(\frac{2\pi ki}{n}\right)\cos\left(\frac{2\pi li}{n}\right) \\
&= \frac{1}{2n}\sum_{i=0}^{n-1} (e^{\imath\frac{2\pi ki}{n}} + e^{-\imath\frac{2\pi ki}{n}})(e^{\imath\frac{2\pi li}{n}} + e^{-\imath\frac{2\pi li}{n}}) \\
&= \frac{1}{2n}\sum_{i=0}^{n-1}(\omega_k^i + \bar{\omega}_k^i)(\omega_l^i + \bar{\omega}_l^i) = \frac{1}{2n}\sum_{i=0}^{n-1} \omega_{k+l}^i + \omega_{k-l}^i + \omega_{l-k}^i + \omega_{-l-k}^i \\
&= \begin{cases} 1, & l = k \\ 0, & l \neq k. \end{cases}
\end{aligned}
$$

Exercise 6.

$$0 = \|f\| \Leftrightarrow \max_{x\in[a,b]} |f(x)| = 0 \Leftrightarrow f(x) = 0.$$

$$\|\alpha f\| = \max_{x\in[a,b]} |\alpha f(x)| = |\alpha| \max_{x\in[a,b]} |f(x)| = |\alpha| \cdot \|f\|.$$

$$\begin{aligned}\|f+g\| &= \max_{x\in[a,b]} |f(x)+g(x)| \le \max_{x\in[a,b]} |f(x)|+|g(x)| \\ &\le \max_{x\in[a,b]} |f(x)| + \max_{x\in[a,b]} |g(x)| = \|f\|+\|g\|.\end{aligned}$$

Exercise 7. Linearity of f is obvious. The only thing, which has to be shown is that it is bounded.

$$|f(x)| = \left|\int_a^b x(t)dt\right| \le \int_a^b |x(t)|dt \le \int_a^b \max_{t\in[a,b]} |x(t)|dt = (b-a)\|x\|.$$

Consequently, $\|f\| \le (b-a)$ Now let $x'(t)=1$.

$$|f(x')| = \left|\int_a^b 1dt\right| = (b-a).$$

Hence

$$\|f\| = \sup_{\|x\|\le 1} |f(x)| \ge |f(x')| = (b-a).$$

Together $\|f\| = (b-a)$ follows.

Exercise 8.

$$|ev_{t_0}(x)| = |x(t_0)| \le \max_{t\in[a,b]} |x(t)| = \|x\|.$$

Hence $\|ev_{t_0}\| \le 1$ holds.

Exercise 9.

$$\begin{aligned}\|x_n - x\| &= \left(\int_{-1}^{1} (x_n(t)-x(t))^2 dt\right)^{1/2} \\ &= \left(\int_{-\frac{1}{n}}^{\frac{1}{n}} (x_n(t)-x(t))^2 dt\right)^{1/2} \\ &= \left(\int_{-\frac{1}{n}}^{0} \left(\frac{2}{n}\left(t+\frac{1}{n}\right)\right)^2 dt + \int_0^{\frac{1}{n}} \left(\frac{2}{n}(t+\frac{1}{n}) - 1\right)^2 dt\right)^{1/2} \\ &= \left(\int_{-\frac{1}{n}}^{0} \frac{4}{n^2}\left(t^2+\frac{2}{n}t+\frac{1}{n^2}\right)dt + \int_0^{\frac{1}{n}} \frac{4}{n^2}\left(t^2+\frac{2-n^2}{n}t+\frac{(2-n^2)^2}{4n^2}\right)dt\right)^{1/2} \\ &= \left(-\frac{4}{n^2}\left(-\frac{1}{3n^3}+\frac{1}{n^3}-\frac{1}{n^3}\right)+\frac{4}{n^2}\left(\frac{1}{3n^3}+\frac{2-n^2}{2}\frac{1}{n^3}+\frac{(2-n^2)^2}{4}\frac{1}{n^3}\right)^{1/2}\right) \\ &= \left(\frac{4}{n^2}\left(\frac{1}{n^3}\left(\frac{2}{3}+\frac{(2-n^2)}{2}+\frac{(2-n^2)^2}{4}\right)\right)\right)^{1/2} \\ &= \left(\frac{8}{3n^5}+2\left(\frac{2}{n^5}-\frac{1}{n^3}\right)+\frac{1}{n}\left(\frac{2}{n^2}-1\right)^2\right)^{1/2}.\end{aligned}$$

Hence

$$\lim_{n\to\infty} \|x_n - x\| = \lim_{n\to\infty} \left(\frac{8}{3n^5} + 2\left(\frac{2}{n^5} - \frac{1}{n^3}\right) + \frac{1}{n}\left(\frac{2}{n^2} - 1\right)^2 \right)^{1/2} = 0.$$

Exercise 10. Let x_n, x be defined as in Exercise 9. Let $y_n := x - x_n$, then

$$|ev_0(y_n)| = \frac{1}{2}.$$

Let M be an arbitrary positive real number. Since $\|y_n\| \to 0$ there is an n_0 such that $\frac{1}{2\|y_n\|} > M$ holds for all $n > n_0$. Consequently, there is no $\|ev_0\| > 0$ with

$$|ev_0(x)| \leq \|ev_0\| \cdot \|x\|.$$

Bibliography

[1] Adam J., K.-P. Schwarz (eds.), Vistas for geodesy in the new millenium. Springer-Verlag, Berlin–Heidelberg–New York 2002.

[2] Antoniadis, A., G. Oppenheim (eds.), Wavelets and statistics. Springer-Verlag, Berlin–Heidelberg–New York 1995.

[3] Bacry, E., J. F. Muzy, A. Arneodo, Singularity spectrum of fractal signals: exact results. J. of Stat. Phys. 70, 635–674, 1993.

[4] Bayer, M., Geomagnetic field modelling from satellite data by first and second generation avelets. PhD thesis, University of Kaiserlautern, Geomathematics Group, Shaker Verlag, Aachen 2000.

[5] Bayer, M., W. Freeden, T. Maier, Wavelet analysis of the geomagnetic field within source regions of ionospheric and magnetospheric currents. Berichte der Arbeitsgruppe Technomathematik, Fachbereich Mathematik, Universität Kaiserslautern Bericht No. 217, 1999.

[6] Bayer, M., W. Freeden, T. Maier, A vector wavelet approach in iono- and magnetospheric geomagnetic satellite data. Journal of Atmospheric and Solar-Terrestrial Physics 63, 581–597, 2001.

[7] Beth, S., Multiscale approximation by vector radial basis functions on the sphere. PhD thesis, University of Kaiserslautern, Geomathematics Group, Shaker Verlag, Aachen 2000.

[8] Beth, S., T. Boos, W. Freeden, On the wavelet determination of scale exponents in energy spectra and structure functions and their application to CCD camera data. In: Schwarz K.-P. (ed.), 221–227, 2000.

[9] Bracewell, R., The fourier transform and its application. McGraw-Hill, New York 1965.

[10] Burt, P. J., E. H. Andelson, The Laplacian pyramid as a compact image code. IEEE Trans. Commun. 31, 20–51, 1983.

[11] Canny, J., A computational approach to edge detection. IEEE Trans. Patt. Anal. and Mach. Intell. 36, 961–1005, 1986.

[12] Cohen, A., I. Daubechies, J. C. Feauveau, Biorthogonal bases of compactly supported wavelets. Comm. Pure Appl. Math. 45, 485–560, 1992.

[13] Collatz, L., Funktionalanalysis und Numerische Mathematik. Springer-Verlag, Berlin–Göttingen–Heidelberg 1964.

[14] Collin F., R. Warnant, Application of wavelet transform for GPS cycle slip correction and comparison with Kalman filter. manuscripta geodaetica 20, 161–172, 1995.

[15] Daubechies I., Ten lectures on wavelets. SIAM, Philadelphia, 1992.

[16] Cui, J., W. Freeden, B. Witte, Gleichmässige Approximation mittels sphärischer Finite-Elemente und ihre Anwendung auf die Geodäsie. Zeitschr. f. Vermess. 117, 266–278, 1992.

[17] Cui, J., W. Freeden, Equidistribution on the sphere. SIAM J. Sci. Stat. Comput. 18, 595–609, 1999.

[18] Dieudonne, J., Grundzüge der modernen Analysis Bd. 1. VEB Deutscher Verlag der Wissenschaften, Berlin 1971.

[19] Donoho, D. L., Wavelet Shrikage and W. V. D., A 10-Minute Tour. In: Proceedings of the International Conference on Wavelet and Applications, Toulouse 1992.

[20] Donoho, D., I. Johnstone, Ideal spatial adaption via wavelet shrinkage. Biometrika 81, 425–455, 1994.

[21] Foufoula-Geourgiou, E., P. Kumar (eds.), Wavelets in geophysics. Academic Press, San Diego 1994.

[22] Freeden, W., On the role of (band limited) harmonic wavelets in Earth's gravitational potential determination. In: Klees R.(ed.), 227–257, 1999.

[23] Freeden, W., Multiscale modelling of spaceborne geodata. Teubner Verlag, Stuttgart Leipzig 1999.

[24] Freeden, W., Multiscale modelling of GOCE-DATA-products. In: ESA-Proceedings of the International GOCE User Workshop, ESTEC, Noordwijk, ESA W-188, 111–126, 2001.

[25] Freeden, W., O. Glockner, R. Litzenberger, Multiscale modelling from geopotential data. In: Haussman et al. (eds.), 151–174, 1999.

[26] Freeden, W., O. Glockner, M. Thalhammer, Multiscale gravitational field recovery from GPS-satellite-to-satellite tracking. Studia Geophysica et Geodaetica 43, 229 264, 1999.

[27] Freeden, W., T. Gerven, M. Schreiner, Constructive approximation on the sphere (with applications to geomathematics). Oxford Science Publications, Claredon 1998.

[28] Freeden, W., E. Groten, V. Michel, A. Kaboodvand, Geoidal reconstruction, decomposition, fast computation, and noise cancellation by harmonic wavelets. Studia Geophysica et Geodeatica 47, 37–73, 2003.

[29] Freeden, W., T. Maier, Spectral and multiscale signal-to-noise thresholding of spherical vector fields. Computational Geosciences 7, 215–250, 2003.

[30] Freeden, W., T. Maier, On multiscale denoising of spherical functions: Basic theory and numerical aspects. Electronic Transactions on Numerical Analysis (ETNA) 14, 40–62, 2002.

[31] Freeden, W., V. Michel, Least squares geopotential approximation by windowed Fourier transform and wavelet transform. Berichte d. Arbeitsgruppe Technomathematik, Fachbereich Mathematik, Univ. Kaiserslautern, Bericht No. 216, 1999.

[32] Freeden, W., V. Michel, Constructive approximation and numerical methods in geodetic research today – an attempt at a categorization based on an uncertainty principle. Berichte d. Arbeitsgruppe Technomathematik, Fachbereich Mathematik, Univ. Kaiserslautern, Bericht No. 208; also in Journal of Geodesy 73, 452 –465, 1999.

[33] Freeden, W., V. Michel, M. Stenger, Multiscale signal-to-noise thresholding. Acta Geodaetica et Geophysica Hungarica 36, 55–86, 2001.

[34] Freeden, W., K. Hesse, On the multiscale solution of satellite problems by use of locally supported kernel functions corresponding to equidistributed data on spherical orbits. Studia Scientiarum Mathematicarum Hungarica 39, 37–74, 2002.

[35] Freeden, W., F. Schneider, An integrated wavelet concept of physical geodesy. Journal of Geodesy 72, 259–281, 1998.

[36] Freeden, W., F. Schneider, Regularization wavelets and multiresolution. Inverse Problems 14, 225–243, 1998.

[37] Freeden, W., M. Schreiner, Satellite gradiometry – a new approach. In: Sanso, F. (ed.), 122–139, 1995.

[38] Ghil, M., R. Benzi, G. Parisi (eds.), North-Holland, Amsterdam, 1985.

[39] Gilbert, A., Wavelet Lösungen geodätischer Integralgleichungen, Diploma thesis, University Stuttgart, 1997.

[40] Gilbert, D., M. Holschneider, J. L. LeMouel, Wavelet Analysis of the Chandler Wobble. JGR, 1998.

[41] Gilbert, A., W. Keller, Deconvolution with wavelets and vaguelettes. Journal of Geodesy 74, 306–320, 2000.

[42] Grafarend E. W., F. W. Krumm, V. S. Schwarze (eds.), Quo vadis geodesy – The challenge of the 3rd millenium, Springer-Verlag, Berlin 2002.

[43] Groupillaud, P., A. Grossmann, J. Morlet, Cycle-octave and related transforms in seismic signal analysis. Geoexploration 23, Elsevier Science Publ., 85–102, 1984/85.

[44] Großmann, Ch., H.-G. Roos, Numerik partieller Differentialgleichungen. B.G. Teubner, Stuttgart 1992.

[45] Hackbusch, W., Multi-grid methods and applications. Springer Series in Computational Mathematics, Berlin 1985.

[46] Hagelberg, C. R., N. K. K. Gamage, Application of structure preserving wavelet decompositions to intermittence turbulence: A case study. In: Foufoula-Geourgiou, E., P. Kumar (eds.), 45–80, 1994.

[47] Haussmann, K. Jetter, M. Reimer (eds.), Advances in multivariate approximation, Mathematical Research, Vol. 107, W. Wiley-VCH, 1999.

[48] Jawert, B., W. Sweldens, An overview of wavelet based multiresolution analysises. SIAM Review 36, 3.

[49] Katul, G. G., J. D. Albertson, C.,R. Chu, M.,B. Parlange, Intermittency in atmospheric surface layer turbulence: The orthonormal wavelet representation. In: Foufoula-Geourgiou E., P. Kumar (eds.), 81–105, 1994.

[50] Kantorovich, L. V., G. P. Akilov, Functional analysis. Pergamon Press, Oxford 1982.

[51] Keller, W., Local geoid determination by wavelet-vaguelette decomposition. Phys. Chem. Earth 21, 147–150, 1996.

[52] Keller, W., A wavelet-vaguelette analysis of geodetic integral formulae. In: Segawa, J., H. Fujimoto, S. Okubo (eds.), 557–565, 1997.

[53] Keller, W., Collocation in reproducing kernel Hilbert spaces of a multiscale analysis. Phys. Chem. Earth 23, 25–29, 1998.

[54] Keller, W., A wavelet approach to non-stationary collocation In : Schwarz, K.-P. (ed.), 208–214, 2000.

[55] Keller, W., A wavelet solution to 1D non-stationary collocation with an extension to 2D. In: Sideris M. G. (ed.), 79–85, 2002.

[56] Keller, W., A wavelet approach for the construction of multi-grid solvers for large linear systems. In: Adam, J., K.-P. Schwarz (eds.), 265–271, 2002.

[57] Kenney, J. F., E. S. Keeping, Mathematics of statistics, Part 1. Van Norstrand, Princeton NJ 1962.

[58] Klees, R. (ed.), Wavelets in the geosciences. Springer-Verlag, Berlin–Heidelberg–New York, 1999.

[59] Kolmogorov, A. N., The local structure of turbulence in incompressible viscous fluid for very large Reynolds numbers. Dokl. Akad. Nauk SSSR 4, 299–303, 1941.

[60] Kolmogorov, A. N., S. Fomin, Reelle Funktionen und Funktionalanalysis. VEB Deutscher Verlag der Wissenschaften, Berlin 1975.

[61] Kronland-Martinet, R., J. Morlet, A. Grossmann, Analysis of sound patterns through wavelet transforms. Int. Journ. of Pattern Recogn. and Artific. Intell. 1, 273–301, 1987.

[62] Kusche, J., Implementation of multi-grid solvers for satellite gravity anomaly recovery. Journal of Geodesy 74, 773–782, 2001.

[63] Kusche, J., S. Rudolph, Satellite gravity anomaly recovery using multi-grid methods. In: Sideris M. G. (ed.), 91–97, 2001.

[64] Little, S. A., P. H. Carter, D. K., Smith. Wavelet analysis of a bathymetric profile reveals anomalous crust. Geophys. Res. Letts. 20, 1915–1918, 1993.

[65] Li, D., J. Shao. The wavelet and its application in image detection. ISPRS Journal of Photogrammetry and Remote Sensing 49, 4–11, 1994.

[66] Li, Q., K. Vasudevan, F. A. Cook, Seismic skeletonization: A new approach to interpretation of seismic reflection data. J. Geophys. Research 102, 8427–8445, 1997.

[67] Little, S. A., Wavelet analysis of seafloor bathymetry: An example. In: Foufoula-Geourgiou E., P. Kumar (eds.), 167–182, 1994.

[68] Louis, A.K., P. Maaß, A. Rieder, Wavelets. B.G. Teubner Verlag, Stuttgart 1994.

[69] Maier, T., Multiscale geomagnetic field modelling from satellite data: Theoretical aspects and numerical applications. PhD thesis, Universität Kaiserslautern, 2002.

[70] Maier, T., C. Mayer, Multiscale downward continuation of CHAMP FGM-data for crustal field modelling. Mathematische Geologie Band 5, C-Press Verlag, Dresden 2002.

[71] Mallat, S., W. L. Hwang, Singularity detection and processing with wavelets. IEEEE Trans INFO. Theory 38, 617–643, 1992.

[72] Mallat, S., A wavelet tour of signal processing. Academic Press San Diego, 1999.

[73] Meyer, Y., Wavelets and operators. Cambridge University Press, 1992.

[74] Mayer, C., T. Maier, Multiscale determination of radial current distribution from CHAMP FGM-data. Mathematische Geologie Band 5, C-Press Verlag Dresden, 2002.

[75] Neumann J. v., Mathematische Grundlagen der Quantenmechanik. Grundlehren der Mathematischen Wissenschaften, Band 38, Springer-Verlag, Heidelberg 1968.

[76] Obuchov, A. M., Statistische Beschreibung stetiger Felder. (Translation from Russian). Sammelband zur statistischen Theorie der Turbulenz, Akademie-Verlag, Berlin 1958.

[77] Parisi, G., U. Frisch, Turbulence and predictability in geophysical fluid dynamics and climate dynamics. In: Ghil, M., R. Benzi, G. Parisi (eds.), 71–88, 1985.

[78] Pike, J. C., Analysis of high resolution marine seismic data using the wavelet Transform. In: Foufoula-Geourgiou E., P. Kumar (eds.), 183–211, 1994.

[79] Polya, G., G. Szegö, Aufgaben und Lehrsätze der Analysis. Vol. II, Springer-Verlag, Berlin 1971.

[80] Rosenfeld, A., M. Thurston, Edge and curve detection for visual scene analysis. IEEE Trans. Comput. C 29, 1971.

[81] Sanso, F. (ed.), International Association of Geodesy Symposia 114, Springer-Verlag, Berlin–Heidelberg–New York 1995.

[82] Sanso, F., M. G. Sideris, On the similarities and differences between system theory and least-squares collocation in physical geodesy. Bolletino di Geodesia e Science Affini LVI, 173–206, 1997.

[83] Schmidt, M., Wavelet analysis of stochatic signals. IERS Technical Note No 28, 65–71, 1997.

[84] Schmidt, M., Ein Beitrag zur zweidimensionalen Wavelet-Analyse von Zufallsprozessen, Zeitschr. f. Vermess. 126, 270–275, 2001.

[85] Schmidt, M., H. Schuh, Wavelet-Analyse der mit VLBI beobachteten Nutationsreihen, Zeitschr. f. Vermess. 124, 24–30, 1999.

[86] Schmidt, M., H. Schuh, Abilities of wavelet analysis for investigating short-period variations of Earth Rotation. IERS Technical Note No 28, 73–80, 2000.

[87] Schmidt, M., Ein Beitrag zur zweidimensionalen Wavelet-Analyse von Zufallsprozessen, Zeitschr. f. Vermess. 126, 270–275, 2001.

[88] Schmidt, M., Grundprinzipien der Wavelet-Analyse und Anwendungen in der Geodäsie, Shaker-Verlag, Reihe Geodäsie Band 9, Aachen 2001.

[89] Schmidt, M., H. Schuh, Wavelet-Analyse von Erdrotationsschwankungen, Zeitschr. f. Vermess. 126, 94–100, 2001.

[90] Schmitz-Hübsch, H., Wavelet-Analysen der Erdrotationsparameter im hochfrequenten Bereich. Schriftenreihe der Deutschen Geodätischen Kommission, Reihe, Heft A 118, München 2002.

[91] Schmitz-Hübsch, H., H. Schuh, Seasonal and short-period fluctuations of Earth Rotation investigated by Wavelet Analysis. In: Grafarend E. W., F. W. Krumm, V. S. Schwarze (eds.), 125–135, 2002.

[92] Schröder, P., W. Sweldens, Pherical wavelets: texture processing. Tech. Rep. 4. Industrial Mathematics Initiative, Department of Mathematics, University of South Carolina 1995.

[93] Schwarz, K. P. (ed.), Geodesy Beyond 2000. The Challenge of the First Decade. International Association of Geodesy Symposia Vol. 121, Springer-Verlag, Berlin–Heidelberg 2000.

[94] Schwarz, K. P., Sideris M.G., R. Forsberg: The use of FFT techniques in physical geodesy. Geophys. J. Int. 100, 485–514, 1990.

[95] Segawa, J., H. Fujimoto, S. Okubo (eds.), Gravity, geoid and marine geodesy. Springer-Verlag, Berlin–Heidelberg–New York 1997.

[96] Sideris, M. G., A fast Fourier transform method for computing terrain corrections, manuscr. geod. 10, 66–73, 1985.

[97] Sideris M. G. (ed.), Gravity, geoid and geodynamics 2000, Springer-Verlag, Berlin–Heidelberg–New York 2002.

[98] Sideris, M. G., K. P. Schwarz, Recent advances in the numerical solution of the linear Molodensky problem. Bull Geod. 62, 521–540, 1988.

[99] Sideris, M. G., I. N. Tziavos, FFT-evaluation and application of gravity field convolution integrals with mean and point data. Bull. Geod. 62, 540–561, 1988.

[100] Sweldens W., The lifting scheme: A custom-design construction of biorthogonal wavelets. Tech. Rep. 7. Industrial Mathematics Initiative, Department of Mathematics, University of South Carolina 1994.

[101] Tsay, J. R., Wavelets für das Facetten-Stereosehen. Veröff. DKG, Reihe C, Nr. 454, München 1996.

[102] Vasudevan, K., F. A. Cook, Empirical mode skeletonization of deep crustal seismic data: Theory and applications. J. Geophys. Research 105, 7845–7855, 2000.

[103] Valdimirov, V. S., Obobcennye fukcii v matematiceskoij fysike. Izdanye Nauka, Moskva 1976.

[104] Wallace, G. K., The JPEG still picture compression standard. Communications of the ACM, 34, 30–44,1991.

Index